Encyclopaedia of

INTEGRAL CALCULUS

Encyclopaedia of
INTEGRAL CALCULUS

Volume 2

R.K. Pandey

ANMOL PUBLICATIONS PVT. LTD.
NEW DELHI - 110 002 (INDIA)

ANMOL PUBLICATIONS PVT. LTD.
H.O.: 4374/4B, Ansari Road, Daryaganj,
New Delhi-110 002 (India)
Ph.: 23278000, 23261597
B.O.: No. 1015, Ist Main Road, BSK IIIrd Stage
IIIrd Phase, IIIrd Block,
Bangalore - 560 085 (India)
Visit us at: www.anmolpublications.com

Encyclopaedia of Integral Calculus

First Edition, 2009

ISBN 978-81-261-3920-0 (Set)

PRINTED IN INDIA

Printed at Mehra Offset Press, Delhi.

Contents

Preface

Encyclopedia of Integral Calculus is a unique Encyclopaedia on Integration, aiming at providing a fairly complete account of the basic concepts required to build a strong foundation for a student endeavoring to study this subject. The *Analytical Approach* to the major concepts makes the book highly self-contained and comprehensive guide that succeeds in making the concepts easily understandable. All the *Elementary Principles and Fundamental Concepts* have been explained Rigorously, leaving no scope for Illusion or Confusion.

The focus throughout the text has been on presenting the subject matter in a well-knit manner and lucid style, so that even a student with average *Mathematical Skill* would find it accessible to himself. In addition, the book provides numerous *well-graded solved examples, generally set in various university and competitive examinations*, which will facilitate easy understanding besides acquainting the students with a variety of questions.

Author

List of Symbols

$\pm$	$\not\subset$	$\subset$	$\supset$	$\subseteq$	$\in$	$\notin$	∂
$\neq$	$\equiv$	$\approx$	$\angle$	∇	Σ	Π	Ω
ϕ	α	β	χ	ε	γ	η	λ
μ	ν	$\perp$	π	θ	ρ	σ	τ
$\Re$	ω	ξ	ψ	ζ	Υ	$\exists$	$\int$
$\iint$	$\iiint$	$\oint$	$\times$	$\sqrt{}$	$\therefore$	$\because$	Δ
$\geq$	$\leq$	$\Rightarrow$	$\Downarrow$	$\Leftrightarrow$	∞	$\propto$	$\mathbb{R}$
$\mathbb{Z}$	$\mathbb{N}$	$\mathbb{C}$	$\mathbb{Q}$				

Chapter 1

Improper Integral

In this section we need to take a look at a couple of different kinds of integrals. Both of these are examples of integrals that are called Improper Integrals. Let's start with the first kind of improper integrals that we're going to take a look at.

INFINITE INTERVAL

In this kind of integrals we are going to take a look at integrals that in which one or both of the limits of integration are infinity. In these cases the interval of integration is said to be over an infinite interval.

Let's take a look at an example that will also show us how we are going to deal with these integrals.

Example: Evaluate the following integral.

$$\int_1^\infty \frac{1}{x^2}\,dx$$

Solution: This is an innocent enough looking integral. However, because infinity is not a real number we can't just integrate as normal and then "plug in" the infinity to get an answer.

To see how we're going to do this integral let's think of this as an area problem. So instead of asking what the integral is, let's instead ask what the area under $f(x)=\frac{1}{x^2}$ on the interval $[1,\infty)$ is.

We still aren't able to do this, however, let's step back a little and instead ask what the area under $f(x)$ is on the interval $[1, t]$ were $t > 1$ and t is finite. This is a problem that we can do.

$$A_t = \int_1^t \frac{1}{x^2}\,dx = -\frac{1}{x}\bigg|_1^t \frac{1}{x^2} = 1-\frac{1}{t}$$

Now, we can get the area under $f(x)$ on $[1, \infty)$ simply by taking the limit of A_t as t goes to infinity.

$$A = \lim_{t\to\infty} A_t \lim_{t\to\infty}\left(1-\frac{1}{t}\right) = 1$$

This is then how we well do the integral itself.

$$\int_1^{\infty} \frac{1}{x^2}dx = \lim_{t\to\infty}\int_1^t \frac{1}{x^2}dx$$

$$= \lim_{t\to\infty}\left(-\frac{1}{x}\right)\Bigg|_1^t \quad 1$$

$$= \lim_{t\to\infty}\left(1-\frac{1}{t}\right) = 1$$

So, this is how we will deal with these kinds of integrals in general. We will replace the infinity with a variable (usually t), do the integral and then take the limit of the result as t goes to infinity. On a side note, notice that the area under a curve on an infinite interval was not infinity as we might have suspected it to be. In fact, it was a surprisingly small number. Of course this won't always be the case, but it is important enough to point out that not all areas on an infinite interval will yield infinite areas.

Let's now get some definitions out of the way. We will call these integrals convergent if the associated limit exists and is a finite number (*i.e.* it's not plus or minus infinity) and divergent if the associated limits either doesn't exist or is (plus or minus) infinity. Let's now formalize up the method for dealing with infinite intervals. There are essentially three cases that we'll need to look at.

1. If $\int_a^t f(x)dx$ exists for every then,

$$\int_a^{\infty} f(x) = \lim_{t\to\infty}\int_a^t f(x)dx$$

provided the limit exists and is finite.

If $\int_{-\infty}^{b} f(x)dx$ exists for every then,

$$\int_{-\infty}^{b} f(x)dx = \lim_{t\to\infty}\int_t^b f(x)dx$$

provided the limits exists and is finite.

If $\int_{-\infty}^{c} f(x)dx$ and $\int_c^{\infty} f(x)dx$ are both convergent then,

$$\int_{-\infty}^{\infty} f(x)dx = \int_{-\infty}^{c} f(x)\,dx + \int_c^{\infty} f(x)\,dx$$

Where c is any number. Note as well that this requires BOTH of the integrals to be convergent in order for this integral to also be convergent. If either of the two integrals is divergent then so is this integral. Let's take a look at a couple more examples.

Example: Determine if the follow integral is convergent or divergent and if it's convergent find its value.

$$\int_1^\infty \frac{1}{x}\,dx$$

Solution: So, the first thing we do is convert the integral to a limit.

$$\int_1^\infty \frac{1}{x}\,dx = \lim_{t\to\infty}\int_1^t \frac{1}{x}\,dx$$

Now, do the integral and the limit.

$$\begin{aligned}\int_1^\infty \frac{1}{x}\,dx &= \lim_{t\to\infty} \ln(x)\Big|_1^t \\ &= \lim_{t\to\infty}\left(\ln(t) - \ln 1\right) \\ &= \infty\end{aligned}$$

So, the limit is infinite and so the integral is divergent.

If we go back to thinking in terms of area notice that the area under $g(x) = \frac{1}{x}$ on the interval $[1,\infty)$ is infinite. This is in contrast to the area under $f(x) = \frac{1}{x^2}$ which was quite small. There really isn't all that much difference between these two functions and yet there is a large difference in the area under them. We can actually extend this out to the following fact.

Fact

If $a > 0$ then

$$\int_a^\infty \frac{1}{x^p}\,dx$$

is convergent if $p > 1$ and divergent if $p \le 1$.

One thing to note about this fact is that it's in essence saying that if an integrand goes to zero fast enough then the integral will converge.

How fast is fast enough? If we use this fact as a guide it looks like integrands that go to zero faster than $\frac{1}{x}$ goes to zero will probably converge.

Let's take a look at a couple more examples.

Example: Determine if the following integral is convergent or divergent. If it is convergent find its value.

$$\int_{-\infty}^0 \frac{1}{\sqrt{3-x}}\,dx$$

Solution: There really isn't much to do with these problems once you know how to do them. We'll convert the integral to a limit/integral pair, evaluate the integral and then the limit.

$$\int_{-\infty}^{0} \frac{1}{\sqrt{3-x}}\,dx = \lim_{t\to-\infty} \int_{t}^{0} \frac{1}{\sqrt{3-x}}\,dx$$
$$= \lim_{t\to-\infty} -2\sqrt{3-x}\Big|_{t}^{0}$$
$$= \lim_{t\to-\infty} \left(-2\sqrt{3}+2\sqrt{3-t}\right)$$
$$= -2\sqrt{3}+\infty$$
$$= \infty$$

So, the limit is infinite and so this integral is divergent.

Example: Determine if the following integral is convergent or divergent. If it is convergent find its value.

$$\int_{-\infty}^{0} xe^{-x^2}\,dx$$

Solution: In this case we've got infinities in both limits and so we'll need to split the integral up into two separate integrals. We can split the integral up at any point, so let's choose $a = 0$ since this will be a convenient point for the evaluate process. The integral is then,

$$\int_{-\infty}^{\infty} xe^{-x^2}\,dx = \int_{-\infty}^{0} xe^{-x^2}\,dx + \int_{0}^{\infty} xe^{-x^2}\,dx$$

We've now got to look at x each of the individual limits.

$$\int_{-\infty}^{0} xe^{-x^2}\,dx = \lim_{t\to-\infty} \int_{t}^{0} xe^{-x^2}\,dx$$
$$= \lim_{t\to-\infty} \left(-\frac{1}{2}e^{-x^2}\right)\Big|_{t}^{0}$$
$$= \lim_{t\to-\infty} \left(-\frac{1}{2}+\frac{1}{2}e^{-t^2}\right)$$
$$= -\frac{1}{2}$$

So, the first integral is convergent. Note that this does NOT mean that the second integral will also be convergent. So, let's take a look at that one.

$$\int_{0}^{\infty} xe^{-x^2}\,dx = \lim_{t\to\infty} \int_{0}^{t} xe^{-x^2}\,dx$$
$$= \lim_{t\to\infty} \left(-\frac{1}{2}e^{-x^2}\right)\Big|_{0}^{t}$$
$$= \lim_{t\to\infty} \left(-\frac{1}{2}e^{-t^2}+\frac{1}{2}\right)$$
$$= \frac{1}{2}$$

This integral is convergent and so since they are both convergent the integral we were actually asked to deal with is also convergent and its value is,

$$\int_{-\infty}^{\infty} xe^{-x^2}\,dx = \int_{-\infty}^{0} xe^{-x^2}\,dx + \int_{0}^{\infty} xe^{-x^2}\,dx = -\frac{1}{2}+\frac{1}{2}=0$$

Example: Determine if the following integral is convergent or divergent. If it is convergent find its value.

$$\int_{-2}^{\infty} \sin x dx$$

Solution: First convert to a limit.

$$\begin{aligned}\int_{-2}^{\infty} \sin x\,dx &= \lim_{t\to\infty}\int_{-2}^{t} \sin x\,dx \\ &= \lim_{t\to\infty}\left(-\cos x\right)\Big|_{-2}^{t} \\ &= \lim_{t\to\infty}\left(\cos 2-\cos t\right)\end{aligned}$$

This limit doesn't exist and so the integral is divergent. In most examples in a Calculus that are worked over infinite intervals the limit either exists or is infinite. However, there are limits that don't exist, as the previous example showed, so don't forget about those.

DISCONTINUOUS INTEGRAND

We now need to look at the second type of improper integrals that we'll be looking at in this section. These are integrals that have discontinuous integrands. The process here is basically the same with one on subtle difference. Here are the general cases that we'll look at for these integrals.

1. If $f(x)$ is continuous on the interval $[a, b]$ and not continuous at $x = b$ then,

$$\int_{a}^{b} f(x)dx = \lim_{t\to b}\int_{a}^{t} f(x)dx$$

provided the limit exists and is finite. Note as well that we do need to use a left hand limit here since the interval of integration is entirely on the left side of the upper limit.

2. If $f(x)$ is continuous on the interval $[a, b]$ and not continuous at then,

$$\int_{a}^{b} f(x)dx = \lim_{t\to b}\int_{t}^{b} f(x)dx$$

provided the limit exists and is finite. In this case we need to use a right hand limit here since the interval of integration is entirely on the right side of the lower limit.

3. If $f(x)$ is not continuous at $x = c$ where $a < c < b$ and $\int_{a}^{c} f(x)dx$ and $\int_{c}^{b} f(x)dx$ are both convergent then,

$$\int_a^b f(x)dx = \int_a^c f(x)dx + \int_c^b f(x)dx$$

As with the infinite interval case this requires BOTH of the integrals to be convergent in order for this integral to also be convergent. If either of the two integrals is divergent then so is this integral.

4. If $f(x)$ is not continuous at $x = a$ and $x = b$ and if $\int_a^c f(x)dx$ and $\int_c^b f(x)dx$ are both convergent then,

$$\int_a^b f(x)dx = \int_a^c f(x)dx + \int_c^b f(x)dx$$

Where c is any number. Again, this requires BOTH of the integrals to be convergent in order for this integral to also be convergent. Note that the limits in these cases really do need to be right or left handed limits. Since we will be working inside the interval of integration we will need to make sure that we stay inside that interval. This means that we'll used one-sided limits to make sure we stay inside the interval.

Let's do a couple of examples of these kinds of integrals.

Example: Determine if the following integral is convergent or divergent. If it is convergent find its value.

$$\int_0^3 \frac{1}{\sqrt{3-x}}dx$$

Solution: The problem point is the upper limit so we are in the first case above.

$$\begin{aligned}\int_0^3 \frac{1}{\sqrt{3-x}}dx &= \lim_{t\to 3^-}\int_0^t \frac{1}{\sqrt{3-x}}dx \\ &= \lim_{t\to 3^-}\left(-2\sqrt{3-x}\right)\Big|_0^t \\ &= \lim_{t\to 3^-}\left(2\sqrt{3}-2\sqrt{3-t}\right) \\ &= 2\sqrt{3}\end{aligned}$$

The limit exists and is finite and so the integral converges and the integrals value is $2\sqrt{3}$.

Example: Determine if the following integral is convergent or divergent. If it is convergent find its value.

$$\int_{-2}^3 \frac{1}{x^3}dx$$

Solution: This integrand is not continuous at $x = 0$ and so we'll need to split the integral up at that point.

$$\int_{-2}^3 \frac{1}{x^3}dx = \int_{-2}^0 \frac{1}{x^3}dx + \int_0^3 \frac{1}{x^3}dx$$

Now we need to look at each of these integrals and see if they are convergent.x

$$\int_{-2}^{0} \frac{1}{x^3}dx = \lim_{t\to 0^-}\int_{-2}^{t}\frac{1}{x^3}dx$$
$$= \lim_{t\to 0^-}\left(-\frac{1}{2x^2}\right)\Bigg|_{-2}^{t}$$
$$= \lim_{t\to 0^-}\left(-\frac{1}{2t^2}+\frac{1}{8}\right)$$
$$= -\infty$$

At this point we're done. One of the integrals is divergent that means the integral that we were asked to look at is divergent. We don't even need to bother with the second integral. Before leaving this section lets note that we can also have integrals that involve both of these cases. Consider the following integral.

Example: Determine if the following integral is convergent or divergent. If it is convergent find its value.

$$\int_0^\infty \frac{1}{x^2}dx$$

Solution: This is an integral over an infinite interval that also contains a discontinuous integrand. To do this integral we'll need to split it up into two integrals. We can split it up anywhere, but pick a value that will be convenient for evaluation purposes.

$$\int_0^\infty \frac{1}{x^2}dx = \int_0^1 \frac{1}{x^2}dx + \int_0^\infty \frac{1}{x^2}dx$$

In order for the integral in the example to be convergent we will need BOTH of these to be convergent. If one or both are divergent then the whole integral will also be divergent. We know that the second integral is convergent by the fact given in the infinite interval portion above. So, all we need to do is check the first integral.x

$$\int_0^1 \frac{1}{x^2}dx = \lim_{t\to 0^+}\int_t^1 \frac{1}{x^2}dx$$
$$= \lim_{t\to 0^+}\left(-\frac{1}{x}\right)\Bigg|_t^1$$
$$= \lim_{t\to 0^+}\left(-1+\frac{1}{t}\right)$$
$$= \infty$$

So, the first integral is divergent and so the whole integral is divergent.

COMPARISON TEST FOR IMPROPER INTEGRALS

Now that we've seen how to actually compute improper integrals we need to address one more topic about them. Often we aren't concerned with the actual value of these integrals. Instead we might only be interested in whether the integral is convergent or divergent. Also, there will be some integrals that we simply won't be able to integrate and yet we would still like to know if they converge or diverge.

To deal with this we've got a test for convergence or divergence that we can use to help us answer the question of convergence for an improper integral. We will give this test only for a sub-case of the infinite interval integral, however versions of the test exist for the other sub-cases of the infinite interval integrals as well as integrals with discontinuous integrands.

COMPARISON TEST

If $f(x) \geq g(x) \geq 0$ on the interval $[a, \infty]$ then,

- If $\int_a^\infty f(x)dx$ converges then so does $\int_a^\infty g(x)dx$.
- If $\int_a^\infty g(x)dx$ diverges then so does $\int_a^\infty f(x)dx$.

Note that if you think in terms of area the Comparison Test makes a lot of sense. $f(x)$ If is larger than $g(x)$ then the area under $f(x)$ must also be larger than the area under $g(x)$.

So, if the area under the larger function is finite (*i.e.* $\int_a^\infty f(x)dx$ converges) then the area under the smaller function must also be finite (*i.e.* $\int_a^\infty g(x)dx$ converges). Likewise, if the area under the smaller function is infinite (*i.e.* $\int_a^\infty g(x)dx$ diverges) then the area under the larger function must also be infinite (*i.e.* $\int_a^\infty f(x)dx$ diverges).

Be careful not to misuse this test. If the smaller function converges there is no reason to believe that the larger will also converge (after all infinity is larger than a finite number...) and if the larger function diverges there is no reason to believe that the smaller function will also converge. Let's work a couple of example using the comparison test. Note that all we'll be able to do is determine the convergence of the integral. We won't be able to determine the value of the integrals and so won't even bother with that.

Example: Determine if the following integral is convergent or divergent.

$$\int_2^\infty \frac{\cos^2}{x^2} dx$$

Solution: Let's take a second and think about how the Comparison Test works. If this integral is convergent then we'll need to find a larger function that also converges on the same interval. Likewise, if this integral is divergent then we'll need to find a smaller function that also diverges. So, it seems like it would be nice to have some idea as to whether the integral converges or diverges ahead of time so we will know whether we will need to look for a larger (and convergent) function or a smaller (and divergent) function.

To get the guess for this function let's notice that the numerator is nice and bounded and simply won't get too large. Therefore, it seems likely that the denominator will determine the convergence/divergence of this integral and we know that

$$\int_2^{\infty} \frac{1}{x^2}\, dx$$

converges since $p = 2 > 1$ by the fact in the previous section. So let's guess that this integral will converge. So we now know that we need to find a function that is larger than

$$\frac{\cos^2 x}{x^2}$$

and also converges. Making a fraction larger is actually a fairly simple process. We can either make the numerator larger or we can make the denominator smaller. In this case can't do a lot about the denominator. However we can use the fact that $0 \le \cos^2 x \le 1$ to make the numerator larger (*i.e.* we'll replace the cosine with something we know to be larger, namely 1). So,

$$\frac{\cos^2 x}{x^2} \le \frac{1}{x^2}$$

Now, as we've already noted

$$\int_2^{\infty} \frac{1}{x^2}\, dx$$

converges and so by the Comparison Test we know that

$$\int_2^{\infty} \frac{\cos^2 x}{x^2}\, dx$$

must also converge.

Example: Determine if the following integral is convergent or divergent.

$$\int_x^{\infty} \frac{1}{x + e^x}\, dx$$

Solution: Let's first take a guess about the convergence of this integral. As noted after the fact in the last section about

$$\int_a^{\infty} \frac{1}{x^p}\, dx$$

if the integrand goes to zero faster than $\frac{1}{x}$ then the integral will probably converge. Now, we've got an exponential in the denominator which is approaching infinity much faster than the x and so it looks like this integral should probably converge. So, we need a larger function that will also converge. In this case we can't really make the numerator larger and so we'll need to make the denominator smaller in order to make the function larger as a whole.

We will need to be careful however. There are two ways to do this and only one, in this case only one, of them will work for us. First, notice that since the lower limit of integration is 3 we can say that x^3 3 > 0 and we know that exponentials are always positive. So, the denominator is the sum of two positive terms and if we were to drop one of them the denominator would get smaller. This would in turn make the function larger.

The question then is which one to drop? Let's first drop the exponential. Doing this gives,

$$\frac{1}{x+e^x} < \frac{1}{x}$$

This is a problem however, since

$$\int_3^{\infty} \frac{1}{x} dx$$

diverges by the fact. We've got a larger function that is divergent. This doesn't say anything about the smaller function. Therefore, we chose the wrong one to drop.

Let's try it again and this time let's drop the x.

$$\frac{1}{x+e^x} < \frac{1}{x^e} = e^x$$

$$\begin{aligned}\int_3^{\infty} \mathbf{e}^{-x}\, dx &= \lim_{t\to\infty} \int_3^t \mathbf{e}^{-x}\, dx \\ &= \lim_{t\to\infty} \left(-\mathbf{e}^{-t} + \mathbf{e}^{-3}\right) \\ &= \mathbf{e}^{-3}\end{aligned}$$

Also,

So, $\int_3^{\infty} e^{-x} dx$ is convergent. Therefore, by the Comparison test

$$\int_3^{\infty} \frac{1}{x+e^x} dx$$

is also convergent.

Example: Determine if the following integral is convergent or divergent.

$$\int_3^{\infty} \frac{1}{x+e^{-x}} dx$$

Solution: This is very similar to the previous example with a couple of very important differences. First. notice that the exponential now goes to zero as x increases instead of growing larger as it did in the previous example (because of the negative in the exponent). Also note that the exponential is now subtracted off the x instead of added onto it.

The fact that the exponential goes to zero means that this time the x in the denominator will probably dominate the term and that means that the integral probably diverges. We will therefore need to find a smaller function that also diverges. Making fractions smaller is pretty much the same as making fractions larger. In this case we'll need to either make the numerator smaller or the denominator larger.

This is where the second change will come into play. As before we know that both x and the exponential are positive. However, this time since we are subtracting the exponential from the x if we were to drop the exponential the denominator will become larger and so the fraction will become smaller. In other words,

$$\frac{1}{x-e^{-x}} > \frac{1}{x}$$

and we know that

$$\int_3^\infty \frac{1}{x}\,dx$$

diverges and so by the Comparison Test we know that

$$\int_3^\infty \frac{1}{x-e^{x}}\,dx$$

must also diverge.

Example: Determine if the following integral is convergent or divergent.

$$\int_1^\infty \frac{1+3\sin^4(2x)}{\sqrt{x}}\,dx$$

Solution: First notice that as with the first example, the numerator in this function is going to be bounded since the sine is never larger than 1. Therefore, since the exponent on the denominator is less than 1 we can guess that the integral will probably diverge. We will need a smaller function that also diverges. We know that. In particular, this term is positive and so if we drop it from the numerator the numerator will get smaller. This gives,

$$\frac{1+3\sin^4(2x)}{\sqrt{x}} > \frac{1}{\sqrt{x}}$$

and

$$\int_1^\infty \frac{1}{\sqrt{x}}\,dx$$

diverges so by the Comparison Test

$$\int_1^\infty \frac{1+3\sin^4(2x)}{\sqrt{x}}\,dx$$

also diverges.

Okay, we've seen a few examples of the Comparison Test now. However, most of them worked pretty much the same way. All the functions were rational and all we did for most of them was add or subtract something from the numerator or denominator to get what we want. Let's take a look at an example that works a little differently so we don't get too locked into these ideas.

Example: Determine if the following integral is convergent or divergent.

$$\int_1^{\infty} \frac{e^{-x}}{x}\, dx$$

Solution: Normally, the presence of just an x in the denominator would lead us to guess divergent for this integral. However, the exponential in the numerator will approach zero so fast that instead we'll need to guess that this integral converges.

To get a larger function we'll use the fact that we know from the limits of integration that $x > 1$. This means that if we just replace the x in the denominator with 1 (which is always smaller than x) we will make the denominator smaller and so the function will get larger.

$$\frac{e^{-x}}{x} < \frac{e^{-x}}{1} = e^{-x}$$

and we can show that

$$\int_1^{\infty} e^{-x}\, dx$$

converges. In fact, we've already done this for a lower limit of 3 and changing that to a 1 won't change the convergence of the integral. Therefore, by the Comparison Test

$$\int_1^{\infty} \frac{e^{-x}}{x}\, dx$$

also converges.

We should also really work an example that doesn't involve a rational function since there is no reason to assume that we'll always be working with rational functions.

Example: Determine if the following integral is convergent or divergent.

$$\int_1^{\infty} e^{-x^2}\, dx$$

Solution: We know that exponentials with negative exponents die down to zero very fast so it makes sense to guess that this integral will be convergent. We need a larger function, but this time we don't have a fraction to work with so we'll need to do something different.

We'll take advantage of the fact that e^{-x} is a decreasing function. This means that

$$x_1 > x_2 \Rightarrow e^{-x_1} < e^{-x_2}$$

In other words, plug in a larger number and the function gets smaller. From the limits of integration we know that and this means that if we square x we will get larger. Or,

$$x^2 > x \text{ provided } x > 1$$

Note that we can only say this since $x > 1$. This won't be true if $x \leq 1$! We can now use the fact that e^{-x} is a decreasing function to get,

$$e^{-x2} < e^{-x}$$

So, e^{-x} is a larger function than e^{-x2} and we know that

$$\int_1^{\infty} e^{-x}\, dx$$

converges so by the Comparison Test we also know that

$$\int_1^{\infty} e^{-x^2}\, dx$$

is convergent.

The last two examples made use of the fact that $x > 1$. Let's take a look at an example to see how do we would have to go about these if the lower limit had been smaller than 1.

Example: Determine if the following integral is convergent or divergent.

$$\int_{\frac{1}{2}}^{\infty} \frac{e^{-x}}{x}\, dx$$

Solution: In this case we can't just replace x with $\frac{1}{2}$ in the denominator and get a larger function for all x in the interval of integration as we did in previous Example above. Remember that we need a function (that is also convergent) that is always larger than the given function..

To see why we can't just replace x with $\frac{1}{2}$ plug in $x = \frac{1}{2}$ into the denominator and compare this to what we would have if we plug in.

$$\frac{e^{-x}}{3/4} = \frac{4}{3} e^{-x} < e^{-x} = \frac{e^{-x}}{1/2}$$

So, for x's in the range $\frac{1}{2} \leq x < 1$ we won't get a larger function as required for use in the Comparison Test. However, this isn't the problem it might at first appear to be. We can always write the integral as follows,

$$\int_{\frac{1}{2}}^{\infty} \frac{\mathbf{e}^{-x}}{x} dx = \int_{\frac{1}{2}}^{1} \frac{\mathbf{e}^{-x}}{x} dx + \int_{1}^{\infty} \frac{\mathbf{e}^{-x}}{x} dx$$

$$= 0.34039 + \int_{1}^{\infty} \frac{\mathbf{e}^{-x}}{x} dx$$

We used Maple to get the value of the first integral. Now, if the second integral converges it will have a finite value and so the sum of two finite values will also be finite and so the original integral will converge. Likewise, if the second integral diverges it will either be infinite or not have a value at all and adding a finite number onto this will not all of a sudden make it finite or exist and so the original integral will diverge. Therefore, this integral will converge or diverge depending only on the convergence of the second integral.

APPROXIMATING DEFINITE INTEGRALS

In this chapter we've spent quite a bit to time on computing the values of integrals. However, not all integrals can be computed. A perfect example is the following definite integral.

$$\int_0^2 e^{x^2} dx$$

We now need to talk a little bit about estimating values of definite integrals. We will look at three different methods, although one should already be familiar to you from your Calculus I days. We will develop all three methods for estimating

$$\int_a^b f(x)dx$$

by thinking of the integral as an area problem and using known shapes to estimate the area under the curve.

Let's get first develop the methods and then we'll try to estimate the integral shown above.

MIDPOINT RULE

This is the rule that you should be somewhat familiar to you. We will divide the interval $[a, b]$ into n subintervals of equal width,

$$\Delta x = \frac{b-a}{n}$$

We will denote each of the intervals as follows,

$$[x_0, x_1], [x_1, x_2], \ldots, [x_{n-1}, x_n], \text{ where } x_0 = a \text{ and } x_n = b$$

Then for each interval let x_i^* be the midpoint of the interval. We then sketch in rectangles for each subinterval with a height of $f(x_i^*)$. Here is a graph showing the set up using $n = 6$.

We can easily find the area for each of these rectangles and so for a general n we get that,

$$\int_a^b f(x)dx \approx \Delta x\,(x_1^*) + \Delta x\, f(x_1^*) + \ldots + \Delta x\, f(x_1^*)$$

Or, upon factoring out a Δx we get the general Mid Point Rule.

$$\int_a^b f(x)\,dx \approx \Delta x\left[f(x_1^*) + f(x_2^*) + \ldots + f(x_n^*)\right]$$

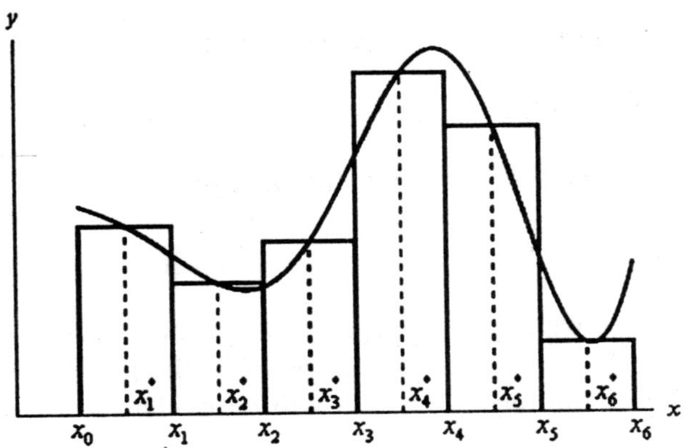

TRAPEZOID RULE

For this rule we will do the same set up as for the Midpoint Rule. We will break up the interval $[a, b]$ into n subintervals of width,

$$\Delta x = \frac{b-a}{n}$$

Then on each subinterval we will approximate the function with a straight line that is equal to the function values at either endpoint of the interval. Here is a sketch of this case for $n = 6$.

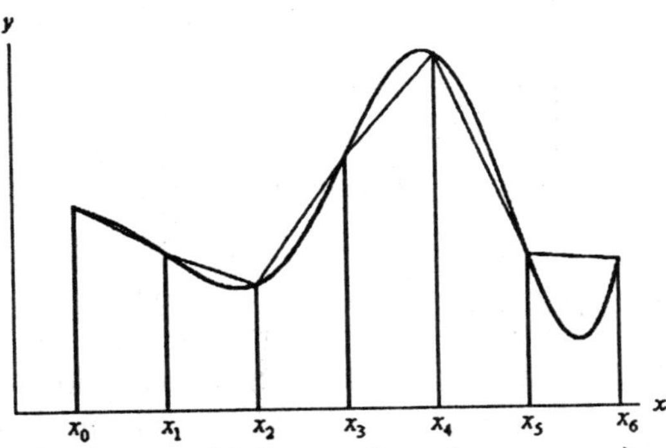

Each of these objects is a trapezoid (hence the rules name…) and as we can see some of them do a very good job of approximating the actual area under the curve and others don't do such a good job.

The area of the trapezoid in the interval $[x_{i-1}, x_i]$ is given by,

$$A_i = \frac{\Delta x}{2}(f(x_{i-1}) + f(x_i))$$

So, if we use n subintervals the integral is approximately,

$$\int_a^b f(x)dx \approx \frac{\Delta x}{2}(f(x_0) + f(x_i)) + \frac{\Delta x}{2}(f(x_1) + f(x_2)) + \cdots + \frac{\Delta x}{2}(f(x_{n-1}) + f(x_n))$$

Upon doing a little simplification we arrive at the general Trapezoid Rule.

$$\int_a^b f(x)dx \approx \frac{\Delta x}{2}[(f(x_0)+2f(x_1)+2f(x_2)+\cdots+2f(f(x_{n-1})+f(x_n)]$$

Note that all the function evaluations, with the exception of the first and last, are multiplied by 2.

SIMPSON'S RULE

This is the final method we're going to take a look at and in this case we will again divide up the interval $[a, b]$ into n subintervals. However unlike the previous two methods we need to require that n be even. The reason for this will be evident in a bit. The width of each subinterval is,

$$\Delta x = \frac{b-a}{n}$$

In the Trapezoid Rule we approximated the curve with a straight line. For Simpson's Rule we are going to approximate the function with a quadratic and we're going to require that the quadratic agree with three of the points from our subintervals. Below is a sketch of this using $n = 6$. Each of the approximations is colored differently so we can see how they actually work.

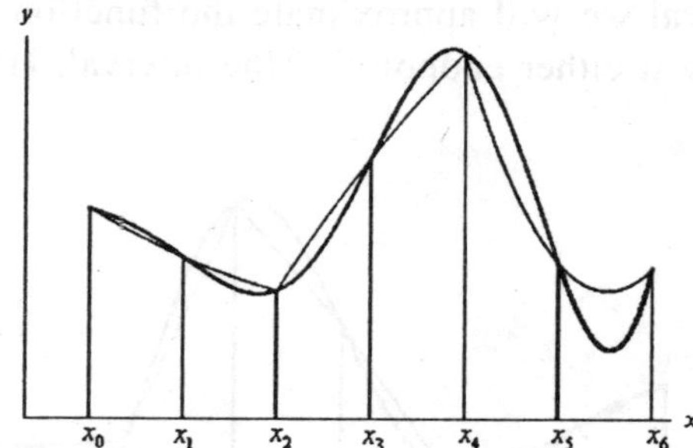

Notice that each approximation actually covers two of the subintervals. This is the reason for requiring n to be even. Some of the approximations look more like a line than a quadratic, but they really are quadratics. Also note that some of the approximations do a better job than others. It can be shown that the area under the approximation on the intervals $[x_{i-1}, x_i]$ and $[x_i, x_{i+1}]$ is,

$$A_i = \frac{\Delta x}{3}(f(x_{i-1})+4f(x_i)+f(x_{i+1}))$$

If we use n subintervals the integral is then approximately,

$$\int f(x)dx \approx \frac{\Delta x}{3}(f(x)+4f(x_1)+f(x_2))+\frac{\Delta x}{3}(f(x_2)+4f(x_3)+f(x_4))$$

$$+\ldots\frac{\Delta x}{3}(f(x_{n-2})+4f(x_{n-1})-f(x_n))$$

Upon simplifying we arrive at the general Simpson's Rule.

$$\int_a^b f(x) \approx \frac{\Delta x}{3}[f(x_0)+4f(x_i)+2f(x_2)+\cdots+2f(x_{n-2})+4f(x_{n-1})+f(x_n)]$$

In this case notice that all the function evaluations at points with odd subscripts are multiplied by 4 and all the function evaluations at points with even subscripts (except for the first and last) are multiplied by 2. If you can remember this, this a fairly easy rule to remember.

Okay, it's time to work an example and see how these rules work.

Example: Using $n = 4$ and all three rules to approximate the value of the following integral.

$$\int_0^2 e^{x^2}\, dx$$

Solution: First, for reference purposes, Maple gives the following value for this integral.

$$\int_0^2 e^{x^2}\, dx = 16.45262776$$

In each case the width of the subintervals will be,

$$\Delta x = \frac{2-0}{4} = \frac{1}{2}$$

and so the subintervals will be,

$$[0, 0.5], [0.5, 1], [1, 1.5], [1.5, 2]$$

Let's go through each of the methods.

Midpoint Rule

$$\int_0^2 e^2 dx \approx \frac{1}{2}\left(e^{(0.25)^2} + e^{(0.75)^2} + e^{(1.25)^2} + e^{(1.75)^2}\right) = 14.48561253$$

Remember that we evaluate at the midpoints of each of the subintervals here! The Midpoint Rule has an error of 1.96701523.

Trapezoid Rule

$$\int_0^2 e^2 dx \approx \frac{1/2}{2}\left(e^{(0)^2} + 2e^{(0.5)^2} + 2e^{(1)^2} + 2e^{(1.5)^2} + e^{(2)^2}\right) = 2064455905$$

The Trapezoid Rule has an error of 4.19193129

Simpson's Rule

$$\int_0^2 e^2 dx \approx \frac{1/2}{2}\left(e^{(0)^2} + 4e^{(0.5)^2} + 2e^{(1)^2} + 4e^{(1.5)^2} + e^{(2)^2}\right) = 17.35362645$$

The Simpson's Rule has an error of 0.90099869.

None of the estimations in the previous example are all that good. The best approximation in this case is from the Simpson's Rule and yet it's still had an error of almost 1. To get a better estimation we would need to use a larger n. So, for completeness sake here are the estimates for some larger value of n.

			Midpoint	*Trapezoid*	*Simpson's*	
n	***Approx.***	***Error***	***Approx.***	***Error***	***Approx.***	***Error***
8	15.9056767	0.5469511	17.5650858	1.1124580	16.5385947	0.0859669
16	16.3118539	0.1407739	16.7353812	0.2827535	16.4588131	0.0061853
32	16.4171709	0.0354568	16.5236176	0.0709898	16.4530297	0.0004019
64	16.4437469	0.0088809	16.4703942	0.0177665	16.4526531	0.0000254
128	16.4504065	0.0022212	16.4570706	0.0044428	16.4526294	0.0000016

In this case we were able to determine the error for each estimate because we could get our hands on the exact value. Often this won't be the case and so we'd next like to look at error bounds for each estimate. These bounds will give the largest possible error in the estimate, but it should also be pointed out that the actual error may be significantly smaller than the bound. The bound is only there so we can say that we know the actual error will be less than the bound. So, suppose that $|f''(x)| \le K$ and $|f^{(4)}(x)| \le M$ for then if E_M, E_T, and E_S are the actual errors for the Midpoint, Trapezoid and Simpson's Rule we have the following bounds,

$$|E_M| \le \frac{K(b-a)^3}{24n^2} \qquad |E_r| \le \frac{K(b-a)^3}{12n^2} \qquad |E_s| \le \frac{M(b-a)^5}{180n^2}$$

Example: Determine the error bounds for the estimations in the last example.

Solution: We already know that, $n = 4$, $a = 0$ and $b = 2$ so we just need to compute K (the largest value of the second derivative) and M (the largest value of the fourth derivative). This means that we'll need the second and fourth derivative of $f(x)$.

$$f''(x) = 2e^{x^2}(1 + 2x^2)$$
$$f^{(4)}(x) = 4e^{x^2}(3 + 12x^2 + 4x^4)$$

Here is a graph of the second derivative.

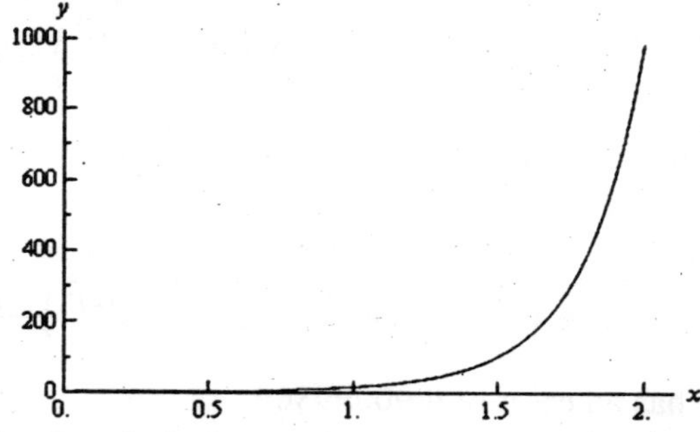

Here is a graph of the fourth derivative.

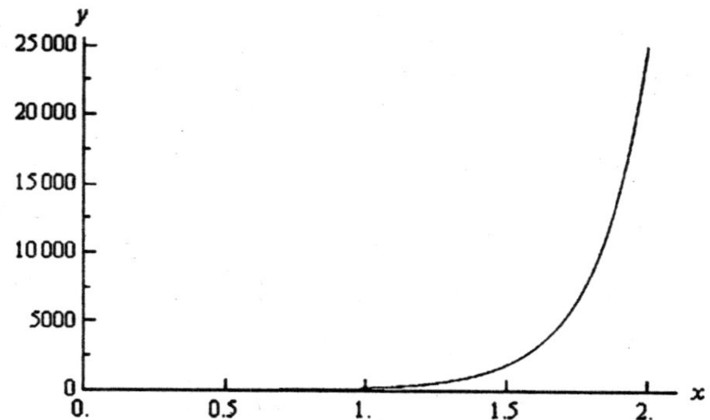

So, from these graphs it's clear that the largest value of both of these are at. So,

$$f''(2) = 982.76667 \quad \Rightarrow \quad K = 983$$

$$f^{(4)}(2) = 25115.14901 \quad \Rightarrow \quad M = 25116$$

We rounded to make the computations simpler.

Here are the bounds for each rule.

$$|E_M| \le \frac{983(2-0)^3}{24(4)^3} = 20.4791666667$$

$$|E_r| \le \frac{983(2-0)^2}{12(4)^2} = 40.9583333333$$

$$|E_S| \le \frac{25116\,(2-0)^5}{180\,(4)^4} = 4.36041666667$$

In each case we can see that the errors are significantly smaller than the actual bounds.

Chapter 2

Line Integral

In mathematics, a line integral sometimes called a path integral or curve integral is an integral where the function to be integrated is evaluated along a curve. Various different line integrals are in use. In the case of a closed curve it is also called a contour integral. The function to be integrated may be a scalar field or a vector field. The value of the line integral is the sum of values of the field at all points on the curve, weighted by some scalar function on the curve (commonly arc length or, for a vector field, the scalar product of the vector field with a differential vector in the curve).

This weighting distinguishes the line integral from simpler integrals defined on intervals. Many simple formulas in physics (for example, $W = \vec{F}.\vec{s}$) have natural continuous analogs in terms of line integrals $\left(W = \int_C \vec{F}.d\vec{s}\right)$. The line integral finds the work done on an object moving through an electric or gravitational field.

VECTOR FIELDS

We need to start this chapter off with the definition of a vector field as they will be a major component of both this chapter and the next. Let's start off with the formal definition of a vector field.

Definition: A vector field on two (or three) dimensionl space is a function $\vec{F}$ that assigns to each point (x, y) (or (x, y, z)) a two (or three dimensional) vector given by $\vec{F}(x, y)$ (or $\vec{F}(x, y, z)$).

That may not make a lot of sense, but most people do know what a vector field is, or at least they've seen a sketch of a vector field. If you've seen a current sketch giving the direction and magnitude of a flow of a fluid or the direction and magnitude of the winds then you've seen a sketch of a vector field.

The standard notation for the function $\vec{F}$ is,

$$\vec{F}(x,y) = P(x,y)\vec{i} + Q(x,y)\vec{j}$$
$$\vec{F}(x,y,z) = P(x,y,z)i + Q(x,y,z)\vec{j} + R(x,y,z)\vec{k}$$

depending on whether or not we're in two or three dimensions. The function P, Q, R (if it is present) are sometimes called scalar functions.

Let's take a quick look at a couple of examples.

Example: Sketch each of the following direction fields.

1. $\vec{F}(x,y) = -y\vec{i} + x\vec{j}$

2. $\vec{F}(x,y,z) = 2x\vec{i} - 2y\vec{j} - 2x\vec{k}$

Solution:

1. $\vec{F}(x,y) = -y\vec{i} + x\vec{j}$

To graph the vector field we need to get some "values" of the function. This means plugging in some points into the function. Here are a couple of evaluations.

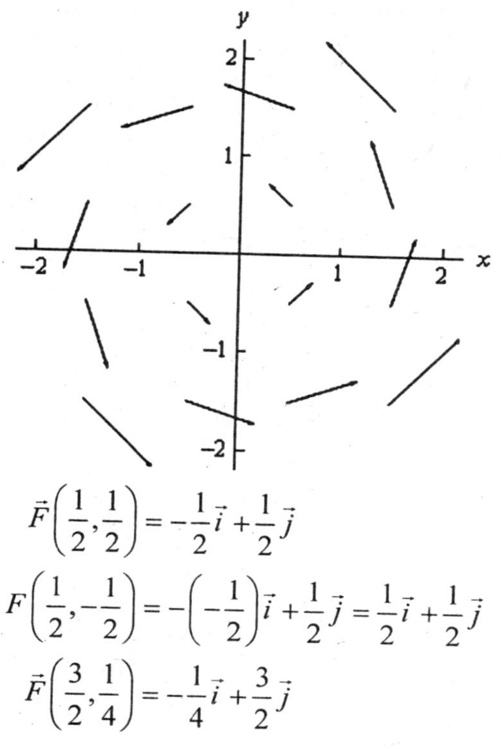

$$\vec{F}\left(\frac{1}{2},\frac{1}{2}\right) = -\frac{1}{2}\vec{i} + \frac{1}{2}\vec{j}$$

$$F\left(\frac{1}{2},-\frac{1}{2}\right) = -\left(-\frac{1}{2}\right)\vec{i} + \frac{1}{2}\vec{j} = \frac{1}{2}\vec{i} + \frac{1}{2}\vec{j}$$

$$\vec{F}\left(\frac{3}{2},\frac{1}{4}\right) = -\frac{1}{4}\vec{i} + \frac{3}{2}\vec{j}$$

So, just what do these evaluations tell us? Well the first one tells us that at the point $\left(\frac{1}{2},\frac{1}{2}\right)$ we will plot the vector $-\frac{1}{2}\vec{i}+\frac{1}{2}\vec{j}$. Likewise, the third evaluation tells us that at the point $\left(\frac{3}{2},\frac{1}{4}\right)$ we will plot the vector $-\frac{1}{4}\vec{i}+\frac{3}{2}\vec{j}$.

We can continue in this fashion plotting vectors for several points and we'll get the following sketch of the vector field.

If we want significantly more points plotted then it is usually best to use a computer aided graphing system such as Maple or Mathematica. Here is a sketch with many more vectors included that was generated with Mathematica.

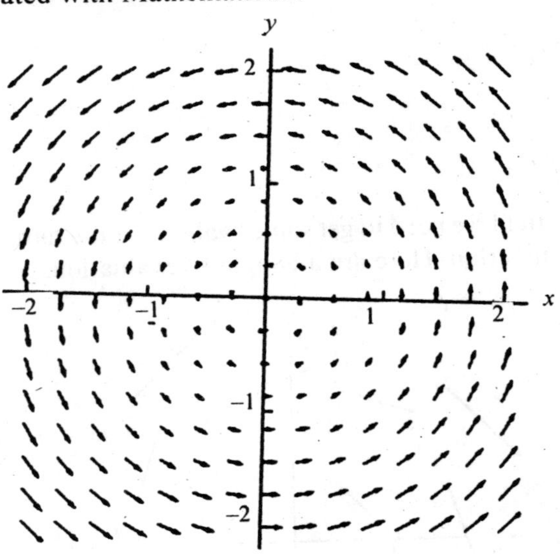

2. $\vec{F}(x,y,z)=2x\vec{i}-2y\vec{j}-2x\vec{k}$

In the case of three dimensional vector fields it is almost always better to use Maple, Mathematica, or some other such tool. Despite that let's go ahead and do a couple of evaluations anyway.

$$\vec{F}(1,-3,2)=2\vec{i}+6\vec{j}-2\vec{k}$$
$$\vec{F}(0,5,3)=-10j$$

Notice that z only affect the placement of the vector in this case and does not affect the direction or the magnitude of the vector. Sometimes this will happen so don't get excited about it when it does. Recall that given a function $f(x,y,z)$ the gradient vector is defined by,

$$\nabla f = \left\langle f_x, f_y, f_z \right\rangle$$

This is a vector field and is often called a gradient vector field.

In these cases the function $f(x, y, z)$ is often called a scalar function to differentiate it from the vector field.

Example: Find the gradient vector field of the following functions.

1. $f(x, y) = x^2 \sin(5y)$
2. $f(x, y, z) = ze^{-xy}$

Solution:

1. $f(x, y) = x^2 \sin(5y)$

Note that we only gave the gradient vector definition for a three dimensional function, but don't forget that there is also a two dimension definition. All that we need to drop off the third component of the vector.

Here is the gradient vector field for this function.

$$\nabla f = \left\langle 2x\sin(5y), 5x^2\cos(5y) \right\rangle$$

2. $f(x, y, z) = ze^{-xy}$

There isn't much to do here other than take the gradient.

$$\nabla f = \left\langle -yze^{-xy}, -xze^{-xy}, e^{-xy} \right\rangle$$

Let's do another example that will illustrate the relationship between the gradient vector field of a function and its contours.

Example: Sketch the gradient vector field for $f(x, y) = x^2 + y^2$ as well as several contours for this function.

Solution: Recall that the contours for a function are nothing more than curves defined by,

$$f(x, y) = k$$

for various values of *k*. So, for our function the contours are defined by the equation,

$$x^2 + y^2 = k$$

and so they are circles centered at the origin with radius $\sqrt{k}$.

Here is the gradient vector field for this function.

$$\nabla f(x, y) = 2xi + 2yj$$

Here is a sketch of several of the contours as well as the gradient vector field.

Notice that the vectors of the vector field are all perpendicular (or orthogonal) to the contours. This will always be the case when we are dealing with the contours of a function as well as its gradient vector field. The k's we used for the graph above were 1.5, 3, 4.5, 6, 7.5, 9, 10.5, 12, and 13.5. Now notice that as we increased k by 1.5 the contour curves get closer together and that as the contour curves get closer together the larger vectors become. In other words, the closer the contour curves are (as k is increased by a fixed amount) the faster the function is changing at that point. Also recall that the direction of fastest change for a function is given by the gradient vector at that point. Therefore, it should make sense that the two ideas should match up as they do here.

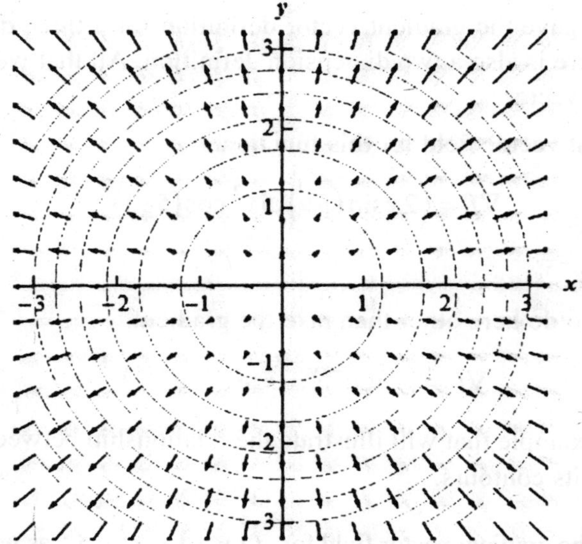

The final topic of this section is that of conservative vector fields. A vector field $\vec{F}$ is called a conservative vector field if there exists a function f such that $\vec{F} = \nabla f$. If $\vec{F}$ is a conservative vector field then the function, f, is called a potential function for $\vec{F}$.

All this definition is saying is that a vector field is conservative if it is also a gradient vector field for some function. For instance the vector field $\vec{F} = y\vec{i} + x\vec{j}$ is a conservative vector field with a potential function of $f(x, y) = xy$ because $\nabla f = \langle y, x \rangle$.

On the other hand, $\vec{F} = -y\vec{i} + x\vec{j}$ is not a conservative vector field since there is no function

f such that $\vec{F} = \nabla f$ .If you're not sure that you believe this at this point be patient, we will be able to prove this in a couple of sections. In that section we will also show how to find the potential function for a conservative vector field.

LINE INTEGRALSPART I

In this section we are now going to introduce a new kind of integra!. However, before we do that it is important to note that you will need to remember how to parameterize equations, or put another way, you will need to be able to write down a set of parametric equations for a given curve. If you need some review you should go back and review some of the basics of parametric equations and curves. Here are some of the more basic curves that we'll need to know how to do as well as limits on the parameter if they are required.

Curve	*Parametric Equations*	
(Ellipse)	Counter-Clockwise	Clockwise
$\frac{x^2}{a^2} + \frac{y^2}{b^2} = 1$	$x = a\cos(t)$ $y = b\sin(t)$ $0 \le t \le 2\pi$	$x = a\cos(t)$ $y = -b\sin(t)$ $0 \le t \le 2\pi$
(Circle)	Counter-Clockwise	Clockwise
$x^2 + y^2 = r^2$	$x = r\cos(t)$ $y = r\sin(t)$ $0 \le t \le 2\pi$	$x = r\cos(t)$ $y = -r\sin(t)$ $0 \le t \le 2\pi$
$y = f(x)$	$x = t$ $y = f(t)$	
$x = g(y)$	$y = g(t)$ $y = t$	
Line Segment From (x_0, y_0, z_0) to (x_1, y_1, z_1)	$\vec{r}(t) = (1-t)\langle x_0, y_0, z_0\rangle + t\langle x_1, y_1, z_1\rangle, 0 \le t \le 1$ or $x = (1-t)x_0 + tx_1$ $y = (1-t)y_0 + ty_1, \quad 0 \le t \le 1$ $z = (1-t)z_0 + tz_1$	

With the final one we gave both the vector form of the equation as well as the parametric form and if we need the two-dimensional version then we just drop the z components. In fact, we will be using the two-dimensional version of this in this section.

For the ellipse and the circle we've given two parameterizations, one tracing out the curve clockwise and the other counter-clockwise. As we'll eventually see the direction that the curve is traced out can, on occasion, change the answer. Also, both of these "start" on the positive x-axis at $t = 0$.

Now let's move on to line integrals $f(x)$. We integrated, a function of a single variable, over an interval $[a,b]$. In this case we were thinking of x as taking all the values in this interval starting at a and ending at b. With line integrals we will start with integrating the function $f(x,y)$, a function of two variables, and the values of x and y that we're going to use will be the points, (x,y), that lie on a curve C.

Note that this is different from the double integrals that we were working with in the previous chapter where the points came out of some two-dimensional region. Let's start with the curve C that the points come from. We will assume that the curve is *smooth* (defined shortly) and is given by the parametric equations,

$$x = h(t) \quad y = g(t) \quad a \le t \le b$$

We will often want to write the parameterization of the curve as a vector function. In this case the curve is given by,

$$\vec{r}(t) = h(t)\vec{i} + g(t)\vec{j} \quad a \le t \le b$$

The curve is called smooth if $\vec{r}(t)$ is continuous and $\vec{r}(t) \neq 0$ for all t. The line integral of $f(x,y)$ along C is denoted by,

$$\int_C f(x,y)\,ds$$

We use a ds here to acknowledge the fact that we are moving along the curve, C, instead of the x-axis (denoted by dx) or the y-axis (denoted by dy). Because of the ds this is sometimes called the line integral of f with respect to arc length. We've seen the notation ds before. When we looked at the arc length of a curve given by parametric equations we found it to be,

$$L = \int_a^b ds, \qquad where\ ds = \sqrt{\left(\frac{dx}{dt}\right)^2 + \left(\frac{dy}{dt}\right)^2}\, dt$$

It is no coincidence that we use ds for both of these problems. The ds is the same for both the arc length integral and the notation for the line integral. So, to compute a line integral we will convert everything over to the parametric equations. The line integral is then,

$$\int_C f(x,y)\,ds = \int_a^b f(h(t),g(t))\sqrt{\left(\frac{dx}{dt}\right)^2 + \left(\frac{dy}{dt}\right)^2}\, dt$$

Don't forget to plug the parametric equations into the function as well. If we use the vector form of the parameterization we can simplify the notation up somewhat by noticing that,

$$\sqrt{\left(\frac{dx}{dt}\right)^2+\left(\frac{dy}{dt}\right)^2}=\|\vec{r}'(t)\|$$

where $\|\vec{r}(t)\|$ is the magnitude or norm of $\vec{r}(t)$. Using this notation the line integral becomes,

$$\int_C f(x,y)\,ds=\int_a^b f(h(t),g(t))\|\vec{r}(t)\|dt$$

Note that as long as the parameterization of the curve C is traced out exactly once at t increases from a to b the value of the line integral will be independent of the parameterization of the curve.

Example: Evaluate $\int_C xy^4\,ds$ where C is the right half of the circle, $x^2+y^2=16$. rotated in the counter clockwise direction.

Solution: We first need a parameterization of the circle. This is given by,

$$x=4\cos t$$

$$y=4\sin t$$

We now need a range of t's that will give the right half of the circle. The following range of t's will do this.

$$-\frac{\pi}{2}\le t\le\frac{\pi}{2}$$

Now, we need the derivatives of the parametric equations and let's compute ds.

$$\frac{dx}{dt}=-4\sin t$$

$$\frac{dy}{dt}=4\cos t$$

$$ds=\sqrt{16\sin^2 t+16\cos^2 t}\,dt=4dt$$

The line integral is then,

$$\int_c xy^4\,ds=\int_{-\pi/2}^{\pi/2}4\cos t(4\sin t)^4(4)dt$$

$$=4096\int_{-\pi/2}^{\pi/2}\cos t\sin^4 t dt$$

$$=\frac{4096}{5}\sin^5 t\Big|_{-\frac{\pi}{2}}^{\frac{\pi}{2}}$$

$$=\frac{8192}{5}$$

Next we need to talk about line integrals over piecewise smooth curves.A piecewise smooth curve is any curve that can be written as the union of a finite number of smooth curves, $C_1, \ldots, C_n$ where the end point of is the starting point of C_t. Below is an illustration of a piecewise smooth curve.

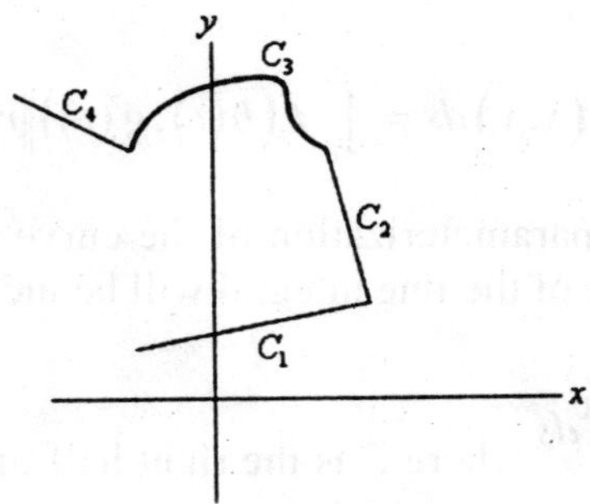

Evaluation of line integrals over piecewise smooth curves is a relatively simple thing to do. All we do is evaluate the line integral over each of the pieces and then add them up. The line integral for some function over the above piecewise curve would be,

$$\int_C f(x,y)\,ds = \int_{C_1} f(x,y)\,ds + \int_{C_2} f(x,y)\,ds + \int_{C_3} f(x,y)\,ds + \int_{C_4} f(x,y)\,ds$$

Let's see an example of this.

Example: Evaluate where $\int_C 4x^3\,ds$ C is the curve shown below.

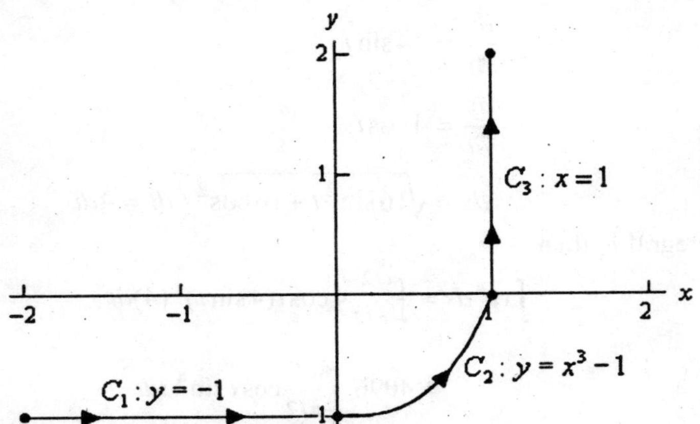

Solution: So, first we need to parameterize each of the curves.

$$C_1 : \; x = t, \; y = -1, \quad -2 \le t \le 0$$

$$C_2 : \; x = t, \; y = t^3 - 1, \quad 0 \le t \le 1$$

$$C_3 : \; x = t, \; y = t, \quad 0 \le t \le 2$$

Now let's do the line integral over each of these curves.

$$\int_{c_1} 4x^3 ds = \int_{-2}^{0} 4t^2\sqrt{(1)^2+(0)^2}\, ds = \int_{-2}^{0} 4t^3 dt = t^4\Big|_{-2}^{0} = -16$$

$$\int_{c_2} 4x^3 ds = \int_0^1 4t^3\sqrt{(1)^2+(3t^2)^2}\, dt$$

$$= \int_0^1 4t^3\sqrt{1+9t^4}\, dt$$

$$= \frac{1}{9}\left(\frac{2}{3}\right)(1+9t^4)\Big|_0^1 = \frac{2}{27}\left(10^{\frac{3}{2}}-1\right) = 2.268$$

Finally, the line integral that we were asked to compute is,

$$\int_c 4x^3 ds = \int_{c_1} 4x^3 ds + \int_{c_2} 4x^3 ds + \int_{c_3} 4x^3 ds$$

$$= -16 + 2.268 + 8$$

$$= -5.732$$

Notice that we put direction arrows on the curve in the above example. The direction of motion along a curve *may* change the value of the line integral as we will see in the next section.

Also note that the curve can be thought of a curve that takes us from the point $(-2,-1)$ to the point $(2,1)$. Let's first see what happens to the line integral if we change the path between these two points.

Example: Evaluate were $\int_c 4x^3 ds$ is the line segment from $(-2,-1)$ to $(2,1)$.

Solution: From the parameterization formulas at the start of this section we know that the line segment start at $(-2,-1)$ and ending at $(2,1)$ is given by,

$$\vec{r}(t) = (1-t)\langle -2,-1\rangle + t\langle 1,2\rangle$$

$$= \langle -2+3t, -1+3t\rangle$$

for $0 \le t \le 1$. This means that the individual parametric equations are,

$$x = -2+3t$$

$$y = -1+3t$$

Using this path the line integral is,

$$\int_C 4x^3 ds = \int_0^1 4(-2+3t)^3 \sqrt{9+9}\, dt$$

$$= 12\sqrt{2}\left(\frac{1}{12}\right)(-2+3t)^4\Big|_0^1$$

$$= 12\sqrt{2}\left(-\frac{5}{4}\right)$$

$$= -15\sqrt{2} = -21.213$$

When doing these integrals don't forget simple substitutions to avoid having to do things like cubing out that term. Cubing it out is not that difficult, but is more work than a simple substitution.

So, the previous two examples seem to suggest that if we change the path between two points then the value of the line integral (with respect to arc length) will change.

While this will happen fairly regularly we can't assume that it will always happen. In a later section we will investigate this idea in more detail

Example: Evaluate $\int_C 4x^3 ds$ were C is the line segment from $(-2,-1)$ to $(2,1)$.

Solution: This one isn't much different, work wise, from the previous example. Here is the parameterization of the curve.

$$\vec{r}(t) = (1-t)\langle 1,2\rangle + t\langle -2,-1\rangle$$
$$= \langle 1-3t, 2-3t\rangle$$

for $0 \le t \le 1$. Remember that we are switch the direction of the curve and this will also change the parameterization so we can make sure that we start/end at the proper point.

Here is the line integral.

$$\int_C 4x^3 ds = \int_0^1 4(1-3t)^3 \sqrt{9+9}$$

$$= 12\sqrt{2}\left(\frac{1}{12}\right)(-2+3t)^4\Big|_0^1$$

$$= 12\sqrt{2}\left(-\frac{5}{4}\right)$$

$$= -15\sqrt{2} = -21.213$$

So, it looks like when we switch the direction of the curve the line integral (with respect to arc length) will not change. This will always be true for these kinds of line integrals. However, there are other kinds of line integrals in which this won't be the case. We will see more examples of this in the next couple of sections so don't get it into your head that changing the direction will never change the value of the line integral.

Before working another example let's formalize this idea up somewhat. Let's suppose that the curve C has the parameterization $x = h(t), y = g(t)$. Let's also suppose that the initial point on the curve is A and the final point on the curve is B. The parameterization $x = h(t), y = g(t)$ will then determine an orientation for the curve where the positive direction is the direction that is traced out as t increases. Finally, let $-C$ be the curve with the same points as C, however in this case the curve has B as the initial point and A as the final point, again t is increasing as we traverse this curve. In other words, given a curve C, the curve $-C$ is the same curve as C except the direction has been reversed.

We then have the following fact about line integrals with respect to arc length.

Fact

$$\int_C f(x,y)\,ds = \int_{-C} f(x,y)\,ds$$

So, for a line integral with respect to arc length we can change the direction of the curve and not change the value of the integral. This is a useful fact to remember as some line integrals will be easier in one direction than the other.

Example: Evaluate for each of the following curves.

1. $C_1 : y = x^2, -1 \le x \le 1$
2. C_2: The line segment from $(-1,1)$ to $(1,1)$.
3. C_3: The line segment from $(1,1)$ to $(-1,1)$.

Solution: Before working any of these line integrals let's notice that all of these curves are paths that connect the points $(-1,1)$ and $(1,1)$. Also notice that $C_3 = -C_2$ and so by the fact above these two should give the same answer. Here is a sketch of the three curves and note that the curves illustrating and have been separated a little to show that they are separate curves in some way even thought they are the same line.

1. $C_1 : y = x^2, -1 \le x \le 1$

Here is a parameterization for this curve.

$$C_1 : x = t, y = t^2, -1 \le t \le 1$$

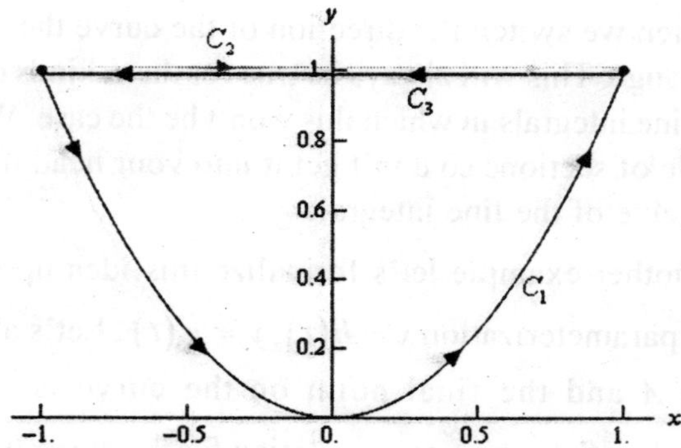

Here is the line integral.

$$\int_C x\,ds = \int_{-1}^{1} t\sqrt{1+4t^2}\,dt = \frac{1}{12}\left(1+4t^2\right)^{\frac{3}{2}}\Bigg|_{-1}^{1} = 0$$

2. C_2: The line segment from $(-1,1)$ to $(1,1)$.

There are two parameterizations that we could use here for this curve. The first is to use the formula we used in the previous couple of examples. That parameterization is,

$$C_2 : \vec{r}(t) = (1-t)\langle -1,1\rangle + t\langle 1,1\rangle$$
$$= \langle 2t-1, 1\rangle$$

for $0 \le t \le 1$.

Sometimes we have no choice but to use this parameterization. However, in this case there is a second (probably) easier parameterization. The second one uses the fact that we are really just graphing a portion of the line $y = 1$. Using this the parameterization is,

$$C_2 : x = t, y = 1, -1 \le t \le 1$$

This will be a much easier parameterization to use so we will use this. Here is the line integral for this curve.

$$\int_C x\,ds = \int_{-1}^{1} t\sqrt{1+0}\,dt = \frac{1}{2}t^2\Bigg|_{-1}^{1} = 0$$

Note that this time, unlike the line integral we worked with in Examples we got the same value for the integral despite the fact that the path is different. This will happen on occasion. We should also not expect this integral to be the same for all paths between these two points. At this point all we know is that for these two paths the line integral will have the same value. It is completely possible that there is another path between these two points that will give a different value for the line integral.

3. C_3: The line segment from $(1,1)$ to $(-1,1)$.

Now, according to our fact above we really don't need to do anything here since we know that $C_3 = -C_2$. The fact tells us that this line integral should be the same as the second part (*i.e.* zero). However, let's verify that, plus there is a point we need to make here about the parameterization.

Here is the parameterization for this curve.

$$\begin{aligned} C_3 : \vec{r}(t) &= (1-t)\langle 1,1\rangle + t\langle -1,1\rangle \\ &= \langle 1-2t, 1\rangle \end{aligned}$$

for $0 \le t \le 1$.

Note that this time we can't use the second parameterization that we used in part (b) since we need to move from right to left as the parameter increases and the second parameterization used in the previous part will move in the opposite direction.

Here is the line integral for this curve.

$$\int_{C_3} x\,ds = \int_{-1}^{1} (1-2t)\sqrt{4+0}\,dt = 2\left(t-t^2\right)\Big|_0^1 = 0$$

Sure enough we got the same answer as the second part.

To this point in this section we've only looked at line integrals over a two-dimensiona curve. However, there is no reason to restrict ourselves like that. We can do line integrals ovei three-dimensional curves as well.

Let's suppose that the three-dimensional curve C is given by the parameterization,

$$\begin{aligned} x &= x(t) \\ y &= y(t) \\ z &= z(t) \\ z &\le t \le b \end{aligned}$$

then the line integral is given by,

$$\int_C f(x,y,z)\,ds = \int_a^b f(x(t),y(t),z(t))\sqrt{\left(\frac{dx}{dt}\right)^2 + \left(\frac{dy}{dt}\right)^2 + \left(\frac{dz}{dt}\right)^2}\,dt$$

Note that often when dealing with three-dimensional space the parameterization will be given as a vector function.

$$\vec{r}(t) = \langle x(t), y(t), z(t)\rangle$$

Notice that we changed up the notation for the parameterization a little. Since we rarely use the function names we simply kept the x, y, and z and added on the (t)part to denote that they may be functions of the parameter.

Also notice that, as with two-dimensional curves, we have,

$$\sqrt{\left(\frac{dx}{dt}\right)^2+\left(\frac{dy}{dt}\right)^2+\left(\frac{dz}{dt}\right)^2}=\left\|\vec{r}'(t)\right\|$$

and the line integral can again be written as,

$$\int_C f(x,y,z)\,ds=\int_a^b f(x(t),y(t),z(t))\left\|\vec{r}'(t)\right\|dt$$

So, outside of the addition of a third parametric equation line integrals in three-dimensional space work the same as those in two-dimensional space. Let's work a quick example.

Example: Evaluate $\int_C xyz\,ds$ where C is the helix given by $\left\|\vec{r}(t)\right\|=\langle\cos(t),\sin(t),3t\rangle$, $0\le t\le 4\pi$.

Solution: Note that we first saw the vector equation for a helix back in the Vector Functions section. Here is a quick sketch of the helix.

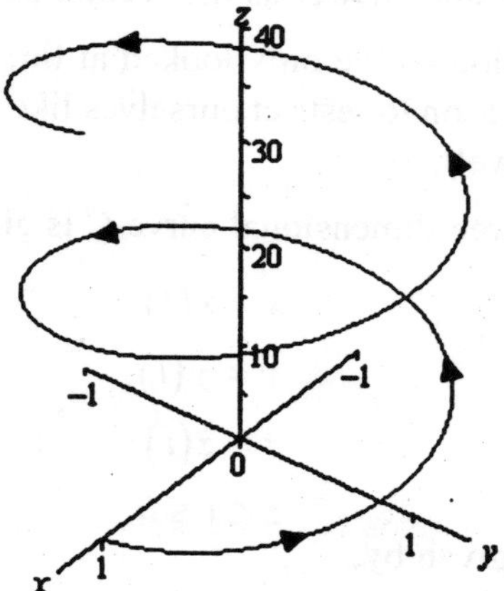

Here is the line integral.

$$\int_C xyz\,ds=\int_0^{4\pi}3t\cos(t)\sin(t)\sqrt{\sin^2 t+\cos^2 t+9}dt$$

$$=\int_0^{4\pi}3t\left(\frac{1}{2}\sin(2t)\right)\sqrt{1+9}$$

$$= \frac{3\sqrt{10}}{2}\left(\frac{1}{4}\sin(2t) - \frac{t}{2}\cos(2t)\right)\Bigg|_0^{4\pi}$$

$$= -3\sqrt{10}\pi$$

You were able to do that integral right? It required integration by parts.

LINE INTEGRALSPART II

In the previous section we looked at line integrals with respect to arc length. In this section we want to look at line integrals with respect to x and/or y. As with the last section we will start with a two-dimensional curve C with parameterization,

$$x = x(t)$$
$$y = y(t)$$
$$a \le t \le b$$

The line integral of f with respect to x is,

$$\int_C f(x, y)\,dx = \int_a^b f(x(t), y(t))\,x'(t)\,dt$$

The line integral of f with respect to y is,

$$\int_C f(x, y)\,dy = \int_a^b f(x(t), y(t))\,y'(t)\,dt$$

Note that the only notational difference between these two and the line integral with respect to arc length (from the previous section) is the differential. These have a dx or dy while the line integral with respect to arc length has a ds. So when evaluating line integrals be careful to first note which differential you've got so you don't work the wrong kind of line integral. These two integral often appear together and so we have the following shorthand notation for these cases.

$$\int_C P\,dx + Q\,dy = \int_a^b P(x, y)\,dx + \int_C Q(x, y)\,dy$$

Let's take a quick look at an example of this kind of line integral.

Example: Evaluate $\int_C \sin(\pi y)\,dy + yx^2\,dx$ where C is the line segment from $(0, 2)$ to $(1, 4)$.

Solution: Here is the parameterization of the curve.

$$\vec{r}(t) = (1-t)\langle 0, 2\rangle + t\langle 1, 4\rangle = \langle t, 2+2t\rangle \qquad 0 \le t \le 1$$

The line integral is,

$$\int_C \sin(\pi y)\,dy + yx^2\,dx = \int_C \sin(\pi y)\,dy + \int_C yx^2\,dx$$

$$-\int_0^1 \sin(\pi(2+2t))(2)\,dt - \int_3^1 (2+2t)(t)^2(1)$$

$$= -\frac{1}{\pi}\cos(2\pi - 2\pi t)\Big|_0^1 + \left(\frac{2}{3}t^3 + \frac{1}{2}t^4\right)\Big|_0^1$$

$$= \frac{7}{6}$$

In the previous section we saw that changing the direction of the curve for a line integral with respect to arc length doesn't change the value of the integral. Let's see what happens with line integrals with respect to x and/or y.

Example: Evaluate $\int_C \sin(\pi y)\,dy + yx^2 dx$ where C is the line segment from $(1,4)$ to $(0,2)$.

Solution: So, we simply changed the direction of the curve. Here is the new parameterization.

$$\vec{r}(t) = (1-t)\langle 1,4\rangle + t\langle 0,2\rangle = \langle 1-t, 4-2t\rangle \quad 0 \le t \le 1$$

The line integral in this case is,

$$\int_C \sin(\pi y)\,dy + yx^2\,dx = \int_C \sin(\pi y)\,dy + \int_C yx^2\,dx$$

$$-\int_0^1 \sin(\pi(4-2t))(-2)\,dt + \int_0^1 (4-2t)(1-t)^2(-1)\,dt$$

$$= -\frac{1}{\pi}\cos(4\pi - 2\pi t)\Big|_0^1 - \left(-\frac{1}{2}t^4 + \frac{8}{3}t^3 - 5t^2 + 4t\right)\Big|_0^1$$

$$= -\frac{7}{6}$$

So, switching the direction of the curve got us a different value or at least the opposite sign of the value from the first example. In fact this will always happen with these kinds of line integrals.

Fact

If C is any curve then,

$$\int_{-C} f(x,y)\,dx = -\int_C f(x,y)\,dx \quad and \quad \int_{-C} f(x,y)\,dx = -\int_C f(x,y)\,dy$$

With the combined form of these two integrals we get,

$$\int_{-C} P\,dx + Q\,dy = -\int_C P\,dx + Q\,dy$$

We can also do these integrals over three-dimensional curves as well. In this case we will pick up a third integral (with respect to z) and the three integrals will be.

$$\int_C f(x,y,z)\,dx = \int_a^b f\left(x(t),y(t),z(t)\right)x'(t)dt$$

$$\int_C f(x,y,z)\,dy = \int_a^b f\left(x(t),y(t),z(t)\right)y'(t)dt$$

$$\int_C f(x,y,z)\,dz = \int_a^b f\left(x(t),y(t),z(t)\right)z'(t)dt$$

where the curve C is parameterized by

$$x = x(t)$$
$$y = y(t)$$
$$z = z(t)$$
$$a \le t \le b$$

As with the two-dimensional version these three will often occur together so the shorthand we'll be using here is,

$$\int_{-C} Pdx + Qdy + Rdz = \int_{-C} P(x,y,z)\,dx + \int_{-C} Q(x,y,z)\,dy + \int_{-C} R(x,y,z)\,dz$$

Let's work an example.

Example: Evaluate $\int_{-C} ydx + xdy + zdz$ where C is given by $x = \cos t, y = \sin t,$ $z = t^2, 0 \le t \le 2\pi$

Solution: So, we already have the curve parameterized so there really isn't much to do other than evaluate the integral.

$$\int_C fdx + xdy + zdz = \int_C fdx + \int_C xdy + \int_C zdz$$

$$= \int_0^{2\pi} \sin t(-\sin t)\,dt + \int_0^{2\pi} \cos t(\cos t)\,dt + \int_0^{2\pi} t^2(2t)\,dt$$

$$- \int_0^{2\pi} \sin^2 t\,dt + \int_0^{2\pi} \cos^2 t dt + \int_0^{2\pi} 2t^3\,dt$$

$$= -\frac{1}{2}\int_0^{2\pi} \left(1 - \cos(2t)\right)dt + \frac{1}{2}\int_0^{2\pi}\left(1 + \cos(2t)\right)dt + \int_0^{2\pi} 2t^3 dt$$

$$= \left(-\frac{1}{2}\left(t - \frac{1}{2}\sin(2t)\right) + \frac{1}{2}\left(t + \frac{1}{2}\sin(2t)\right) + \frac{1}{2}t^4\right)\Bigg|_0^{2\pi}$$

$$= 8\pi^4$$

LINE INTEGRALS OF VECTOR FIELDS

In the previous two sections we looked at line integrals of functions. In this section we are going to evaluate line integrals of vector fields. We'll start with the vector field,

$$\vec{F}(x,y,z) = P(x,y,z)\vec{i} + Q(x,y,z)j + R(x,y,z)\vec{k}$$

and the three-dimensional, smooth curve given by

$$\vec{r}(t) = x(t)\vec{i} + y(\vec{j}) + z(t)\vec{k} \quad a \le t \le b$$

The line integral of $\vec{F}$ along C is

$$\int_C \vec{F} \bullet d\vec{r} = \int_a^b \vec{F}(\vec{r}(t)) \bullet \vec{r}(t)\,dt$$

Note the notation in the left side. That really is a dot product of the vector field and the differential and the differential really is a vector. Also, $\vec{F}(\vec{r}(t))$ is a shorthand for,

$$\vec{F}(\vec{r}(t)) = \vec{F}(x(t), y(t), z(t))$$

We can also write line integrals of vector fields as a line integral with respect to arc length as follows,

$$\int_C \vec{F} \bullet d\vec{r} = \int_a^b \vec{F} \bullet \vec{T}\,ds$$

here $\vec{T}(t)$ is the unit tangent vector and is given by,

$$\vec{T}(t) = \frac{\vec{r}'(t)}{\|\vec{r}'(t)\|}$$

If we use our knowledge on how to compute line integrals with respect to arc length we can see that this second form is equivalent to the first form given above.

$$\begin{aligned}\int_C \vec{F} \bullet d\vec{r} &= \int_C \vec{F} \bullet \vec{T}\,ds \\ &= \int_a^b \vec{F}(\vec{r}(t)) \bullet \frac{\vec{r}'(t)}{\|\vec{r}'(t)\|}\|\vec{r}'(t)\|\,dt \\ &= \int_a^b \vec{F}(\vec{r}(t)) \bullet \vec{r}'(t)\,dt\end{aligned}$$

In general we use the first form to compute these line integrals as it is usually much easier to use. Let's take a look at a couple of examples.

Example: Evaluate $\int\limits_C \vec{F} \bullet d\vec{r}$ where $\vec{F}(x,y,z) = 8x^2yz\vec{i} + 5z\vec{i} - 4xy\vec{k}$ and C is the curve given by $\vec{r}(t) = t\vec{i} + t^2\vec{j} + t^3\vec{k}, 0 \le t \le 1$.

Solution: We first need the vector field evaluated along the curve.

$$\vec{F}(\vec{r}(t)) = 8t^2(t^2)(t^3)\vec{i} + 5t^3\vec{j} - 4t(t^2)\vec{k} = 8t^7\vec{i} + 5t^3\vec{j} - 4t^3\vec{k}$$

Next we need the derivative of the parameterization.

$$\vec{r}(t) = \vec{i} + 2t\,\vec{j} + 3t^2\vec{k}$$

Finally, let's get the dot product taken care of.

$$\vec{F}(\vec{r}(t)) \bullet \vec{r}'(t) = 8t^7 + 10t^4 - 12t^5$$

The line integral is then,

$$\int\limits_C \vec{F} \bullet d\vec{r} = \int_0^1 8t^7 + 10t^4 - 12t^5 \, dt$$

$$= \left(t^8 + 2t^5 - 2t^6\right)\Big|_0^1$$

$$= 1$$

Example: Evaluate $\int\limits_C \vec{F} \bullet d\vec{r}$ where $\vec{F}(x,y,z) = xz\vec{i} - yz\vec{k}$ and C is the line segment from $(-1,2,0)$ and $(3,0,1)$.

Solution: We'll first need the parameterization of the line segment. We saw how to get the parameterization of line segments in the first section on line integrals.

We've been using the two dimensional version of this over the last couple of sections. Here is the parameterization for the line.

$$\vec{r}(t) = (1-t)\langle -1,2,0\rangle + t\langle 3,0,1\rangle$$

$$= \langle 4t-1, 2-2-2t, t\rangle, \ 0 \le t \le 1$$

So, let's get the vector field evaluated along the curve.

$$\vec{F}(\vec{r}(t)) = (4t-1)(t)\vec{i} - (2-2t)(t)\vec{k}$$

$$= \left(4t^2 - t\right)\vec{i} - \left(2t - 2t^2\right)\vec{k}$$

Now we need the derivative of the parameterization.

$$\vec{r}(t) = \langle 4, -2, 1 \rangle$$

The dot product is then,

$$\vec{F}(\vec{r}(t)) \cdot \vec{r}'(t) = 4(4t^2 - t) - (2t - 2t^2) = 18t^2 - 1t$$

The line integral becomes,

$$\int_C \vec{F} \cdot d\vec{r} = \int_0^1 18t^2 - 6t dt$$

$$= \left(6t^2 - 3t^2\right)\Big|_0^1$$

$$= 3$$

Let's close this section out by doing one of these in general to get a nice relationship between line integrals of vector fields and line integrals with respect to *x*, *y*, and *z*.

Given the vector field $\vec{F}(x, y, z) - P\vec{i} + Q\vec{j} + R\vec{k}$ and the curve C parameterized by $\vec{r}(t) = x(t)\vec{i} + y(t)\vec{j} + z(t)\vec{k}, a \le t \le b$ the line integral is,

$$\int_C \vec{F} \cdot d\vec{r} = \int_a^b \left(P\vec{i} + Q\vec{j} + R\vec{k}\right) \cdot \left(x'\vec{i} + y'\vec{j} + z'\vec{k}\right) dt$$

$$= \int_a^b Px' + Qy' + Rz' \, dt$$

$$= \int_a^b Px' \, dt + \int_a^b Qy' \, dt + \int_a^b Rz' \, dt$$

$$= \int_C Pdx + \int_C Qdy + \int_C Rdz$$

$$= \int_C Pdx + Qdy + Rdz$$

So, we see that,

$$\int_C \vec{F} \cdot d\vec{t} = \int_C Pdx + Qdy + R\,dz$$

Note that this gives us another method for evaluating line integrals of vector fields.

This also allows us to say the following about reversing the direction of the path with line integrals of vector fields.

Fact

$$\int_C \vec{F} \cdot d\vec{r} = -\int_C \vec{F} \cdot d\vec{r}$$

This should make some sense given that we know that this is true for line integrals with respect to x, y, and/or z and that line integrals of vector fields can be defined in terms of line integrals with respect to x, y, and z.

FUNDAMENTAL THEOREM FOR LINE INTEGRALS

We had the Fundamental Theorem of Calculus that told us how to evaluate definite integrals. This told us,

$$\int_a^b F'(x)\,dx = F(b) - F(a)$$

It turns out that there is a version of this for line integrals over certain kinds of vector fields. Here it is.

Theorem: Suppose that C is a smooth curve given by $\vec{r}(t), a \le t \le b$. Also suppose that f is a function whose gradient vector,, is continuous on C. Then,

$$\int_C \nabla f \bullet d\vec{r} = f(\vec{r}(b)) - f(\vec{r}(a))$$

Note that $\vec{r}(a)$ represents the initial point on C while $\vec{r}(b)$ represents the final point on C. Also, we did not specify the number of variables for the function since it is really immaterial to the theorem. The theorem will hold regardless of the number of variables in the function.

Proof: This is a fairly straight forward proof.

For the purposes of the proof we'll assume that we're working in three dimensions, but it can be done in any dimension.

Let's start by just computing the line integral.

$$\begin{aligned}\int_C \nabla f \bullet d\vec{r} &= \int_a^b \nabla f(\vec{r}(t)) \bullet \vec{r}'(t) \\ &= \int_a^b \left(\frac{\partial f}{\partial x}\frac{dx}{dt} + \frac{\partial f}{\partial y}\frac{\partial y}{\partial t} + \frac{\partial f}{\partial z}\frac{dz}{dt} \right) dt\end{aligned}$$

Now, at this point we can use the Chain Rule to simplify the integrand as follows,

$$\begin{aligned}\int_C \nabla f \bullet d\vec{r} &= \int_a^b \left(\frac{\partial f}{\partial x}\frac{dx}{dt} + \frac{\partial f}{\partial y}\frac{\partial y}{\partial t} + \frac{\partial f}{\partial z}\frac{dz}{dt} \right) dt \\ &= \int_a^b \frac{d}{dt}\left[f(\vec{r}(t)) \right] dt\end{aligned}$$

To finish this off we just need to use the Fundamental Theorem of Calculus for single integrals.

$$\int_C \nabla f \bullet d\vec{r} = f(\vec{r}(b)) - f(\vec{r}(a))$$

Let's take a quick look at an example of using this theorem.

Example: Evaluate $\int\limits_C \nabla f \bullet d\vec{r}$ where $f(x,y,z)=\cos(\pi x)+\sin(\pi y)-xyz$ and C is *any* path that starts $\left(1,\frac{1}{2},2\right)$ at and ends at $(2,1,-1)$.

Solution: First let's notice that we didn't specify the path for getting from the first point to the second point. The reason for this is simple. The theorem above tells us that all we need are the initial and final points on the curve in order to evaluate this kind of line integral.

So, let $\vec{r}(t), a \le t \le b$ be any path that starts at $\left(1,\frac{1}{2},2\right)$ and ends at $(2,1,-1)$. Then,

$$\vec{r}(a)=\left\langle 1,\frac{1}{2},2\right\rangle$$
$$\vec{r}(b)=\langle 2,1,-1\rangle$$

The integral is then,

$$\begin{aligned}\int\limits_C \nabla f \bullet d\vec{r} &= f(2,1,-1)-f\left(1,\frac{1}{2},2\right)\\ &=\cos(2\pi)+\sin\pi-2(1)(-1)-\left(\cos\pi+\sin\left(\frac{\pi}{2}\right)-1\left(\frac{1}{2}\right)(2)\right)\\ &=4\end{aligned}$$

Notice that we also didn't need the gradient vector to actually do this line integral. However, for the practice of finding gradient vectors here it is,

$$\nabla f=\langle -\pi sin(\pi x)-yz, \pi\cos(\pi y)-xz,-xy\rangle$$

The most important idea to get from this example is not how to do the integral as that's pretty simple, all we do is plug the final point and initial point into the function and subtract the two results. The important idea from this example (and hence about the Fundamental Theorem of Calculus) is that, for these kinds of line integrals, we didn't really need to know the path to get the answer. In other words, we could use any path we want and we'll always get the same results.

In the first section on line integrals (even though we weren't looking at vector fields) we saw that often when we change the path we will change the value of the line integral. We now have a type of line integral for which we know that changing the path will NOT change the value of the line integral.

Let's formalize this idea up a little. Here are some definitions. The first one we've already seen before, but it's been a while and it's important in this section so we'll give it again. The remaining definitions are new.

Definitions: First suppose that is a continuous vector field in some domain *D*.

1. $\vec{F}$ is a conservative vector field if there is a function *f* such that $\vec{F} = \nabla f$. The function *f* is called a potential function for the vector field. We first saw this definition in the first section of this chapter.
2. $\int_C \vec{F} \cdot d\vec{r}$ is independent of path if $\int_{C_1} \vec{F} \cdot d\vec{r} = \int_{C_2} \vec{F} \cdot d\vec{r}$ for any two paths C_1 and C_2 in *D* with the same initial and final points.
3. A path *C* is called closed if its initial and final points are the same point. For example a circle is a closed path.
4. A path *C* is simple if it doesn't cross itself. A circle is a simple curve while a figure 8 type curve is not simple.
5. A region *D* is open if it doesn't contain any of its boundary points.
6. A region *D* is connected if we can connect any two points in the region with a path that lies completely in *D*.
7. A region *D* is simply-connected if it is connected and it contains no holes. We won't need this one until the next section, but it fits in with all the other definitions given here so this was a natural place to put the definition.

With these definitions we can now give some nice facts.

Facts

1. $\int_C \nabla f \cdot d\vec{r}$ is independent of path.

This is easy enough to prove since all we need to do is look at the theorem above. The theorem tells us that in order to evaluate this integral all we need are the initial and final points of the curve. This in turn tells us that the line integral must be independent of path.

2. $\vec{F}$ If is a conservative vector field then $\int_C \vec{F} \cdot d\vec{r}$ is independent of path.

This fact is also easy enough to prove. If $\vec{F}$ is conservative then it has a potential function, *f*, and so the line integral becomes $\int_C \vec{F} \cdot d\vec{r} = \int_C \nabla f \cdot d\vec{r}$ .Then using the first fact we know that this line integral must be independent of path.

3. If $\vec{F}$ is a continuous vector field on an open connected region D and if $\int_C \vec{F} \cdot d\vec{r}$ is independent of path (for any path in D) then $\vec{F}$ is a conservative vector field on D.

4. If $\int_C \vec{F} \cdot d\vec{r}$ is independent of path then $\int_C \vec{F} \cdot d\vec{r} = 0$ for every closed path C.

5. If $\int_C \vec{F} \cdot d\vec{r} = 0$ for every closed path C then $\int_C \vec{F} \cdot d\vec{r}$ is independent of path.

These are some nice facts to remember as we work with line integrals over vector fields. Also notice that 2 & 3 and 4 & 5 are converses of each other.

CONSERVATIVE VECTOR FIELDS

In the previous section we saw that if we knew that the vector field $\vec{F}$ was conservative then $\int_C \vec{F} \cdot d\vec{r}$ was independent of path. This in turn means that we can easily evaluate this line integral provided we can find a potential function for $\vec{F}$.

In this section we want to look at two questions. First, given a vector field $\vec{F}$ is there any way of determining if it is a conservative vector field? Secondly, if we know that $\vec{F}$ is a conservative vector field how do we go about finding a potential function for the vector field?

The first question is easy to answer at this point if we have a two-dimensional vector field. For higher dimensional vector fields we'll need to wait until the final section in this chapter to answer this question. With that being said let's see how we do it for two-dimensional vector fields.

Theorem: Let $\vec{F} = P\vec{i} + Q\vec{j}$ be a vector field on an open and simply-connected region D. Then if P and Q have continuous first order partial derivatives in D and

$$\frac{\partial P}{\partial y} = \frac{\partial P}{\partial y}$$

the vector field $\vec{F}$ is conservative.

Let's take a look at a couple of examples.

Example: Determine if the following vector fields are conservative or not.

1. $\vec{F}(x, y) = \left(x^2 - yx\right)\vec{i} + \left(y^2 - xy\right)\vec{j}$

2. $\vec{F}(x,y) = \left(2xe^{xy} + x^2ye^{xy}\right)\vec{i} + \left(x^3e^{xy} + 2yxy\right)\vec{j}$

Solution: There really isn't too much to these. All we do is identify P and Q then take a couple of derivatives and compare the results.

1. $\vec{F}(x,y) = \left(x^2 - yx\right)\vec{i} + \left(y^2 - xy\right)\vec{j}$

In this case here is P and Q and the appropriate partial derivatives.

$$P = x^2 - yx \qquad \frac{\partial P}{\partial x} = -x$$

$$Q = y^2 - xy \qquad \frac{\partial Q}{\partial x} = -y$$

So, since the two partial derivatives are not the same this vector field is NOT conservative.

2. $\vec{F}(x,y) = \left(2xe^{xy} + x^2ye^{xy}\right)\vec{i} + \left(x^3e^{xy} + 2yxy\right)\vec{j}$

Here is P and Q as well as the appropriate derivatives.

$$P = 2xe^{xy} + x^2ye^{xy} \qquad \frac{\partial P}{\partial y} = 2x^2e^{xy} + x^2e^{xy} + x^3ye^{xy} = 3x^2e^{xy} + x^3ye^{xy}$$

$$Q = x^3e^{xy} + 2y \qquad \frac{\partial Q}{\partial x} = 3x^2e^{xy} + x^3ye^{xy}$$

The two partial derivatives are equal and so this is a conservative vector field.

Now that we know how to identify if a two-dimensional vector field is conservative we need to address how to find a potential function for the vector field. This is actually a fairly simple process. First, let's assume that the vector field is conservative and so we know that a potential function, $f(x,y)$ exists. We can then say that,

$$\nabla f = \frac{\partial f}{\partial x}\vec{i} + \frac{\partial f}{\partial y}\vec{j} = P\vec{i} + Q\vec{j} = \vec{F}$$

Or by setting components equal we have,

$$\frac{\partial f}{\partial x} = P \text{ and } \frac{\partial f}{\partial y} = Q$$

By integrating each of these with respect to the appropriate variable we can arrive at the following two equations.

$$f(x,y) = \int P(x,y)\,dx \text{ or } f(x,y) = \int Q(x,y)\,dy$$

We saw this kind of integral briefly at the end of the section on iterated integrals in the previous chapter.

It is usually best to see how we use these two facts to find a potential function in an example or two.

Example: Determine if the following vector fields are conservative and find a potential function for the vector field if it is conservative.

1. $\vec{F} = \left(2x^3y^4 + x\right)\vec{i} + \left(2x^4y^3 + y\right)\vec{j}$

2. $\vec{F}(x, y) = \left(2xe^{xy} + x^2ye^{xy}\right)\vec{i} + \left(x^3e^{xy} + 2y\right)\vec{j}$

Solution:

1. $\vec{F} = \left(2x^3y^4 + x\right)\vec{i} + \left(2x^4y^3 + y\right)\vec{j}$

Let's first identify P and Q and then check that the vector field is conservative..

$$P = 2x^3y^4 + x \qquad \frac{\partial P}{\partial y} = 8x^3y^3$$

$$Q = 2x^4y^3 + y \qquad \frac{\partial Q}{\partial x} = 8x^3y^3$$

So, the vector field is conservative. Now let's find the potential function. From the first fact above we know that,

$$\frac{\partial f}{\partial x} = 2x^3y^4 + x \qquad \frac{\partial f}{\partial y} = 2x^4y^3 + y$$

From these we can see that

$$f(x, y) = \int 2x^3y^4 + x\,dx \quad \text{or} \quad f(x, y) = \int 2x^4y^3 + y\,dy$$

We can use either of these to get the process started. Recall that we are going to have to be careful with the "constant of integration" which ever integral we choose to use. For this example let's work with the first integral and so that means that we are asking what function did we differentiate with respect to x to get the integrand.

This means that the "constant of integration" is going to have to be a function of y since any function consisting only of y and/or constants will differentiate to zero when taking the partial derivative with respect to x.

Here is the first integral.

$$f(x, y) = \int 2x^3y^4 + x\,dx$$

$$= \frac{1}{2}x^4y^4 + \frac{1}{2}x^2 + h(y)$$

where $h(y)$is the "constant of integration".

We now need to now determine $h(y)$. This is easier that it might at first appear to be. To get to this point we've used the fact that we knew P, but we will also need to use the fact that we know Q to complete the problem. Recall that Q is really the derivative of f with respect to y. So, if we differentiate our function with respect to y we know what it should be.

So, let's differentiate f (including the $h(y)$) with respect to y and set it equal to Q since that is what the derivative is supposed to be.

$$\frac{\partial f}{\partial y} = 2x^4y^3 + h'(y) = 2x^4y^3 + y = Q$$

From this we can see that,

$$h'(y) = y$$

Notice that since $h'(y)$ is a function only of y so if there are any x's in the equation at this point we will know that we've made a mistake. At this point finding $h(y)$ is simple.

$$h(y) \int h'(y)\,dy = \int y\,dx = \frac{1}{2}y^2 + c$$

So, putting this all together we can see that a potential function for the vector field is,

$$f(x, y) = \frac{1}{2}x^4y^4 + \frac{1}{2}x^2 + \frac{1}{2}y^2 + c$$

Note that we can always check our work by verifying that $\nabla f \bullet \vec{F}$. Also note that because the c can be anything there are an infinite number of possible potential functions, although they will only vary by an additive constant.

2. $\vec{F}(x, y) = \left(2xe^{xy} + x^2ye^{xy}\right)\vec{i} + \left(x^3e^{xy} + 2y\right)\vec{j}$

This one will go a lot faster since we don't need to go through as much explanation. We've already verified that this vector field is conservative in the first set of examples so we won't bother redoing that.

Let's start with the following,

$$\frac{\partial f}{\partial x} = 2xe^{xy} + x^2ye^{xy}$$

$$\frac{\partial f}{\partial y} = x^2e^{xy} + 2y$$

This means that we can do either of the following integrals,

$$f(x, y) = \int 2xe^{xy} + x^2ye^{xy}\,dx \quad \text{or} \quad f(x, y) = \int x^3e^{xy} + 2y\,dy$$

While we can do either of these the first integral would be somewhat unpleasant as we would need to do integration by parts on each portion. On the other hand the second integral is fairly simple since the second term only involves y's and the first term can be done with the substitution $u = xy$. So, from the second integral we get,

$$f(x, y) = x^2 e^{xy} + y^2 + h(x)$$

Notice that this time the "constant of integration" will be a function of x. If we differentiate this with respect to x and set equal to P we get,

$$\frac{\partial f}{\partial x} = 2xe^{xy} + x^2 ye^{xy} + h'(x)$$

$$= 2xe^{xy} + x^2 ye^{xy} = P$$

So, in this case it looks like,

$$h'(x) = 0 \quad \Rightarrow \quad h(x) = c$$

So, in this case the "constant of integration" really was a constant. Sometimes this will happen and sometimes it won't.

Here is the potential function for this vector field.

$$f(x, y) = x^2 e^{xy} + y^2 + c$$

Now, as noted above we don't have a way (yet) of determining if a three-dimensional vector field is conservative or not. However, if we are given that a three-dimensional vector field is conservative finding a potential function is similar to the above process, although the work will be a little more involved.

In this case we will use the fact that,

$$\nabla f = \frac{\partial f}{\partial x}\vec{i} + \frac{\partial f}{\partial y}\vec{j} + \frac{\partial f}{\partial z}\vec{k} = P\vec{i} + Q\vec{j} + R\vec{k} = \vec{F}$$

Let's take a quick look at an example.

Example: Find a potential function for the vector field,

$$\vec{F} = 2xy^3z^4\vec{i} + 3x^2y^3z^3\vec{j} + 4x^2y^3z^3\vec{k}$$

Solution: We'll start off with the following equalities.

$$\frac{\partial f}{\partial x} = 2xy^3z^4$$

$$\frac{\partial f}{\partial y} = 3x^2y^2z^4$$

$$\frac{\partial f}{\partial z} = 4x^2y^3z^3$$

To get started we can integrate the first one with respect to x, the second one with respect to y, or the third one with respect to z. Let's integrate the first one with respect to x.

$$f(x,y,z) = \int 2xy^3z^4\,dx = x^2y^3z^4 + g(y,z)$$

Note that this time the "constant of integration" will be a function of both y and z since differentiating anything of that form with respect to x will differentiate to zero.

Now, we can differentiate this with respect to y and set it equal to Q. Doing this gives,

$$\frac{\partial f}{\partial y} = 3x^2y^2z^4 + g_y(y,z) = 3x^2y^2z^4 = Q$$

Of course we'll need to take the partial derivative of the constant of integration since it is a function of two variables. It looks like we've now got the following,

$$g_y(y,z) = 0$$
$$g(y,z) = h(z)$$

Since differentiating $g(y,z)$ with respect to y gives zero then $g(y,z)$ could at most be a function of z. This means that we now know the potential function must be in the following form.

$$f(x,y,z) = x^2y^3z^4 + h(z)$$

To finish this out all we need to do is differentiate with respect to z and set the result equal to R.

$$\frac{\partial f}{\partial z} = 4x^2y^3z^3 + h'(z) = 4x^2y^3z^3 = R$$

So,

$$h'(z) = 0 \quad \Rightarrow \quad h(z) = c$$

The potential function for this vector field is then,

$$f(x,y,z) = x^2y^3z^4 + c$$

Note that to keep the work to a minimum we used a fairly simple potential function for this example. It might have been possible to guess what the potential function was based simply on the vector field. However, we should be careful to remember that this usually won't be the case and often this process is required.

Also, there were several other paths that we could have taken to find the potential function. Each would have gotten us the same result.

Let's work one more slightly (and only slightly) more complicated example.

Example: Find a potential function for the vector field,

$$\vec{F} = \left(2x\cos(y) - 2z^3\right)\vec{i} + \left(3 + 2ye^z - x^2\sin(y)\right)\vec{j} + \left(y^2e^2 - 6xz^2\right)\vec{k}$$

Solution: Here are the equalities for this vector field.

$$\frac{\partial f}{\partial x} = 2x\cos(y) - 2z^3$$

$$\frac{\partial f}{\partial y} = 3 + 2ye^z \sin(y)$$

$$\frac{\partial f}{\partial z} = y^2e^2 - 6xz^2$$

For this example let's integrate the third one with respect to z.

$$f(x,y,z) = \int y^2e^z - 6xz^2 dz$$

$$= y^2e^z - 2xz^3 + g(x,y)$$

The "constant of integration" for this integration will be a function of both x and y. Now, we can differentiate this with respect to x and set it equal to P. Doing this gives,

$$\frac{\partial f}{\partial x} = -2z^3 + g_x(x,y)$$

$$= 2x\cos(y) - 2z^3 = P$$

So, it looks like we've now got the following,

$$g_x(x,y) = 2x\cos(y) \quad \Rightarrow \quad g(x,y) = x^2\cos(y) + h(y)$$

The potential function for this problem is then,

$$f(x,y,z) = y^2e^z - 2xz^3 + x^2\cos(y) + h(y)$$

To finish this out all we need to do is differentiate with respect to y and set the result equal to Q.

$$\frac{\partial f}{\partial y} = 2ye^z - x^2\sin(y) + h'(y)$$

$$= 3 + 2ye^z - x^2\sin(y) = Q$$

So,

$$h'(y) = 3 \quad \Rightarrow \quad h(y) = 3y + c$$

The potential function for this vector field is then,

$$f(x,y,z) = y^2e^z - 2xz^3 + x^2\cos(y) + 3y + c$$

So, a little more complicated than the others and there are again many different paths that we could have taken to get the answer.

We need to work one final example in this section.

Example: Evaluate $\int\limits_C \vec{F} \cdot d\vec{r}$ where $\vec{F} = \left(2x^3y^4 + x\right)\vec{i} + \left(2x^4y^3 + y\right)\vec{j}$ and C is given by $\vec{r}(t) = \left(t\cos(\pi t) - 1\right)\vec{i} + \sin\left(\frac{\pi t}{2}\right)\vec{j}$, $0 \le t \le 1$

Solution: Now, we could use the techniques we discussed when we first looked at line integrals of vector fields however that would be particularly unpleasant solution.

Instead, let's take advantage of the fact that we know from Example above this vector field is conservative and that a potential function for the vector field is,

$$f(x,y) = \frac{1}{2}x^4y^4 + \frac{1}{2}x^4 + \frac{1}{2}y^2 + c$$

Using this we know that integral must be independent of path and so all we need to do is use the theorem from the previous section to do the evaluation.

$$\int\limits_C \vec{F} \cdot d\vec{r} = \int\limits_C \nabla f \cdot d\vec{r} = f\left(\vec{r}(1)\right) - f\left(\vec{r}(0)\right)$$

where,

$$\vec{r}(1) = \langle -2, 1 \rangle$$
$$\vec{r}(0) = \langle -1, 0 \rangle$$

So, the integral is,

$$\begin{aligned} \int\limits_C \vec{F} \cdot d\vec{r} &= f(-2,1) - f(-1,0) \\ &= \left(\frac{21}{2} + c\right) - \left(\frac{1}{2} + c\right) \\ &= 10 \end{aligned}$$

GREEN'S THEOREM

In this section we are going to investigate the relationship between certain kinds of line integrals (on closed paths) and double integrals.

Let's start off with a simple (recall that this means that it doesn't cross itself) closed curve C and let D be the region enclosed by the curve. Here is a sketch of such a curve and region.

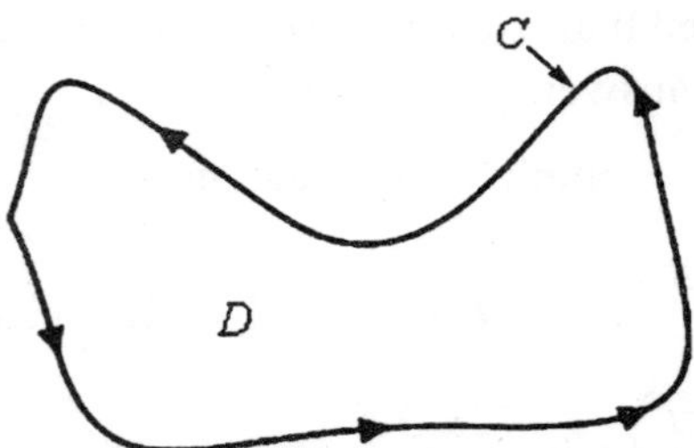

First, notice that because the curve is simple and closed there are no holes in the region D. Also notice that a direction has been put on the curve. We will use the convention here that the curve C has a positive orientation if it is traced out in a counter-clockwise direction. Another way to think of a positive orientation (that will cover much more general curves as well see later) is that as we traverse the path following the positive orientation the region D must always be on the left.

Given curves/regions such as this we have the following theorem.

GREEN'S THEOREM

Let C be a positively oriented, piecewise smooth, simple, closed curve and let D be the region enclosed by the curve. If P and Q have continuous first order partial derivatives on D then,

$$\int_C P dx + Q dy = \iint_D \left(\frac{\partial Q}{\partial x} - \frac{\partial P}{\partial y} \right) dA$$

Before working some examples there are some alternate notations that we need to acknowledge. When working with a line integral in which the path satisfies the condition of Green's Theorem we will often denote the line integral as,

$$\oint_C P dx + Q dy \quad \text{or} \quad \oint_C P dx + Q dy$$

Both of these notations do assume that C satisfies the conditions of Green's Theorem so be careful in using them. Also, sometimes the curve C is not thought of as a separate curve but instead as the boundary of some region D and in these cases you may see C denoted as ∂D.

Let's work a couple of examples.

Example: Use Green's Theorem to evaluate $\oint_C xy\,dx + x^2 y^2 dy$ where C is the triangle with vertices $(0,0)(1,0),(1,2)$ with positive orientation.

Solution: Let's first sketch C and D for this case to make sure that the conditions of Green's Theorem are met for C and will need the sketch of D to evaluate the double integral.

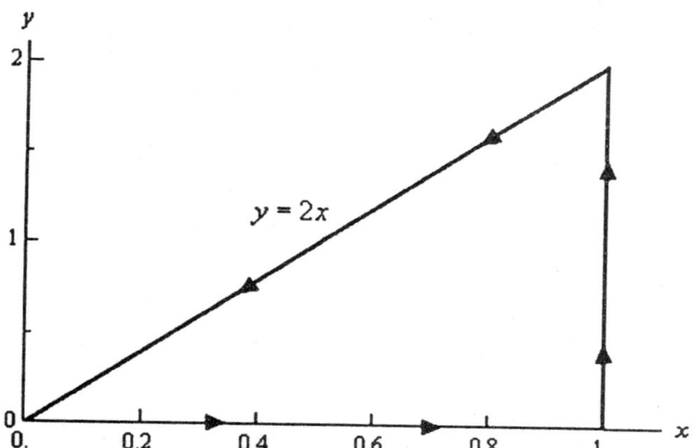

So, the curve does satisfy the conditions of Green's Theorem and we can see that the following inequalities will define the region enclosed.

$$0 \le x \le 1 \qquad 0 \le y \le 2x$$

We can identify P and Q from the line integral. Here they are.

$$P = xy$$

$$Q = x^2 y^2$$

So, using Green's Theorem the line integral becomes,

$$\oint_C xy\,dx + x^2 y^3 dy = \iint_D 2xy^3 - x dA$$

$$= \int_0^1 \int_0^{2\pi} 2xy^3 - x dy\,dx$$

$$= \int_0^1 \left(\frac{1}{2}xy^4 - xy\right)\Bigg|_0^{2x} dx$$

$$= \int_0^1 8x^5 - 2x^2 dx$$

$$= \left(\frac{4}{3}x^6 - \frac{2}{3}x^3\right)\Bigg|_0^1$$

$$= \frac{2}{3}$$

Example: Evaluate $\oint_C y^3\,dx - x^3 dy$ where C is the positively oriented circle of radius 2 centered at the origin.

Solution: A circle will satisfy the conditions of Green's Theorem since it is closed and simple and so there really isn't a reason to sketch it.

Let's first identify P and Q from the line integral.

$$P = y^3$$

$$Q = -x^3$$

Be careful with the minus sign on Q!

Now, using Green's theorem on the line integral gives,

$$\oint_C y^3\,dx - x^3 y^2 dy = \iint_D -3x^2 - 3y^2\,dA$$

where D is a disk of radius 2 centered at the origin.

Since D is a disk it seems like the best way to do this integral is to use polar coordinates. Here is the evaluation of the integral.

$$\begin{aligned}\oint_C y^3\,dx - x^3 dy &= -3\iint_D \left(x^2 + y^2\right) dA \\ &= -3\int_0^{2\pi}\int_0^2 r^3\,dr\,d\theta \\ &= -3\int_0^{2\pi} \frac{1}{4}r^4\Big|_0^2\,d\theta \\ &= -3\int_0^{2\pi} 4d\theta \\ &= -24\pi\end{aligned}$$

So, Green's theorem, as stated, will not work on regions that have holes in them. However, many regions do have holes in them. So, let's see how we can deal with those kinds of regions.

Let's start with the following region. Even though this region doesn't have any holes in it the arguments that we're going to go through will be similar to those that we'd need for regions with holes in them, except it will be a little easier to deal with and write down.

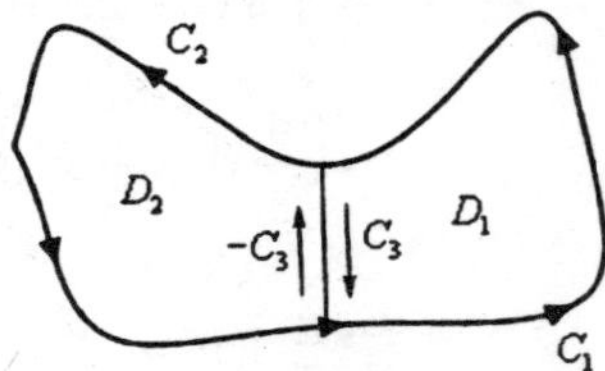

The region D will be $D_1 \cup D_2$ and recall that the symbol is called the union and means that we'll D consists of both D_1 and D_2. The boundary of D_1 is $C_1 \cup C_3$ while the boundary of D_2 is and notice that both of these boundaries are positively oriented. As we traverse each boundary the corresponding region is always on the left. Finally, also note that we can think of the whole boundary, C, as,

$$C = (C_1 \cup C_3) \cup (C_2 \cup (-C_3)) = C_1 \cup C_2$$

ince both $C_3$and $-C_3$will "cancel" each other out.

Now, let's start with the following double integral and use a basic property of double integrals to break it up.

$$\iint_D (Q_x - P_y)\, dA = \iint_{D_1 \cup D_2} (Q_x - P_y)\, dA$$

$$= \iint_{D_1} (Q_x - P_y)\, dA + \iint_{D2} (Q_x - P_y)\, dA$$

Next, use Green's theorem on each of these and again use the fact that we can break up line integrals into separate line integrals for each portion of the boundary.

$$\iint_D (Q_x - P_y)\, dA = \iint_{D_1} (Q_x - P_y)\, dA + \iint_{D_2} (Q_x - P_y)\, dA$$

$$= \oint_{C_1 \cup C_3} Pdx + Qdy + \oint_{C_2 \cup (-C_3)} Pdx + Qdy$$

$$= \oint_{C_1} Pdx + Qdy + \oint_{C_3} Pdx + Qdy + \oint_{C_2} Pdx + Qdy + \oint_{-C_3} Pdx + Qdy$$

Next, we'll use the fact that,

$$\oint_{-C_3} Pdx + Qdy = -\oint_{C_3} Pdx + Qdy$$

Recall that changing the orientation of a curve with line integrals with respect to x and/or y will simply change the sign on the integral. Using this fact we get,

$$\iint_D (Q_x - P_y)\, dA = \oint_{C_1} Pdx + Qdy + \oint_{C_3} Pdx + Qdy + \oint_{C_2} Pdx + Qdy + \oint_{C_3} Pdx + Qdy$$

$$= \oint_{C_1} Pdx + Qdy + \oint_{C_2} Pdx + Qdy$$

Finally, put the line integrals back together and we get,

$$\iint_D (Q_x - P_y)\, dA = \oint_{C_1} Pdx + Qdy + \oint_{C_2} Pdx + Qdy$$

$$= \oint_{C_1 \cup C_2} Pdx + Qdy$$

$$= \oint_C Pdx + Qdy$$

So, what did we learn from this? If you think about it this was just a lot of work and all we got out of it was the result from Green's Theorem which we already knew to be true. What this exercise has shown us is that if we break a region up as we did above then the portion of the line integral on the pieces of the curve that are in the middle of the region (each of which are in the opposite direction) will cancel out. This idea will help us in dealing with regions that have holes in them. To see this let's look at a ring.

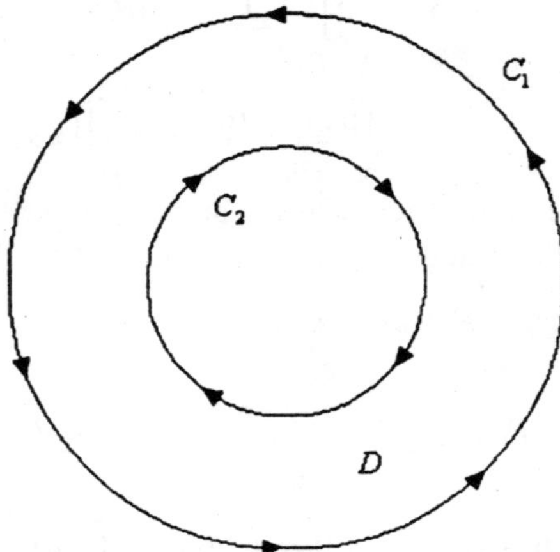

Notice that both of the curves are oriented positively since the region D is on the left side as we traverse the curve in the indicated direction. Note as well that the curve C_2 seems to violate the original definition of positive orientation.

We originally said that a curve had a positive orientation if it was traversed in a counter-clockwise direction. However, this was only for regions that do not have holes. For the boundary of the hole this definition won't work and we need to resort to the second definition that we gave above. Now, since this region has a hole in it we will apparently not be able to use Green's Theorem on any line integral with the curve $C = C_1 \cup C_2$. However, if we cut the disk in half and rename all the various portions of the curves we get the following sketch.

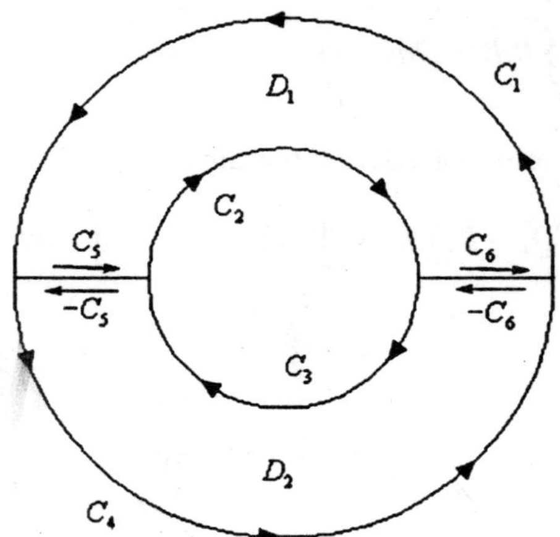

The boundary of the upper portion (D_1)of the disk is $C_1 \cup C_2 \cup C_5 \cup C_6$ and the boundary on the lower portion (D_2)of the disk is $C_3 \cup C_4 \cup (-C_5) \cup (-C_6)$.

Also notice that we can use Green's Theorem on each of these new regions since they don't have any holes in them. This means that we can do the following,

$$\iint_D (Q_x - P_y)\, dA = \iint_{D_1} (Q_x - P_y)\, dA + \iint_{D_2} (Q_x - P_y)\, dA$$

$$= \oint_{C_1 \cup C_2 \cup C_5 \cup C_6} P\,dx + Q\,dy + \oint_{C_3 \cup C_4 \cup (-C_5) \cup (-C_6)} P\,dx + Q\,dy$$

Now, we can break up the line integrals into line integrals on each piece of the boundary. Also recall from the work above that boundaries that have the same curve, but opposite direction will cancel. Doing this gives,

$$\iint_D (Q_x - P_y)\, dA = \iint_{D_1} (Q_x - P_y)\, dA + \iint_{D_2} (Q_x - P_y)\, dA$$

$$= \oint_{C_1} P\,dx + Q\,dy + \oint_{C_2} P\,dx + Q\,dy + \oint_{C_3} P\,dx + Q\,dy + \oint_{C_4} P\,dx + Q\,dy$$

But at this point we can add the line integrals back up as follows,

$$\iint_D (Q_x - P_y)\, dA = \oint_{C_1 \cup C_2 \cup C_3 \cup C_4} P\,dx + Q\,dy$$

$$= \oint_C P\,dx + Q\,dy$$

The end result of all of this is that we could have just used Green's Theorem on the disk from the start even though there is a hole in it. This will be true in general for regions that have holes in them.

Let's take a look at an example.

Example: Evaluate $\oint_C y^3\, dx - x^3 dy$ where C are the two circles of radius 2 and radius 1 centered at the origin with positive orientation.

Solution: Notice that this is the same line integral as we looked at in the second example and only the curve has changed. In this case the region D will now be the region between these two circles and that will only change the limits in the double integral so we'll not put in some of the details here. Here is the work for this integral.

$$\oint_C y^3 dx - x^3 dy = -3\iint_D \left(x^2 + y^2\right) dA$$

$$= -3\int_0^{2\pi}\int_1^2 r^3\, dr\, d\theta$$

$$= -3\int_0^{2\pi} \frac{1}{4}r^4\Big|_1^2 d\theta$$

$$= -3\int_0^{2\pi} \frac{15}{4} d\theta$$

$$= -\frac{45\pi}{2}$$

We will close out this section with an interesting application of Green's Theorem. Recall that we can determine the area of a region D with the following double integral.

$$A = \iint_D dA$$

Let's think of this double integral as the result of using Green's Theorem. In other words, let's assume that

$$Q_x - P_y = 1$$

And see if we can get some functions P and Q that will satisfy this. There are many functions that will satisfy this. Here are some of the more common functions.

$$\begin{array}{c|c|c} P = 0 & P = -y & P = -\dfrac{y}{2} \\ Q = x & Q = 0 & Q = \dfrac{x}{2} \end{array}$$

Then, if we use Green's Theorem in reverse we see that the area of the region D can also be computed by evaluating any of the following line integrals.

$$A = \oint_C x\, dy = -\oint_C y\, dx = \frac{1}{2}\oint_C x\, dy - y dx$$

where C is the boundary of the region D.

Let's take a quick look at an example of this.

Example: Use Green's Theorem to find the area of a disk of radius a.

Solution: We can use either of the integrals above, but the third one is probably the easiest. So,

$$A = \frac{1}{2}\oint_C x\, dy - y\, dx$$

where C is the circle of radius a. So, to do this we'll need a parameterization of C. This is,

$$x = a\cos t$$
$$y = a\sin t$$
$$0 \le t \le 2\pi$$

The area is then,

$$A = \frac{1}{2}\oint_C x\,dy - y\,dx$$
$$= \frac{1}{2}\left(\int_0^{2\pi} a\cos t(a\cos t)\,dt - \int_0^{2\pi} a\sin t(-a\sin t)\,dt\right)$$
$$= \frac{1}{2}\int_0^{2\pi} a^2\cos^2 t + a^2\sin^2 t\,dt$$
$$= \frac{1}{2}\int_0^{2\pi} a^2\,dt$$
$$= \pi a^2$$

CURL AND DIVERGENCE

In this section we are going to introduce a couple of new concepts, the curl and the divergence of a vector.

Let's start with the curl. Given the vector field $\vec{F} = P\vec{i} + Q\vec{j} + R\vec{k}$ the curl is defined to be,

$$\text{curl}\vec{F} = (R_y - Q_z)\vec{i} + (P_z - R_x)\vec{j} + (Q_x - P_y)\vec{k}$$

There is another (potentially) easier definition of the curl of a vector field. We use it we will first need to define the *operator*. This is defined to be,

$$\nabla = \frac{\partial}{\partial x}\vec{i} + \frac{\partial}{\partial y}\vec{j} + \frac{\partial}{\partial z}\vec{k}$$

We use this as if it's a function in the following manner.

$$\nabla f = \frac{\partial f}{\partial x}\vec{i} + \frac{\partial f}{\partial y}\vec{j} + \frac{\partial f}{\partial z}\vec{k}$$

whatever function is listed after the $\nabla$is substituted into the partial derivatives. Note as well that when we look at it in this light we simply get the gradient vector.

Using the ∇we can define the curl as the following cross product,

$$\text{curl}\,\vec{F} = \nabla \times \vec{F} = \begin{vmatrix} \vec{i} & \vec{j} & \vec{k} \\ \frac{\partial}{\partial x} & \frac{\partial}{\partial y} & \frac{\partial}{\partial z} \\ P & Q & R \end{vmatrix}$$

We have a couple of nice facts that use the curl of a vector field.

Facts

1. If $f(x, y, z)$ has continuous second order partial derivatives then $\text{curl}(\nabla f) = \vec{0}$. This is easy enough to check by plugging into the definition of the derivative so we'll leave it to you to check.

2. If $\vec{F}$ is a conservative vector field then $\text{curl}\vec{F} = \vec{0}$. This is a direct result of what it means to be a conservative vector field and the previous fact.

3. If $\vec{F}$ is defined on all of $\mathbb{R}^3$ whose components have continuous first order partial derivative and then is a conservative vector field. This is not so easy to verify and so we won't try.

Example: Determine if $\vec{F} = x^2 y\vec{i} + xyz\vec{j} - x^2 y^2 \vec{k}$ is a conservative vector field.

Solution: So all that we need to do is compute the curl and see if we get the zero vector or not.

$$\begin{aligned}
\text{curl}\,\vec{F} &= \begin{vmatrix} \vec{i} & \vec{j} & \vec{k} \\ \dfrac{\partial}{\partial x} & \dfrac{\partial}{\partial y} & \dfrac{\partial}{\partial z} \\ x^2 y & xyz & -x^2 y^2 \end{vmatrix} \\
&= -2x^2 y\vec{i} + yz\vec{k} - \left(-2xy^2\vec{j}\right) - xy\vec{i} - x^2\vec{k} \\
&= -\left(2x^2 y + xy\right)\vec{i} + 2xy^2\vec{j} + \left(yz - x^2\right)\vec{k} \\
&\neq \vec{0}
\end{aligned}$$

So, the curl isn't the zero vector and so this vector field is not conservative.

Next we should talk about a physical interpretation of the curl. Suppose that $\vec{F}$ is the velocity field of a flowing fluid. Then $\text{curl}\vec{F}$ represents the tendency of particles at the point (x, y, z) to rotate about the axis that points in the direction of $\text{curl}\vec{F} = \vec{0}$. If then the fluid is called irrotational. Let's now talk about the second new concept in this section. Given the vector field $\vec{F} = P\vec{i} + Q\vec{j} + R\vec{k}$ the divergence is defined to be,

$$\text{div}\vec{F} = \frac{\partial P}{\partial x} + \frac{\partial Q}{\partial y} + \frac{\partial R}{\partial z}$$

There is also a definition of the divergence in terms of the ∇operator. The divergence can be defined in terms of the following dot product.

$$\text{div}\vec{F} = \nabla \bullet \vec{F}$$

Example: Compute $\text{div}\vec{F}$ for $\vec{F} = x^2 y\vec{i} + xyz\vec{j} - x^2 y^2 \vec{k}$

Solution: There really isn't much to do here other than compute the divergence.

$$\begin{aligned}\text{div}\vec{F} &= \frac{\partial}{\partial x}\left(x^2 y\right) + \frac{\partial}{\partial y}(zyz) + \frac{\partial}{\partial z}\left(-x^2 y^2\right) \\ &= 2xy + xz\end{aligned}$$

We also have the following fact about the relationship between the curl and the divergence.

$$\text{div}\left(\text{curl}\,\vec{F}\right) = 0$$

Example: Verify the above fact for the vector field $\vec{F} = yz^2\,\vec{i} + xy\,\vec{j} + yz\vec{k}$.

Solution: Let's first compute the curl.

$$\begin{aligned}\text{curl}\,\vec{\text{F}} &= \begin{vmatrix} \vec{i} & \vec{j} & \vec{k} \\ \frac{\partial}{\partial x} & \frac{\partial}{\partial y} & \frac{\partial}{\partial z} \\ yz^2 & xy & yz \end{vmatrix} \\ &= z\vec{i} + 2yz\,\vec{j} + y\vec{k} - z^2\vec{k} \\ &= z\vec{i} + 2yz\,\vec{j} + \left(y - z^2\right)\vec{k}\end{aligned}$$

Now compute the divergence of this.

$$\begin{aligned}\text{div}\left(\text{curl}\,\vec{F}\right) &= \frac{\partial}{\partial x}(z) + \frac{\partial}{\partial y}(2yz) + \frac{\partial}{\partial x}\left(y - z^2\right) \\ &= 2z - 2z = 0\end{aligned}$$

We also have a physical interpretation of the divergence. If we again think of $\vec{F}$ as the velocity field of a flowing fluid then $\text{div}\vec{F}$ represents the net rate of change of the mass of the fluid flowing from the point (x, y, z) per unit volume.

This can also be thought of as the tendency of a fluid to diverge from a point. If $\text{div}\vec{F} = 0$ then the $\vec{F}$ is called incompressible.

The next topic that we want to briefly mention is the Laplace operator. Let's first take a look at,

$$\operatorname{div}(\nabla f) = \nabla \cdot \nabla f = f_{xx} + f_{yy} + f_{zz}$$

The Laplace operator is then defined as,

$$\nabla^2 = \nabla \cdot \nabla$$

The Laplace operator arises naturally in many fields including heat transfer and fluid flow.

The final topic in this section is to give two vector forms of Green's Theorem. The first form uses the curl of the vector field and is,

$$\oint_C \vec{F} \cdot d\vec{r} = \iint_D (\operatorname{curl} \vec{F}) \cdot \vec{k}\, dA$$

Where $\vec{k}$ is the standard unit vector in the positive z direction.

The second form uses the divergence. In this case we also need the outward unit normal to the curve C. If the curve is parameterized by

$$\vec{r}(t) = x(t)\vec{i} + y(t)\vec{j}$$

then the outward unit normal is given by,

$$\vec{n} = \frac{y'(t)}{\|\vec{r}(t)\|}\vec{i} - \frac{x'(t)}{\|\vec{r}(t)\|}\vec{j}$$

Here is a sketch illustrating the outward unit normal for some curve C at various points.

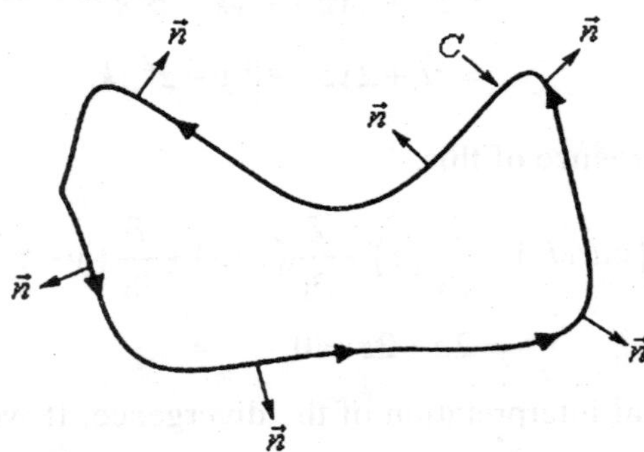

The vector form of Green's Theorem that uses the divergence is given by,

$$\oint_C \vec{F} \cdot \vec{n}\, ds = \iint_D \operatorname{div} \vec{F}\, dA.$$

Chapter 3

Surface Integral

In mathematics, a surface integral is a definite integral taken over a surface which may be a curved set in space; it can be thought of as the double integral analog of the line integral. Given a surface, one may integrate over it scalar fields that is, functions which return numbers as values, and vector fields that is, functions which return vectors as values. Surface integrals have applications in physics, particularly with the classical theory of electromagnetism.

PARAMETRIC SURFACES

Before we get into surface integrals we first need to talk about how to parameterize a surface. When we parameterized a curve we took values of t from some interval $[a, b]$ and plugged them into

$$\vec{r}(t) = x(t)\vec{i} + y(t)\vec{j} + z(t)\vec{k}$$

and the resulting set of vectors will be the position vectors for the points on the curve. With surfaces we'll do something similar. We will take points, (u, v), out of some two-dimensional space D and plug them into

$$\vec{r}(u, v) = x(u, v)\vec{i} + y(u, v)\vec{j} + z(u, v)\vec{k}$$

and the resulting set of vectors will be the position vectors for the points on the surface S that we are trying to parameterize. This is often called the parametric representation of the parametric surface S. We will sometimes need to write the parametric equations for a surface. There are really nothing more than the components of the parametric representation explicitly written down.

$$x = x(u, v)$$
$$y = y(u, v)$$
$$z = z(u, v)$$

Example: Determine the surface given by the parametric representation

$$\vec{r}(u,v) = u\,\vec{i} + u\cos v\,\vec{j} + u\sin v\vec{k}$$

Let's first write down the parametric equations.

$$x = u \quad y = u\cos v \quad z = u\sin v$$

Now if we square y and z and then add them together we get,

$$\begin{aligned} y^2 + z^2 &= u^2\cos^2 v + u^2\sin^2 v \\ &= u^2\left(\cos^2 v + \sin^2 v\right) \\ &= u^2 = x^2 \end{aligned}$$

So, we were able to eliminate the parameters and the equation in x, y, and z is given by,

$$x^2 = y^2 + z^2$$

From the Quadric Surfaces section notes we can see that this is a cone that opens along the x-axis. We are much more likely to need to be able to write down the parametric equations of a surface than identify the surface from the parametric representation so let's take a look at some examples of this.

Example: Give parametric representations for each of the following surfaces.

1. The elliptic paraboloid $x = 5y^2 + 2z^2 - 10$
2. The elliptic paraboloid $x = 5y^2 + 2z^2 - 10$ that is in front of the yz-plane.
3. The sphere $x = y^2 + z^2 = 30$.
4. The cylinder $y^2 + z^2 = 25$.

Solution:

1. The elliptic paraboloid $x = 5y^2 + 2z^2 - 10$.

This one is probably the easiest one of the four to see how to do. Since the surface is in the form we can quickly write down a set of parametric equations as follows,

$$x = 5y^2 + 2z^2 - 10 \quad y = y \quad z = z$$

The last two equations are just there to acknowledge that we can choose y and z to be anything we want them to be. The parametric representation is then,

$$\vec{r}(y,z) = \left(5y^2 + 2z^2 - 10\right)\vec{i} + y\vec{j} + z\vec{k}$$

2. The elliptic paraboloid $x = 5y^2 + 2z^2 - 10$ that is in front of the yz-plane.

This is really a restriction on the previous parametric representation. The parametric representation stays the same.

$$\vec{r}(y,z) = \left(5y^2 + 2z^2 - 10\right)\vec{i} + y\vec{j} + z\vec{k}$$

However, since we only want the surface that lies in front of the yz-plane we also need to require that $x \geq 0$. This is equivalent to requiring,

$$5y^2 + 2z^2 - 10 \geq 0 \text{ or } 5y^2 + 2z^2 \geq 10$$

3. The sphere $x^2 + y^2 + z^2 = 30$.

This one can be a little tricky until you see how to do it. In spherical coordinates we know that the equation of a sphere of radius a is given by,

$$\rho = a$$

and so the equation of this sphere (in spherical coordinates) is $\rho = \sqrt{30}$. Now, we also have the following conversion formulas for converting Cartesian coordinates into spherical coordinates.

$$x = \rho \sin\varphi \cos\theta$$
$$y = \rho \sin\varphi \cos\theta$$
$$z = \rho \cos\varphi$$

However, we know what $\rho$is for our sphere and so if we plug this into these conversion formulas we will arrive at a parametric representation for the sphere. Therefore, the parametric representation is,

$$\vec{r}(\theta,\varphi) = \sqrt{30}\sin\varphi\cos\theta\vec{i} + \sqrt{30}\sin\varphi\cos\theta\vec{j} + \sqrt{30}\cos\varphi\vec{k}$$

All we need to do now is come up with some restriction on the variables. First we know that we have the following restriction.

$$0 \leq \varphi \leq \pi$$

This is enforced upon us by choosing to use spherical coordinates. Also, to make sure that we only trace out the sphere once we will also have the following restriction.

$$0 \leq \theta \leq 2\pi$$

4. The cylinder $y^2 + z^2 = 25$.

As with the last one this can be tricky until you see how to do it. In this case it makes some sense to use cylindrical coordinates since they can be easily used to write down the equation of a cylinder. In cylindrical coordinates the equation of a cylinder of radius a is given by

$$r = a$$

and so the equation of the cylinder in this problem is $r = 5$.

Next, we have the following conversion formulas.

$$x = x \quad y = r\sin\theta \quad z = r\cos\theta$$

Notice that they are slightly different from those that we are used to seeing. We needed to change them up here since the cylinder was centered upon the x-axis. Finally, we know what r is so we can easily write down a parametric representation for this cylinder.

$$\vec{r}(x,\theta) = x\vec{i} + 5\sin\theta\vec{j} + 5\cos\theta\vec{k}$$

We will also need the restriction $0 \le \theta \le 2\pi$ to make sure that we don't retrace any portion of the cylinder. Since we haven't put any restrictions on the "height" of the cylinder there won't be any restriction on x. In the first part of this example we used the fact that the function was in the form $x = f(y,z)$ to quickly write down a parametric representation. This can always be done for functions that are in this basic form.

$$z = f(x,y) \quad \Rightarrow \quad \vec{r}(x,y) = x\vec{i} + y\vec{j} + f(x,y)\vec{k}$$
$$x = f(y,z) \quad \Rightarrow \quad \vec{r}(x,y) = f(y,z)\vec{i} + y\vec{j} + z\vec{k}$$
$$y = f(x,z) \quad \Rightarrow \quad \vec{r}(x,y) = x\vec{i} + f(x,z)\vec{j} + z\vec{k}$$

We have practiced writing down some parametric representations for some surfaces let's take a quick look at a couple of applications. Let's take a look at finding the tangent plane to the parametric surface S given by,

$$\vec{r}(u,v) = x(u,v)\vec{i} + y(u,v)\vec{j} + z\left(u,v\vec{k}\right)$$

First, define

$$\vec{r}_u(u,v) = \frac{\partial x}{\partial u}(u,v)\vec{i} + \frac{\partial y}{\partial u}(u,v)\vec{j} + \frac{\partial z}{\partial u}(u,v)\vec{k}$$
$$\vec{r}_v(u,v) = \frac{\partial x}{\partial v}(u,v)\vec{i} + \frac{\partial y}{\partial v}(u,v)\vec{j} + \frac{\partial z}{\partial v}(u,v)\vec{k}$$

Now, provided $\vec{r}_u \times \vec{r}_v \ne \vec{0}$ it can be shown that the vector $\vec{r}_u \times \vec{r}_v$ will be orthogonal to the surface S. This means that it can be used for the normal vector that we need in order to write down the equation of a tangent plane. This is an important idea that will be used many times throughout the next couple of sections.

Let's take a look at an example.

Example: Find the equation of the tangent plane to the surface given by

$$\vec{r}(u,v) = u\vec{i} + 2v^2\vec{j} + \left(u^2 + v\right)\vec{k}$$

at the point $(2,2,3)$.

Solution: Let's first compute $\vec{r}_u \times \vec{r}_v$. Here are the two individual vectors.

$$\vec{r}_u(u,v) = \vec{i} + 2uk$$
$$\vec{r}_v(u,v) = 4v\vec{j} + \vec{k}$$

Now the cross product (which will give us the normal vector $\vec{n}$) is,

$$\vec{n} = \vec{r}_u \times \vec{t}_v = \begin{vmatrix} \vec{i} & \vec{j} & \vec{k} \\ 1 & 0 & 2u \\ 0 & 4v & 1 \end{vmatrix} = -8uv\vec{i} - \vec{j} + 4v\vec{k}$$

Now, this is all fine, but in order to use it we will need to determine the value of u and v that will give us the point in question.

We can easily do this by setting the individual components of the parametric representation equal to the coordinates of the point in question. Doing this gives,

$$2 = u \quad \Rightarrow \quad u = 2$$
$$2 = 2v^2 \quad \Rightarrow \quad v = \pm 1$$
$$3 = u^2 + v$$

Now, as shown, we have the value of u, but there are two possible values of v. To determine the correct value of v let's plug u into the third equation and solve for v. This should tell us what the correct value is.

$$3 = 4 + v \quad \Rightarrow \quad v = -1$$

We know that we'll be at the point in question when $u = -1$ and $v = -1$. At this point the normal vector is,

$$\vec{n} = 16\vec{i} - \vec{j} - 4\vec{k}$$

The tangent plane is then,

$$16(x-2) - (y-2) - 4(z-3) = 0$$
$$16x - y - 4z = 18$$

You do remember how to write down the equation of a plane, right?

The second application that we want to take a quick look at is the surface area of the parametric surface S given by,

$$\vec{r}(u,v) = x(u,v)\vec{i} + y(u,v)\vec{j} + z(u,v)\vec{k}$$

and as we will see it again comes down to needing the vector $\vec{r}_u \times \vec{r}_v$.

So, provided S is traced out exactly once as (u,v) ranges over the points in D the surface area of S is given by,

$$A = \iint_D \left\|\vec{r}_u \times \vec{r}_v\right\| dA$$

Let's take a look at an example.

Example: Find the surface area of the portion of the sphere of radius 4 that lies inside the cylinder $x^2 + y^2 = 12$ and above the *xy*-plane.

Solution: First we need the parameterization of the sphere. We parameterized a sphere earlier in this section so there isn't too much to do at this point. Here is the parameterization.

$$\vec{r}(\theta,\varphi) = 4\sin\varphi\cos\theta\vec{i} + 4\sin\varphi\sin\theta\vec{j} + 4\cos\varphi\vec{k}$$

Next we need to determine D. Since we are not restricting how far around the z-axis we are rotating with the sphere we can take the following range for θ.

$$0 \le \theta \le 2\pi$$

Now, we need to determine a range for φ $_{\varphi\varphi}$. This will take a little work, although it's not too bad. First, let's start with the equation of the sphere.

$$x^2 + y^2 + z^2 = 16$$

Now, if we substitute the equation for the cylinder into this equation we can find the value of z where the sphere and the cylinder intersect.

$$x^2 + y^2 + z^2 = 16$$
$$12 + z^2 = 16$$
$$z^2 = 4 \quad \Rightarrow \quad z = \pm 2$$

Now, since we also specified that we only want the portion of the sphere that lies above the *xy*-plane we know that we need $z = 2$. We also know that $\rho = 4$. Plugging this into the following conversion formula we get,

$$z = \rho\cos\varphi$$
$$2 = 4\cos\varphi$$
$$\cos\varphi = \frac{1}{2} \quad \Rightarrow \quad \varphi = \frac{\pi}{3}$$

So, it looks like the range of φ will be,

$$0 \le \varphi \le \frac{\pi}{3}$$

Finally, we need to determine $\vec{r}_\theta \times \vec{r}_\varphi$ .Here are the two individual vectors.

$$\vec{r}_\theta(\theta,\varphi) = -4\sin\varphi\sin\theta\vec{i} - 4\sin\varphi\cos\theta\vec{j}$$

$$\vec{r}_\varphi(\theta,\varphi) = 4\cos\varphi\cos\theta\vec{i} + 4\cos\varphi\sin\theta\vec{j} - 4\sin\varphi\vec{k}$$

Now let's take the cross product.

$$\vec{r}_\theta \times \vec{r}_\varphi = \begin{vmatrix} \vec{i} & \vec{j} & \vec{k} \\ -4\sin\varphi\sin\theta & 4\sin\varphi\cos\theta & 0 \\ 4\cos\varphi\cos\theta & 2\cos\varphi\sin\theta & -4\sin\varphi \end{vmatrix}$$

$$= -16\sin^2\varphi\cos\theta\vec{i} - 16\sin\varphi\cos\varphi\sin^2\theta\vec{k} - 16\sin^2\varphi\sin\theta\vec{j} - 16\sin\varphi\cos\varphi\cos^2\theta\vec{k}$$

$$= -16\sin^2\varphi\cos\theta\vec{i} - 16\sin^2\varphi\sin\theta\vec{j} - 16\sin\varphi\cos\varphi\left(\sin^2\theta + \cos^2\theta\right)\vec{k}$$

$$= -16\sin^2\varphi\cos\theta\vec{i} - 16\sin^2\varphi\sin\theta\vec{j} - 16\sin\varphi\cos\varphi\vec{k}$$

We now need the magnitude of this,

$$\left\|\vec{r}_\theta \times \vec{r}_\varphi\right\| = \sqrt{256\sin^4\varphi\cos^2\theta + 256\sin^4\varphi\sin^2\theta + 256\sin^2\varphi cos^2\varphi}$$

$$= \sqrt{256\sin^4\varphi\left(\cos^2\theta + \sin^2\theta\right) + 256\sin^2\varphi cos^2\varphi}$$

$$= \sqrt{256\sin^4\varphi\left(\sin^2\varphi + \cos^2\varphi\right)}$$

$$= 16\sqrt{\sin^2\varphi}$$

$$= 16\left|\sin\varphi\right|$$

$$= 16\sin\varphi$$

We can drop the absolute value bars in the sine because sine is positive in the range of φthat we are working with. We can finally get the surface area.

$$A = \iint_D 16\sin\varphi dA$$

$$= \int_0^{2\pi}\int_0^{\frac{\pi}{3}} 16\sin\varphi d\varphi d\theta$$

$$= \int_0^{2\pi} -16\cos\varphi\Big|_0^{\pi/3} d\theta$$

$$= \int_0^{2\pi} 8 d\theta$$

$$= 16\pi$$

SURFACE INTEGRALS

It is now time to think about integrating functions over some surface, S, in three-dimensional space. Let's start off with a sketch of the surface S since the notation can get a little confusing once we get into it. Here is a sketch of some surface S.

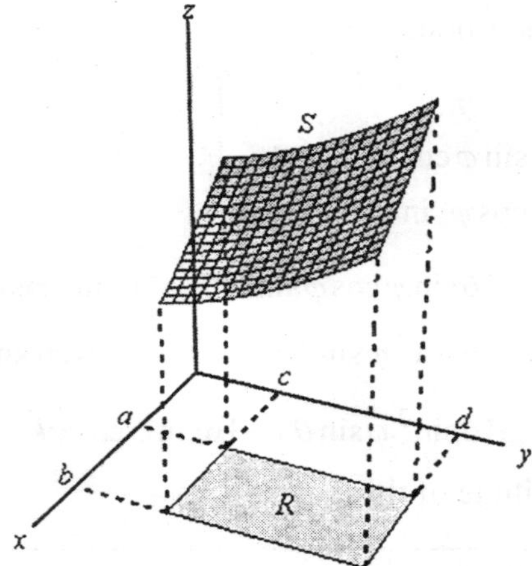

The region S will lie above (in this case) some region D that lies in the xy-plane. We used a rectangle here, but it doesn't have to be of course. Also note that we could just as easily looked at a surface S that was in front of some region D in the yz-plane or the xz-plane. Do not get so locked into the xy-plane that you can't do problems that have regions in the other two planes. Now, how we evaluate the surface integral will depend upon how the surface is given to us. There are essentially two separate methods here, although as we will see they are really the same.

First, let's look at the surface integral in which the surface S is given by $z = g(x, y)$. In this case the surface integral is,

$$\iint_S f(x,y,z)\,ds = \iint_D f(x,y,g(x,y))\sqrt{\left(\frac{\partial g}{\partial x}\right)^2 + \left(\frac{\partial g}{\partial y}\right)^2 + 1}\,dA$$

Now, we need to be careful here as both of these look like standard double integrals. In fact the integral on the right is a standard double integral.

The integral on the left however is a surface integral. The way to tell them apart is by looking at the differentials. The surface integral will have a dS while the standard double integral will have a dA.

In order to evaluate a surface integral we will substitute the equation of the surface in for z in the integrand and then add on the often messy square root. After that the integral is a standard double integral and by this point we should be able to deal with that. Note as well that there are similar formulas for surfaces given by $y = g(x,z)$ (with D in the xz-plane) and $y = g(x,z)$ (with D in the yz-plane). We will see one of these formulas in the examples and we'll leave the other to you to write down. The second method for evaluating a surface integral is for those surfaces that are given by the parameterization,

$$\vec{r}(u,v) = x(u,v)\vec{i} + y(u,v)\vec{j} + z(u,v)\vec{k}$$

In these cases the surface integral is,

$$\iint_S f(x,y,z)\,dS = \iint_S f(\vec{r}(u,v))\|\vec{r}_u \times \vec{r}_v\|\,dA$$

where D is the range of the parameters that trace out the surface S. Before we work some examples let's notice that since we can parameterize a surface given by $z = g(x,y)$ as,

$$\vec{r}(x,y) = x\vec{i} + y\vec{j} + g(x,y)\vec{k}$$

we can always use this form for these kinds of surfaces as well. In fact it can be shown that,

$$\|\vec{r}_x \times \vec{r}_y\| = \sqrt{\left(\frac{\partial g}{\partial x}\right)^2 + \left(\frac{\partial g}{\partial y}\right)^2 + 1}$$

for these kinds of surfaces. You might want to verify this for the practice of computing these cross products.

Let's work some examples.

Example: Evaluate $\iint_S 6xy\,dS$ where S is the portion of the plane $x + y + z = 1$ that lies in front of the yz-plane.

Solution: We are looking for the portion of the plane that lies in front of the yz-plane we are going to need to write the equation of the surface in the form $x = g(y,z)$. This is easy enough to do.

$$x = 1 - y - z$$

Next we need to determine just what D is. Here is a sketch of the surface S.

Here is a sketch of the region D.

Notice that the axes are labeled differently than we are used to seeing in the sketch of D. This was to keep the sketch consistent with the sketch of the surface. We arrived at the equation of the hypotenuse by setting x equal to zero in the equation of the plane and solving for z. Here are the ranges for y and z.

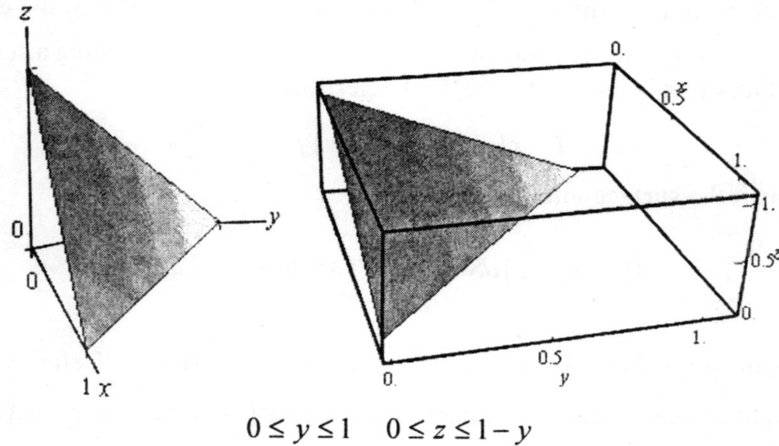

$$0 \le y \le 1 \quad 0 \le z \le 1 - y$$

Now, because the surface is not in the form $z = g(x, y)$ we can't use the formula above. However, as noted above we can modify this formula to get one that will work for us. Here it is,

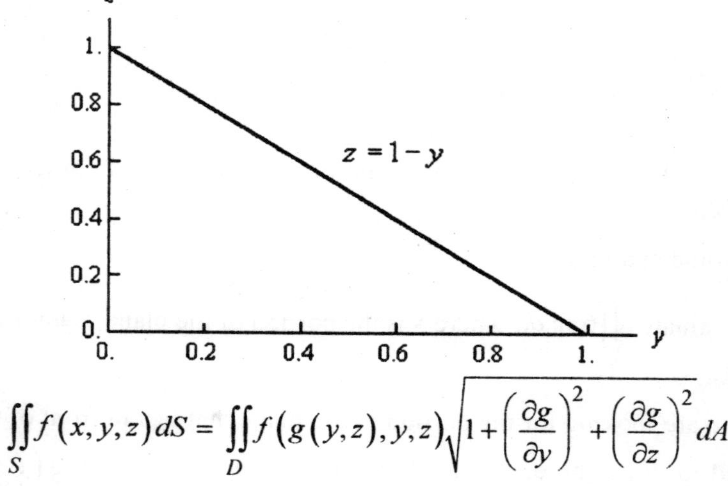

$$\iint\limits_S f(x, y, z)\, dS = \iint\limits_D f(g(y, z), y, z)\sqrt{1 + \left(\frac{\partial g}{\partial y}\right)^2 + \left(\frac{\partial g}{\partial z}\right)^2}\, dA$$

The changes made to the formula should be the somewhat obvious changes. So, let's do the integral.

$$\iint\limits_S 6xy\, dS = \iint\limits_D 6(1 - y - z)\, y\sqrt{1 + (-1)^2 + (-1)^2}\, dA$$

Notice that we plugged in the equation of the plane for the x in the integrand. At this point we've got a fairly simple double integral to do. Here is that work.

$$\iint_S 6xy\,dS = \sqrt{3}\iint_D 6\left(y - y^2 - zy\right)dA$$

$$= 6\sqrt{3}\int_0^1\int_0^{1-y} y - y^2 - zy\,dz\,dy$$

$$= 6\sqrt{3}\int_0^1\left(yz - zy^2 - \frac{1}{2}z^2y\right)\Bigg|_0^{1-y} dy$$

$$= 6\sqrt{3}\int_0^1 \frac{1}{2}y - y^2 + \frac{1}{2}y^3\,dy$$

$$-6\sqrt{3}\left(\frac{1}{4}y^2 - \frac{1}{3}y^3 + \frac{1}{8}y^4\right)\Bigg|_0^1 - \frac{\sqrt{3}}{4}$$

Example: Evaluate $\iint_S z\,dS$ where S is the upper half of a sphere of radius 2.

Solution: We gave the parameterization of a sphere in the previous section. Here is the parameterization for this sphere.

$$\vec{r}(\theta,\varphi) = 2\sin\varphi\cos\theta\vec{i} + 2\sin\varphi\sin\theta\vec{j} + 2\cos\varphi\vec{k}$$

Since we are working on the upper half of the sphere here are the limits on the parameters.

$$0 \le \theta \le 2\pi \qquad 0 \le \varphi \le \frac{\pi}{2}$$

Next, we need to determine $\vec{r}_\theta \times \vec{r}_\varphi$. Here are the two individual vectors.

$$\vec{r}_\theta(\theta,\varphi) = -2\sin\varphi\sin\theta\vec{i} + 2\sin\varphi\cos\theta\vec{j}$$

$$\vec{r}_\varphi(\theta,\varphi) = 2\cos\varphi\cos\theta\vec{i} + 2\cos\varphi\sin\theta\vec{j} - 2\sin\varphi\vec{k}$$

Now let's take the cross product.

$$\vec{r}_\theta \times \vec{r}_\varphi = \begin{vmatrix} \vec{i} & \vec{j} & \vec{k} \\ -2\sin\varphi\sin\theta & 2\sin\varphi\cos\theta & 0 \\ 2\cos\varphi\cos\theta & 2\cos\varphi\cos\theta & -2\sin\varphi \end{vmatrix}$$

$$= -4\sin^2\varphi\cos\theta\vec{i} - 4\sin\varphi\cos\varphi\sin^2\theta\vec{k} - 4\sin^2\varphi\sin\theta\vec{j} - 4\sin\varphi\cos\varphi\cos^2\theta\vec{k}$$

$$= -4\sin^2\varphi\cos\theta\vec{i} - 4\sin^2\varphi\sin\theta\vec{j} - 4\sin\varphi\cos\varphi\left(\sin^2\theta + \cos^2\theta\right)\vec{k}$$

$$= -4\sin^2\varphi\cos\theta\vec{i} - 4\sin^2\varphi\sin\theta\vec{j} - 4\sin\varphi\cos\varphi\vec{k}$$

Finally, we need the magnitude of this,

$$\|\vec{r}_\theta \times \vec{r}_\varphi\| = \sqrt{16\sin^4\varphi\cos^2\theta + 16\sin^4\varphi\sin^2\theta + 16\sin^2\varphi cos^2\varphi}$$
$$= \sqrt{16\sin^4\varphi\left(\cos^2\theta + \sin^2\theta\right) + 16\sin^2\varphi cos^2\varphi}$$
$$= \sqrt{16\sin^2\varphi\left(\sin^2\varphi + \cos^2\varphi\right)}$$
$$= 4\sqrt{\sin^2\varphi}$$
$$= 4|\sin\varphi|$$
$$= 4\sin\varphi$$

We can drop the absolute value bars in the sine because sine is positive in the range of φ that we are working with. The surface integral is then,

$$\iint_S z\, dS = \iint_D 2\cos\varphi(4\sin\varphi)\,dA$$

Don't forget that we need to plug in for x, y and/or z in these as well, although in this case we just needed to plug in z. Here is the evaluation for the double integral.

$$\iint_S z\,dS = \int_0^{2\pi}\int_0^{\frac{\pi}{2}} 4\sin(2\varphi)\,d\varphi\,d\theta$$
$$= \int_0^{2\pi}\left(-2\cos(2\varphi)\right)\Big|_0^{\frac{\pi}{2}}\,d\theta$$
$$= \int_0^{2\pi} 4d\theta$$
$$= 8\pi$$

Example: Evaluate $\iint_S y\,dS$ where S is the portion of the cylinder $x^2 + y^2 = 3$ that lies between $z = 0$ and $z = 6$.

Solution: We parameterized up a cylinder in the previous section. Here is the parameterization of this cylinder.

$$\vec{r}(z,\theta) = \sqrt{3}\cos\theta\vec{i} + \sqrt{3}\sin\theta\vec{j} + z\vec{k}$$

The ranges of the parameters are,

$$0 \le z \le 6 \quad 0 \le \theta \le 2\pi$$

Now we need $\vec{r}_z \times \vec{r}_\theta$. Here are the two vectors.

$$\vec{r}_z(z,\theta) = \vec{k}$$
$$\vec{r}_\theta(z,\theta) = -\sqrt{3}\sin\theta\vec{i} + \sqrt{3}\cos\theta\vec{j}$$

Here is the cross product.

$$\vec{r}_z \times \vec{r}_\theta = \begin{vmatrix} \vec{i} & \vec{j} & \vec{k} \\ 0 & 0 & 1 \\ -\sqrt{3}\sin\theta & \sqrt{3}\cos\theta & 0 \end{vmatrix}$$
$$= -\sqrt{3}\cos\theta\vec{i} - \sqrt{3}\sin\theta\vec{j}$$

The magnitude of this vector is,

$$\|\vec{r}_z \times \vec{r}_\theta\| - \sqrt{3\cos^2\theta + 3\sin^2\theta} = \sqrt{3}$$

The surface integral is then,

$$\iint_S y\,dS = \iint_D \sqrt{3}\sin\theta\left(\sqrt{3}\right)dA$$
$$= 3\int_0^{2\pi}\int_0^6 \sin\theta\,dz\,d\theta$$
$$= 3\int_0^{2\pi} 6\sin\theta\,d\theta$$
$$= \left(-18\cos\theta\right)\Big|_0^{2\pi}$$
$$= 0$$

Example: Evaluate $\iint_S y + zdS$ where S is the surface whose side is the cylinder $x^2 + y^2 = 3$, whose bottom is the disk $x^2 + y^2 \le 3$ in the *xy*-plane and whose top is the plane $z = 4 - y$.

Solution: There is a lot of information that we need to keep track of here. First, we are using pretty much the same surface (the integrand is different however) as the previous example. However, unlike the previous example we are putting a top and bottom on the surface this time. Let's first start out with a sketch of the surface.

Actually we need to be care.ul here. There is more to this sketch than the actual surface itself. We're going to let S_1be the portion of the cylinder that goes from the *xy*-plane to the plane. In other words, the top of the cylinder will be at an angle.

We'll call the portion of the plane that lies inside (*i.e.* the cap on the cylinder) S_2 $_{S_2 S_1}$.

Finally, the bottom of the cylinder (not shown here) is the disk of radius $\sqrt{3}$ in the *xy*-plane and is denoted byS_3.

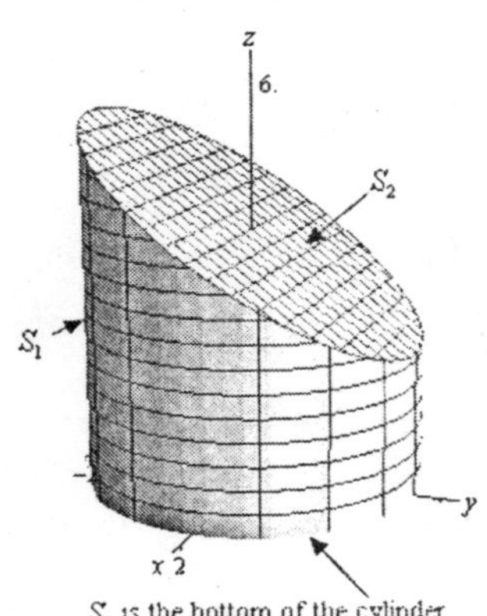

S_3 is the bottom of the cylinder

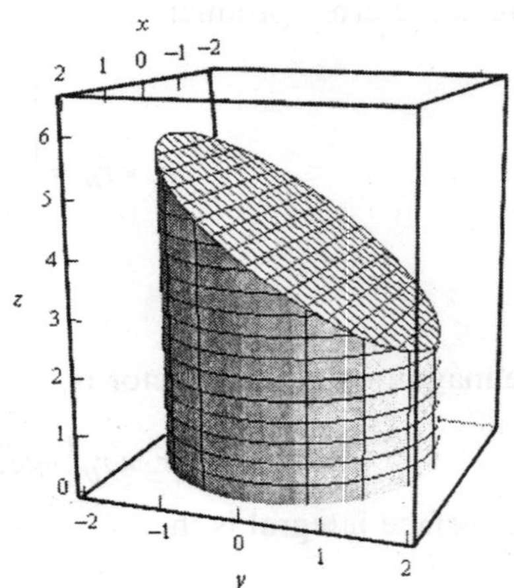

In order to do this integral we'll need to note that just like the standard double integral, if the surface is split up into pieces we can also split up the surface integral. So, for our example we will have,

$$\iint_S y+z\,dS = \iint_{S_1} y+z\,dS + \iint_{S_2} y+z\,dS + \iint_{S_3} y+z\,dS$$

We're going to need to do three integrals here. However, we've done most of the work for the first one in the previous example so let's start with that.

S_1: THE CYLINDER

The parameterization of the cylinder and $\|\vec{r}_z \times \vec{r}_\theta\|$ is,

$$\vec{r}(z,\theta) = \sqrt{3}\cos\theta\vec{i} + \sqrt{3}\sin\theta\vec{j} + z\vec{k}$$

$$\|\vec{r}_z \times \vec{r}_\theta\| = \sqrt{3}$$

The difference between this problem and the previous one is the limits on the parameters. Here they are.

$$0 \le \theta \le 2\pi$$

$$0 \le z \le 4 - y = 4 - \sqrt{3}\sin\theta$$

The upper limit for the z's is the plane so we can just plug that in. However, since we are on the cylinder we know what y is from the parameterization so we will also need to plug that in. Here is the integral for the cylinder.

$$\iint_{S_1} y+z\,ds = \iint_D \left(\sqrt{3}\sin\theta + z\right)\left(\sqrt{3}\right)dA$$

$$= \sqrt{3}\int_0^{2\pi}\int_0^{4-\sqrt{3}\sin\theta} \sqrt{3}\sin\theta + z\,dz\,d\theta$$

$$= \sqrt{3} \int_0^{2\pi} \sqrt{3} \sin\theta \left(4 - \sqrt{3}\sin\theta\right) + \frac{1}{2}\left(4 - \sqrt{3}\sin\theta\right)^2 d\theta$$
$$= \sqrt{3} \int_0^{2\pi} 8 - \frac{3}{2}\sin^2\theta \, d\theta$$
$$= \sqrt{3} \int_0^{2\pi} 8 - \frac{3}{4}\left(1 - \cos(2\theta)\right) d\theta$$
$$= \sqrt{3}\left(\frac{29}{4}\theta + \frac{3}{4}\sin(2\theta)\right)\Big|_0^{2\pi}$$
$$= \frac{29\sqrt{3}\pi}{2}$$

S_2: PLANE ON TOP OF THE CYLINDER

In this case we don't need to do any parameterization since it is set up to use the formula that we gave at the start of this section. Remember that the plane is given by $z = 4 - y$. Also note that, for this surface, D is the disk of radius $\sqrt{3}$ centered at the origin.

Here is the integral for the plane.

$$\iint_{S_2} y + z \, dS = \iint_D (y + 4 - y)\sqrt{(0)^2 + (-1)^2 + 1} \, dA$$
$$= \sqrt{2} \iint_D 4 dA$$

Don't forget that we need to plug in for z! Now at this point we can proceed in one of two ways. Either we can proceed with the integral or we can recall that $\iint_D dA$ is nothing more than the area of D and we know that D is the disk of radius and so there is no reason to do the integral. Here is the remainder of the work for this problem.

$$\iint_{S_2} y + z \, dS = 4\sqrt{2} \iint_D dA$$
$$= 4\sqrt{2}\left(\pi\left(\sqrt{3}\right)^2\right)$$
$$= 12\sqrt{2}\pi$$

S_3: BOTTOM OF THE CYLINDER

Again, this is set up to use the initial formula we gave in this section once we realize that the equation for the bottom is given by $g(x, y) = 0$ and D is the disk of radius $\sqrt{3}$ centered at the origin. Also, don't forget to plug in for z.

Here is the work for this integral.

$$\iint_{S_3} y+z\,ds = \iint_D (y+0)\sqrt{(0)^2+(0)^2+(1)^2}\,dA$$
$$= \iint_D y\,dA$$
$$= \int_0^{2\pi}\int_0^{\sqrt{3}} r^2 \sin\theta\,dr\,d\theta$$
$$= \int_0^{2\pi}\left(\frac{1}{3}r^3\sin\theta\right)\Bigg|_0^{\sqrt{3}} d\theta$$
$$= \sqrt{3}\cos\theta\Big|_0^{2\pi}$$
$$= 0$$

We can now get the value of the integral that we are after.

$$\iint_S y+z\,dS = \iint_{S_1} y+z\,dS + \iint_{S_2} y+z\,dS + \iint_{S_3} y+z\,dS$$
$$= \frac{29\sqrt{3}\pi}{2} + 12\sqrt{2}\pi + 0$$
$$= \frac{\pi}{2}\left(29\sqrt{3}+24\sqrt{2}\right)$$

SURFACE INTEGRALS OF VECTOR FIELDS

Just as we did with line integrals we now need to move on to surface integrals of vector fields. Recall that in line integrals the orientation of the curve we were integrating along could change the answer. The same thing will hold true with surface integrals. So, before we really get into doing surface integrals of vector fields we first need to introduce the idea of an oriented surface.

Let's start off with a surface that has two sides (while this may seem strange, recall that the Mobius Strip is a surface that only has one side!) that has a tangent plane at every point (except possibly along the boundary). Making this assumption means that every point will have two unit normal vectors, and $\vec{n}_2 = -\vec{n}_1$. This means that every surface will have two sets of normal vectors. The set that we choose will give the surface an orientation.

There is one convention that we will make in regards to certain kinds of oriented surfaces. First we need to define a closed surface. A surface *S* is closed if it is the boundary of some solid region *E*. A good example of a closed surface is the surface of a sphere. We say that the closed surface *S* has a positive orientation if we choose the set of unit normal vectors that point outward from the region E while the negative orientation will be the set of unit normal vectors that point in towards the region E.

Note that this convention is only used for closed surfaces.

In order to work with surface integrals of vector fields we will need to be able to write down a formula for the unit normal vector corresponding to the orientation that we've chosen to work with. We have two ways of doing this depending on how the surface has been given to us.

First, let's suppose that the function is given by $z = g(x, y)$. In this case we first define a new function,

$$f(x, y, z) = z - g(x, y)$$

In terms of our new function the surface is then given by the equation $f(x, y, z)$. Now, recall that ∇f ∇f will be orthogonal (or normal) to the surface given by $f(x, y, z) = 0$. This means that we have a normal vector to the surface.

The only potential problem is that it might not be a unit normal vector. That isn't a problem since we also know that we can turn any vector into a unit vector by dividing the vector by its length. In or case this is,

$$\vec{n} = \frac{\nabla f}{\|\nabla f\|}$$

In this case it will be convenient to actually compute the gradient vector and plug this into the formula for the normal vector. Doing this gives,

$$\vec{n} = \frac{\nabla f}{\|\nabla f\|} = \frac{-g_x\vec{i} - g_y\vec{j} + \vec{k}}{\sqrt{(g_x)^2 + (g_y)^2 + 1}}$$

Now, from a notational standpoint this might not have been so convenient, but it does allow us to make a couple of additional comments. First, notice that the component of the normal vector in the z-direction (identified by the $\vec{k}$ $\vec{kk}$ in the normal vector) is always positive and so this normal vector will generally point upwards. It may not point directly up, but it will have an upwards component to it.

This will be important when we are working with a closed surface and we want the positive orientation. If we know that we can then look at the normal vector and determine if the "positive" orientation should point upwards or downwards. Remember that the "positive" orientation must point out of the region and this may mean downwards in places. Of course if it turns out that we need the downward orientation we can always take the negative of this unit vector and we'll get the one that we need.

Again, remember that we always have that option when choosing the unit normal vector. Before we move onto the second method of giving the surface we should point out that we only did this for surfaces in the form $z = g(x, y)$. We could just as easily done the above work for surfaces in the form $y = g(x, z)$ (so $f(x, y, z) = y - g(x, y)$) or for surfaces in the form $z = g(x, y)$ (so $f(x, y, z) = y - g(x, y)$).

Now, we need to discuss how to find the unit normal vector if the surface is given parametrically as,

$$\vec{r}(u, v) = x(u, v)\vec{i} + y(u, v)\vec{j} + z(u, v)\vec{k}$$

In this case recall that the vector $\vec{r}_u \times \vec{r}_v$ will be normal to the tangent plane at a particular point. But if the vector is normal to the tangent plane at a point then it will also be normal to the surface at that point. So, this is a normal vector. In order to guarantee that it is a unit normal vector we will also need to divide it by its magnitude. So, in the case of parametric surfaces one of the unit normal vectors will be,

$$\vec{n} = \frac{\vec{r}_u \times \vec{r}_v}{\left\| \vec{r}_u \times \vec{r}_v \right\|}$$

As with the first case we will need to look at this once it's computed and determine if it points in the correct direction or not. If it doesn't then we can always take the negative of this vector and that will point in the correct direction. Finally, remember that we can always parameterize any surface given by $z = g(x, y)$ (or $y = g(x, z)$ or $x = g(y, z)$) easily enough and so if we want to we can always use the parameterization formula to find the unit normal vector. Now look at oriented surfaces and their associated unit normal vectors we can actually give a formula for evaluating surface integrals of vector fields.

Given a vector field $\vec{F}$ with unit normal vector $\vec{n}$ then the surface integral of $\vec{F}$ over the surface S is given by,

$$\iint_S \vec{F} \bullet d\vec{S} = \iint_S \vec{F} \bullet \vec{n}\, dS$$

where the right hand integral is a standard surface integral. This is sometimes called the flux of $\vec{F}$ across S.

Before we work any examples let's notice that we can substitute in for the unit normal vector to get a *somewhat* easier formula to use. We will need to be careful with each of the following formulas however as each will assume a certain orientation and we may have to change the normal vector to match the given orientation.

Let's first start by assuming that the surface is given by $z = g(x, y)$ .In this case let's also assume that the vector field is given by $\vec{F} = P\vec{i} + Q\vec{j} + R\vec{k}$ and that the orientation that we are after is the "upwards" orientation. Under all of these assumptions the surface integral of $\vec{F}$ over S is,

$$\iint_S \vec{F} \bullet +d\vec{S} = \iint_S \vec{F} \bullet \vec{n}\, dS$$

$$= \iint_D \left(P\vec{i} + Q\vec{j} + R\vec{k}\right) \bullet \left(\frac{-g_x\vec{i} - g_y\vec{j} + \vec{k}}{\sqrt{\left(g_x\right)^2 + \left(g_y\right)^2 + 1}} \right) \sqrt{\left(g_x\right)^2 + \left(g_y\right)^2 + 1}\, dA$$

$$= \iint_D \left(P\vec{i} + Q\vec{j} + R\vec{k}\right) \bullet \left(-g_x\vec{i} - g_y\vec{j} + \vec{k}\right) dA$$

$$= \iint_D -Pg_x - Qg_y + R\, dA$$

Now, remember that this assumed the "upward" orientation. If we'd needed the "downward" orientation then we would need to change the signs on the normal vector. This would in turn change the signs on the integrand as well. So, we really need to be careful here when using this formula. In general it is best to rederive this formula as you need it. When we've been given a surface that is not in parametric form there are in fact 6 possible integrals here.

Two for each form of the surface $z = g(x, y), y = g(x, y)$ and $x = g(y, z)$. Given each form of the surface there will be two possible unit normal vectors and we'll need to choose the correct one to match the given orientation of the surface.

However, the derivation of each formula is similar to that given here and so shouldn't be too bad to do as you need to. Notice as well that because we are using the unit normal vector the messy square root will always drop out.

This means that when we do need to derive the formula we won't really need to put this in. All we'll need to work with is the numerator of the unit vector. We will see at least one more of these derived in the examples below. It should also be noted that the square root is nothing more than,

$$\sqrt{\left(g_x\right)^2 + \left(g_y\right)^2 + 1} = \left\|\nabla f\right\|$$

so in the following work we will probably just use this notation in place of the square root when we can to make things a little simpler. Let's now take a quick look at the formula for the surface integral when the surface is given parametrically by $\vec{r}(u, v)$. In this case the surface integral is,

$$\iint_S \vec{F} \bullet d\vec{S} = \iint_S \vec{F} \bullet \vec{n}\, dS$$

$$= \iint_D \vec{F} \bullet \left(\frac{\vec{r}_u \times \vec{r}_y}{\| \vec{r}_u \times \vec{r}_v \|} \right) \| \vec{r}_u \times \vec{r}_v \| \, dA$$

$$= \iint_D \vec{F} \bullet (\vec{r}_u \times \vec{r}_v) dA$$

Again note that we may have to change the sign on $\vec{r}_u \times \vec{r}_v$ to match the orientation of the surface and so there is once again really two formulas here. Also note that again the magnitude cancels in this case and so we won't need to worry that in these problems either.

Note as well that there are even times when we will used the definition, $\iint_S \vec{F} \bullet d\vec{S} = \iint_S \vec{F} \bullet \vec{n}\, dS$, directly. We will see an example of this below.

Let's now work a couple of examples.

Example: Evaluate $\iint_S \vec{F} \bullet d\vec{S}$ where $\vec{F} = y\vec{j} - z\vec{k}$ and S is the surface given by the paraboloid $y = x^2 + z^2, 0 \le y \le 1$ and the disk $x^2 + z^2 \le 1$ at $y = 1$. Assume that S has positive orientation.

Solution: The disk is really nothing more than the cap on the paraboloid. This means that we have a closed surface. This is important because we've been told that the surface has a positive orientation and by convention this means that all the unit normal vectors will need to point outwards from the region enclosed by *S*. Let's first get a sketch of *S* so we can get a feel for what is going on and in which direction we will need to unit normal vectors to point.

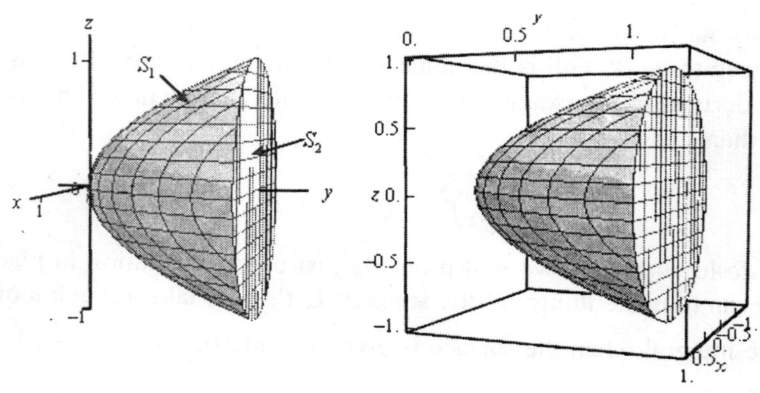

As noted in the sketch we will denote the paraboloid by S_1 and the disk by S_2. Also note that in order for unit normal vectors on the paraboloid to point away from the region they will all need to point generally in the negative y direction.

On the other hand, unit normal vectors on the disk will need to point in the positive y direction in order to point away from the region. Since S is composed of the two surfaces we'll need to do the surface integral on each and then add the results to get the overall surface integral.

Let's start with the paraboloid. In this case we have the surface in the form $y = g(x,z)$ so we will need to derive the correct formula since the one given initially wasn't for this kind of function. This is easy enough to do however. First define,

$$f(x,y,z) = y - g(x,z) = y - x^2 - z^2$$

We will next need the gradient vector of this function.

$$\nabla f = \langle -2x, 1, --2z \rangle$$

Now, the y component of the gradient is positive and so this vector will generally point in the positive y direction. However, as noted above we need the normal vector point in the negative y direction to make sure that it will be pointing away from the enclosed region. This means that we will need to use

$$\vec{n} = \frac{-\nabla f}{\|-\nabla f\|} = \frac{\langle 2x, -1, 2z \rangle}{\|\nabla f\|}$$

Let's note a couple of things here before we proceed. We don't really need to divide this by the magnitude of the gradient since this will just cancel out once we actually do the integral. So, because of this we didn't bother computing it. Also, the dropping of the minus sign is not a typo. When we compute the magnitude we are going to square each of the components and so the minus sign will drop out.

S_1: THE PARABOLOID

The surface integral in this case.

$$\begin{aligned}\iint_S \vec{F} \bullet +d\vec{S} &= \iint_D \left(y\vec{j} - z\vec{k} \right) \bullet \left(\frac{\langle 2x, -1, 2z \rangle}{\|\nabla f\|} \right) \|\nabla f\| \, dA \\ &= \iint_D -y - 2z^2 \, dA \\ &= \iint_D -\left(x^2 + z^2\right) - 2z^2 \, dA \\ &= -\iint_D x^2 + 3z^2 \, dA\end{aligned}$$

Don't forget that we need to plug in the equation of the surface for y before we actually compute the integral. In this case D is the disk of radius 1 in the xz-plane and so it makes sense to use polar coordinates to complete this integral. Here are polar coordinates for this region.

$$x = r\cos\theta \quad z = r\sin\theta$$
$$0 \le \theta \le 2\pi \quad 0 \le r \le 1$$

Note that we kept the x conversion formula the same as the one we are used to using for x and let z be the formula that used the sine. We could have done it any order, however in this way we are at least working with one of them as we are used to working with.

Here is the evaluation of this integral.

$$\iint_{S_1} \vec{F} \bullet +d\vec{S} = -\iint_D x^2 + 3z^2 dA$$

$$= -\int_0^{2\pi}\int_0^1 \left(r^2\cos^2\theta + 3r^2\sin^2\theta\right) r dr\, d\theta$$

$$= -\int_0^{2\pi}\left(\frac{1}{2}(1+\cos(2\theta)) + \frac{3}{2}(1-\cos(2\theta))\right)\left(\frac{1}{4}r^4\right)\Bigg|_0^1 d\theta$$

$$= -\frac{1}{8}\int_0^{2\pi} 4 - 2\cos(2\theta)\, d\theta$$

$$= -\frac{1}{8}(4\theta - \sin(2\theta))\Bigg|_0^{2\pi}$$

$$= -\pi$$

S_2: THE CAP OF THE PARABOLOID

We can now do the surface integral on the disk (cap on the paraboloid). This one is actually fairly easy to do and in fact we can use the definition of the surface integral directly. First let's notice that the disk is really just the portion of the plane $y = 1$ that is in front of the disk of radius 1 in the xz-plane. Now we want the unit normal vector to point away from the enclosed region and since it must also be orthogonal to the plane $y = 1$ then it must point in a direction that is parallel to the y-axis, but we already have a unit vector that does this. Namely,

$$\vec{n} = \vec{j}$$

the standard unit basis vector. It also points in the correct direction for us to use. Because we have the vector field and the normal vector we can plug directly into the definition of the surface integral to get,

$$\iint_{S_2} \vec{F} \bullet d\vec{S} = \iint_{S_2} \left(y\vec{j} - z\vec{k}\right) \bullet \left(\vec{j}\right) dS = \iint_{S_2} y\, dS$$

At this point we need to plug in for y (since S_2 is a portion of the plane $y = 1$ we do know what it is) and we'll also need the square root this time when we convert the surface integral over to a double integral. In this case since we are using the definition directly we won't get the canceling of the square root that we saw with the first portion. To get the square root well need to acknowledge that

$$y = 1 = g(x, z)$$

and so the square root is,

$$\sqrt{(g_x)^2 + 1 + (g_z)^2}$$

The surface integral is then,

$$\iint_{S_2} \vec{F} \bullet d\vec{S} = \iint_{S_2} \vec{F} \bullet d\vec{S}$$

$$= \iint_D 1\sqrt{0+1+0}\,dA = \iint_D dA$$

At this point we can acknowledge that D is a disk of radius 1 and this double integral is nothing more than the double integral that will give the area of the region D so there is no reason to compute the integral. Here is the value of the surface integral.

$$\iint_{S_2} \vec{F} \bullet d\vec{S} = \pi$$

Finally, to finish this off we just need to add the two parts up. Here is the surface integral that we were actually asked to compute.

$$\iint_S \vec{F} \bullet d\vec{S} = \iint_{S_1} \vec{F} \bullet d\vec{S} + \iint_{S_2} \vec{F} \bullet d\vec{S} = -\pi + \pi = 0$$

Example: Evaluate $\iint_S \vec{F} \bullet d\vec{S}$ where $\vec{F} = x\vec{i} + y\vec{j} + z^4\vec{k}$ and S is the upper half the sphere $x^2 + y^2 + z^2 = 9$ and the disk $x^2 + y^2 \le 9$ in the plane $z = 0$. Assume that S has the positive orientation.

Solution: So, as with the previous problem we have a closed surface and since we are also told that the surface has a positive orientation all the unit normal vectors must point away from the enclosed region. To help us visualize this here is a sketch of the surface.

We will call S_1 the hemisphere and S_2 will be the bottom of the hemisphere (which isn't shown on the sketch). Now, in order for the unit normal vectors on the sphere to point away from enclosed region they will all need to have a positive z component.

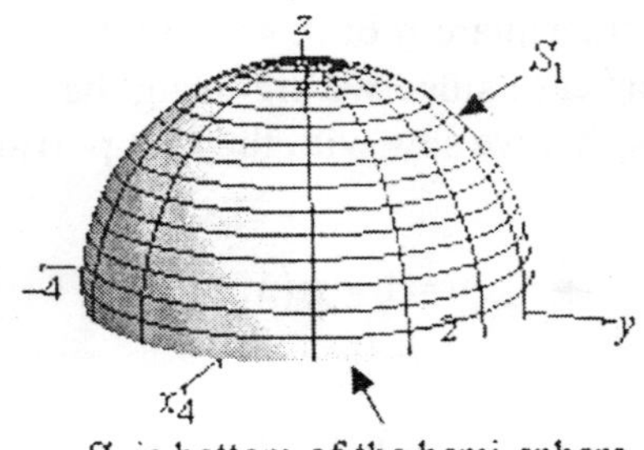

S_2 is bottom of the hemi-sphere

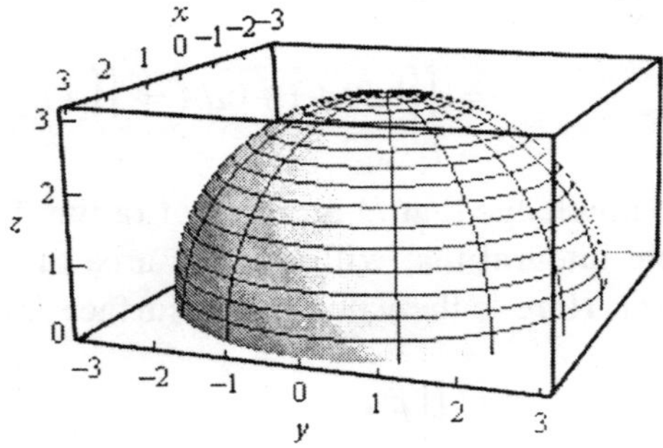

Remember that the vector must be normal to the surface and if there is a positive z component and the vector is normal it will have to be pointing away from the enclosed region. On the other hand, the unit normal on the bottom of the disk must point in the negative z direction in order to point away from the enclosed region.

S_1: THE SPHERE

Let's do the surface integral on first. In this case since the surface is a sphere we will need to use the parametric representation of the surface. This is,

$$\vec{r}(\theta,\varphi) = 3\sin\varphi\cos\theta\vec{i} + 3\sin\varphi\sin\theta\vec{j} + 3\cos\varphi\vec{k}$$

Since we are working on the hemisphere here are the limits on the parameters that we'll need to use.

$$0 \le \theta \le 2\pi \quad 0 \le \varphi \le \frac{\pi}{2}$$

Next, we need to determine $\vec{r}_\theta \times \vec{r}_\varphi$. Here are the two individual vectors and the cross product.

$$\vec{r}_\theta(\theta,\varphi) = -3\sin\varphi\sin\theta\vec{i} + 3\sin\varphi\cos\theta\vec{j}$$
$$\vec{r}_\varphi(\theta,\varphi) = 3\cos\varphi\cos\theta\vec{i} + 3\cos\varphi\sin\theta\vec{j} - 3\sin\varphi\vec{k}$$

$$\vec{r}_\theta \times \vec{r}_\varphi = \begin{vmatrix} \vec{i} & \vec{j} & \vec{k} \\ -3\sin\varphi\sin\theta & 3\sin\varphi\cos\theta & 0 \\ 3\cos\varphi\cos\theta & 3\cos\varphi sin\theta & -3\sin\varphi \end{vmatrix}$$
$$= -9\sin^2\varphi\cos\theta\vec{i} - 9\sin\varphi\cos\varphi\sin^2\theta\vec{k} - 9\sin^2\varphi\sin\theta\vec{j} - 9\sin\varphi\cos\varphi\cos^2\theta\vec{k}$$
$$= -9\sin^2\varphi\cos\theta\vec{i} - 9\sin^2\varphi\sin\theta\vec{j} - 9\sin\varphi\cos\theta\left(\sin^2\theta + \cos^2\theta\right)\vec{k}$$
$$-9\sin^2\varphi\cos\theta\vec{i} - 9\sin^2\varphi\sin\theta\vec{j} - 9\sin\varphi\cos\varphi\vec{k}$$

Note that we won't need the magnitude of the cross product since that will cancel out once we start doing the integral. Notice that for the range of φthat we've got both sine and cosine are positive and so this vector will have a negative z component and as we noted above in order for this to point away from the enclosed area we will need the z component to be positive. Therefore we will need to use the following vector for the unit normal vector.

$$\vec{n} = -\frac{\vec{r}_\theta \times \vec{r}_\varphi}{\left\|\vec{r}_\theta \times \vec{r}_\varphi\right\|} = \frac{9\sin^2\varphi\cos\theta\vec{i} + 9\sin^2\theta\sin\theta\vec{j} + 9\sin\varphi\cos\varphi\vec{k}}{\left\|\vec{r}_\theta \times \vec{r}_\varphi\right\|}$$

Again, we will drop the magnitude once we get to actually doing the integral since it will just cancel in the integral.

Next we'll need

$$\vec{F}\left(\vec{r}(\theta,\varphi)\right) = 3\sin\varphi\cos\theta\vec{i} + 3\sin\varphi\sin\theta\vec{j} + 81\cos^4\varphi\vec{k}$$

Remember that in this evaluation we are just plugging in the x component of $\vec{r}(\theta,\varphi)$ into the vector field *etc.* We also may as well get the dot product out of the way that we know we are going to need.

$$\vec{F}\left(\vec{r}(\theta,\varphi)\right)\cdot\left(\vec{r}_\theta \times \vec{r}_\varphi\right) = 27\sin^3\varphi\cos^2\theta + 27\sin^3\varphi\sin^2\theta + 729\sin\varphi\cos^5\varphi$$
$$= 27\sin^3\varphi + 729\sin\varphi\cos^5\varphi$$

Now we can do the integral.

$$\iint_{S_1} \vec{F}\cdot d\vec{S} = \iint_D \vec{F}\cdot\left(\frac{\vec{r}_u \times \vec{r}_v}{\left\|\vec{r}_\theta \times \vec{r}_\varphi\right\|}\right)\left\|\vec{r}_\theta \times \vec{r}_\varphi\right\| dA$$

$$= \int_0^{2\pi} \int_0^{\frac{\pi}{2}} 27\sin^3\varphi\left(1 \cos^2\varphi\right) + 729\sin\varphi\cos^5\varphi d\varphi d\theta$$

$$= -\int_0^{\pi 2}\left(27\left(\cos\varphi - \frac{1}{3}\cos^3\varphi\right) + \frac{729}{6}\cos^6\varphi\right)\Bigg|_0^{\frac{\pi}{2}} d\theta$$

$$= \int_0^{\pi 2} \frac{279}{2} d\theta$$

$$= 279\pi$$

S_1: THE BOTTOM OF THE HEMI-SPHERE

Now, we need to do the integral over the bottom of the hemisphere. In this case we are looking at the disk $x^2 + z^2 \le 9$ that lies in the plane $z = 0$ and so the equation of this surface is actually $z = 0$. The disk is really the region D that tells us how much of the surface we are going to use. This also means that we can use the definition of the surface integral here with

$$\vec{n} = -\vec{k}$$

We need the negative since it must point away from the enclosed region. The surface integral in this case is,

$$\iint_{S_2} \vec{F} \bullet d\vec{S} = \iint_{S_2} \left(x\vec{i} + y\vec{j} + z^4\vec{k}\right) \bullet \left(-\vec{k}\right) dS$$

$$= \iint_{S_2} -z^4 d\vec{S}$$

Remember, however, that we are in the plane given by $z = 0$ and so the surface integral becomes,

$$\iint_{S_2} \vec{F} \bullet d\vec{S} = \iint_{S_2} -z^4 dS$$

$$= \iint_{S_2} 0 \, dS = 0$$

The last step is to then add the two pieces up. Here is surface integral that we were asked to look at.

$$\iint_{S} \vec{F} \bullet d\vec{S} = \iint_{S1} \vec{F} \bullet dS + \iint_{S_2} \vec{F} \bullet dS$$

$$= 279\pi + 0 = 279\pi$$

We will leave this section with a quick interpretation of a surface integral over a vector field. If is the velocity field of a fluid then the surface integral

$$\iint_S \vec{v} \cdot d\vec{S}$$

represents the net amount of fluid flowing through S.

STOKES' THEOREM

In this section we are going to take a look at a theorem that is a higher dimensional version of Green's Theorem. In Green's Theorem we related a line integral to a double integral over some region. In this section we are going to relate a line integral to a surface integral. However, before we give the theorem we first need to define the curve that we're going to use in the line integral.

Let's start off with the following surface with the indicated orientation.

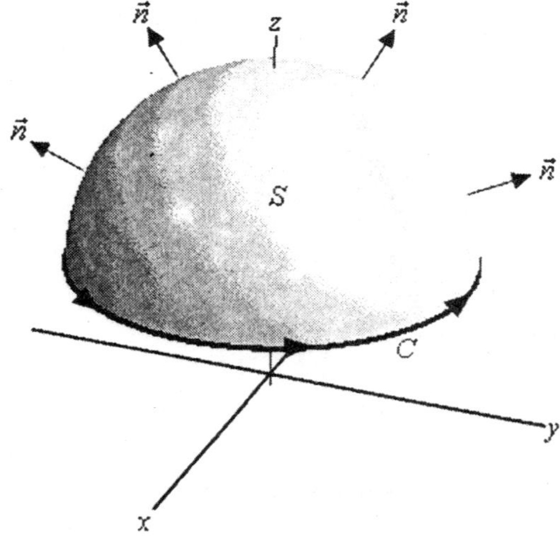

Around the edge of this surface we have a curve C. This curve is called the boundary curve. The orientation of the surface S will induce the positive orientation of C. To get the positive orientation of C think of yourself as walking along the curve. While you are walking along the curve if your head is pointing in the same direction as the unit normal vectors while the surface is on the left then you are walking in the positive direction on C.

STOKES' THEOREM

Let S be an oriented smooth surface that is bounded by a simple, closed, smooth boundary curve C with positive orientation. Also let $\vec{F}$ be a vector field then,

$$\int_C \vec{F} \cdot d\vec{r} = \iint_S \text{curl } \vec{F} \cdot d\vec{S}$$

In this theorem note that the surface S can actually be any surface so long as its boundary curve is given by C. This is something that can be used to our advantage to simplify the surface integral on occasion.

Let's take a look at a couple of examples.

Example: Use Stokes' Theorem to evaluate $\iint_S \text{curl}\, \vec{F} \bullet d\vec{S}$ where $\vec{F} = z^2\, \vec{i} - 3xy\, \vec{j} + x^3y^3\vec{k}$ and S is the part of $z = 5 - x^2 - y^2$ above the plane $z = 1$. Assume that S is oriented upwards.

Solution: Let's start this off with a sketch of the surface.

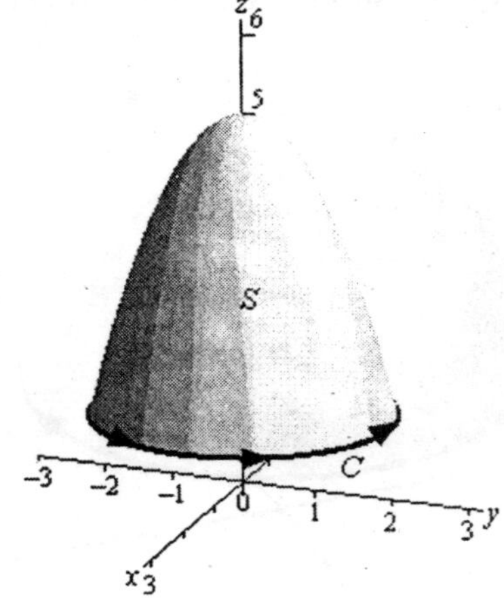

In this case the boundary curve C will be where the surface intersects the plane and so will be the curve

$$1 = 5 - x^2 - y^2$$
$$x^2 + y^2 = 4 \text{ at } z = 1$$

So, the boundary curve will be the circle of radius 2 that is in the plane. The parameterization of this curve is,

$$\vec{r}(t) 2\cos t\, \vec{i} + 2\sin t\, \vec{j} + \vec{k}, \quad 0 \le t \le 2\pi$$

The first two components give the circle and the third component makes sure that it is in the plane z = 1.

Using Stokes' Theorem we can write the surface integral as the following line integral.

$$\iint_S \text{curl}\, \vec{F} \bullet d\vec{S} = \int_C \vec{F} \bullet d\vec{r} = \int_0^{2\pi} (\vec{r}(t)) \bullet \vec{r}(t)\, dt$$

So, it looks like we need a couple of quantities before we do this integral. Let's first get the vector field evaluated on the curve. Remember that this is simply plugging the components of the parameterization into the vector field.

$$\vec{F}(\vec{r}(t)) = (1)^2\,\vec{i} - 3(2\cos t)(2\sin t)\,\vec{j} + (2\cos t)^3(2\sin t)^3\,\vec{k}$$
$$= \vec{i} - 12\cos t\sin y\;\vec{j} + 64\cos^3 t\sin^3 t\sin^3 t\vec{k}$$

Next, we need the derivative of the parameterization and the dot product of this and the vector field.

$$\vec{r}'(t) = -2\sin y\vec{i} + 2\cos t\,\vec{j}$$
$$\vec{F}(\vec{r}(t))\cdot\vec{r}(t) = -2\sin t - 24\sin t\cos^2 t$$

We can now do the integral.

$$\iint_S \operatorname{curl}\vec{F}\cdot d\vec{S} = \int_0^{2\pi} -2\sin t - 24\sin t\cos^2$$
$$= \left(2\cos t + 8\cos^3 t\right)\Big|_0^{2\pi}$$
$$= 0$$

Example: Use Stokes' Theorem to evaluate $\int_C \vec{F}\cdot d\vec{r}$ where $\vec{F} = z^2\,\vec{i} + y^2\,\vec{j} + x\,\vec{k}$ and C is the triangle with vertices $(1,0,0), (0,0,1)$ and $(0,0,1)$ with counter-clockwise rotation.

Solution: We are going to need the curl of the vector field eventually so let's get that out of the way first.

$$\operatorname{curl}\vec{F} = \begin{vmatrix} \vec{i} & \vec{j} & \vec{k} \\ \dfrac{\partial}{\partial z} & \dfrac{\partial}{\partial y} & \dfrac{\partial}{\partial z} \\ z^2 & y^2 & x \end{vmatrix} = 2z\;\vec{j} - \vec{j} = (2z-1)\,\vec{j}$$

Now, all we have is the boundary curve for the surface that we'll need to use in the surface integral. However, as noted above all we need is any surface that has this as its boundary curve. So, let's use the following plane with upwards orientation for the surface.

Since the plane is oriented upwards this induces the positive direction on C as shown. The equation of this plane is,

$$x + y + z = 1 \quad\Rightarrow\quad z = g(x,y) = 1 - x - y$$

Now, let's use Stokes' Theorem and get the surface integral set up.

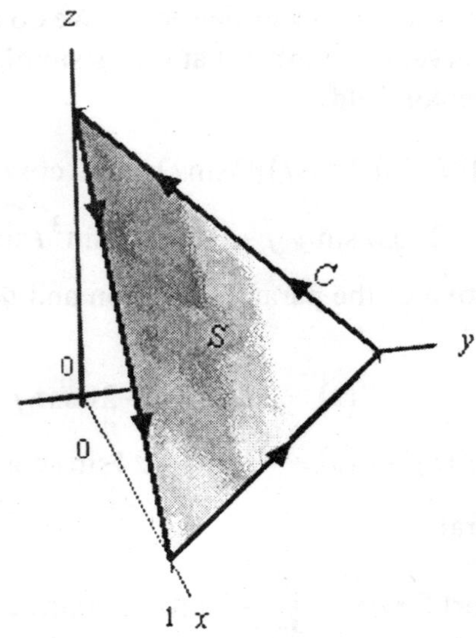

$$\iint_C \vec{F} \bullet d\vec{r} = \iint_S \text{curl } \vec{F} \bullet d\vec{S}$$

$$= \iint_S (2z-1)\,\vec{j} \bullet d\vec{S}$$

$$= \iint_D (2z-1)\,\vec{j} \bullet \frac{\nabla f}{\|\nabla f\|} \|\nabla f\| \, dA$$

First let's get the gradient. Recall that this comes from the function of the surface.

$$f(x,y,z) = z - g(x,y) = z - 1 + x + y$$

$$\nabla f = \vec{i} + \vec{j} + \vec{k}$$

Note as well that this also points upwards and so we have the correct direction.

Now, D is the region in the xy-plane shown below,

We get the equation of the line by plugging in $z = 0$into the equation of the plane. So based on this the ranges that define D are,

$$0 \le x \le 1 \quad 0 \le y \le -x+1$$

The integral is then,

$$\int_C \vec{F} \bullet d\vec{r} = \iint_D (2z-1)\,\vec{j} \bullet \left(\vec{i} + \vec{j} + \vec{k}\right)$$

$$= \int_0^1 \int_0^{-x+1} 2(1-x-y) - 1 dy\, dx$$

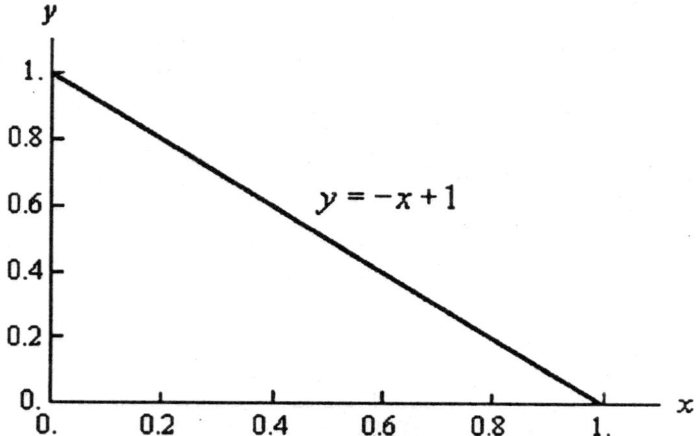

Don't forget to plug in for z since we are doing the surface integral on the plane. Finishing this out gives,

$$\iint_C \vec{F} \cdot d\vec{r} = \int_0^1 \int_0^{-x+1} 1 - 2x - 2y \, dy \, dx$$

$$= \int_0^1 \left(y - 2xy - y^2\right)\Big|_0^{1x+1} dx$$

$$= \int_0^1 x^2 - x \, dx$$

$$= \left(\frac{1}{3}x^3 - \frac{1}{2}x^2\right)\Bigg|_0^1$$

$$= -\frac{1}{6}$$

DIVERGENCE THEOREM

In this section we are going to relate surface integrals to triple integrals. We will do this with the Divergence Theorem.

DIVERGENCE THEOREM

Let E be a simple solid region and S is the boundary surface of E with positive orientation. Let $\vec{F}$ be a vector field whose components have continuous first order partial derivatives. Then,

$$\iint_S \vec{F} \cdot dS = \iiint_E \text{div } \vec{F} \, dV$$

Let's see an example of how to use this theorem.

Example: Use the divergence theorem to evaluate $\iint_S \vec{F} \cdot dS$ where $\vec{F} = xy\vec{i} - \frac{1}{2}$

$y^2\vec{j} + z\vec{k}$ and the surface consists of the three surfaces, $z = 4 - 3x^2 - 3y^2, 1 \le z \le 4$ on the top, $x^2 + y^2 = 1, 0 \le z \le 1$ on the sides and z = 0on the bottom.

Solution: Let's start this off with a sketch of the surface.

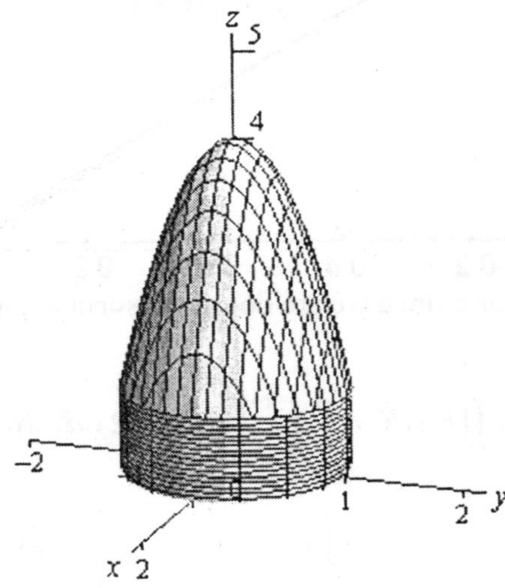

The region E for the triple integral is then the region enclosed by these surfaces. Note that cylindrical coordinates would be a perfect coordinate system for this region. If we do that here are the limits for the ranges.

$$0 \le z \le 4 - 3r^2$$
$$0 \le r \le 1$$
$$0 \le \theta \le 2\pi$$

We'll also need the divergence of the vector field so let's get that.

$$\operatorname{div}\vec{F} = y - y + 1 = 1$$

The integral is then,

$$\iint_S \vec{F} \bullet d\vec{S} = \iiint_E \operatorname{div} \vec{F}\, dV$$
$$= \int_0^{2\pi}\int_0^1\int_0^{4-3r^2} r\, dz\, dr\, d\theta$$
$$= \int_0^{2\pi}\int_0^1 4r - 3r^3\, dr\, d\theta$$
$$= \int_0^{2\pi}\left(2r^2 - \frac{3}{4}r^4\right)\Bigg|_0^1 d\theta$$
$$= -\int_0^{2\pi}\frac{5}{4}d\theta = \frac{5}{2}\pi$$

Chapter 4

Multiple Integral

The multiple integral is a type of *definite* integral extended to functions of more than one real variable, for example, $f(x,y)$ or $f(x,y,z)$. Just as the definite integral of a positive function of one variable represents the area of the region between the graph of the function and the x-axis, the double integral of a positive function of two variables represents the volume of the region between the surface defined by the function and the plane which contains its domain. (Note that the same volume can be obtained via the triple integral — the integral of a function in three variables — of the constant function $f(x, y, z) = 1$ over the above-mentioned region between the surface and the plane.) If there are more variables, a multiple integral will yield hypervolumes of multi-dimensional functions.

Multiple integration of a function in n variables: $\int(x_1,x_2,...,x_n)$ over a domain D is most commonly represented by nesting integral signs in the reverse order of execution (the leftmost integral sign is computed last) proceeded by the function and integrand arguments in proper order (the rightmost argument is computed last). The domain of integration is either represented symbolically for every integrand over each integral sign, or is often abbreviated by a variable at the rightmost integral sign:

$$\int\int...\int_D f(x_1,x_2,...,x_n)dx_1dx_2...dx_n$$

Since it is impossible to calculate the antiderivative of a function of more than one variable, *indefinite* multiple integrals do not exist. Therefore all multiple integrals are *definite* integrals.

DOUBLE INTEGRALS

Before starting on double integrals let's do a quick review of the definition of a definite integrals for functions of single variables. First, when working with the integral,

$$\int_a^b f(x)\,dx$$

we think of x's as coming from the interval $a \le x \le b$. For these integrals we can say that we are integrating over the interval $a \leq x \leq b$. Note that this does assume that $a \leq b$, however, if we have $b \leq a$ then we can just use the interval $b \leq x \leq a$.

Now, when we derived the definition of the definite integral we first thought of this as an area problem. We first asked what the area under the curve was and to do this we broke up the interval $a \leq x \leq b$ into n subintervals of width ∇x and choose a point, x_i^*, from each interval as shown below,

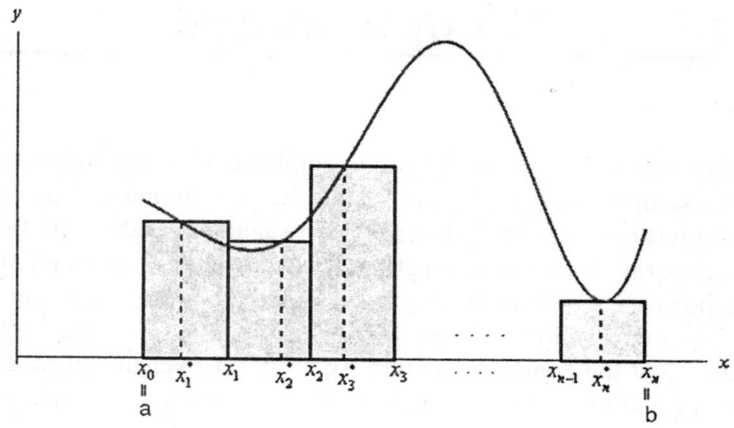

Each of the rectangles has height of $f\left(x_i^*\right)$ and we could then use the area of each of these rectangles to approximate the area as follows.

$$Ak+ \approx f\left(x_0^*\right)\Delta x + f\left(x_1^*\right)\Delta x + \cdots + f\left(x_1^*\right)\Delta x + \cdots + f\left(x_n^*\right)\Delta x$$

To get the exact area we then took the limit as n goes to infinity and this was also the definition of the definite integral.

$$\int_a^b f(x)\,dx = \lim_{n \to \infty} \sum_{i=1}^{n} f\left(x_i^*\right)\Delta x$$

In this section we want to integrate a function of two variables, $f(x, y)$. With functions of one variable we integrated over an interval (*i.e.* a one-dimensional space) and so it makes some sense then that when integrating a function of two variables we will integrate over a region of $\mathbb{R}^2$ (two-dimensional space).

We will start out by assuming that the region in $\mathbb{R}^2$ is a rectangle which we will denote as follows,

$$R = |a, b| \times |c, d|$$

This means that the ranges for x and y are $a \leq x \leq b$ and $c \leq y \leq d$.

Also, we will initially assume that $f(x, y) \geq 0$ although this doesn't really have to be the case. Let's start out with the graph of the surface S give by graphing $f(x, y)$ over the rectangle R.

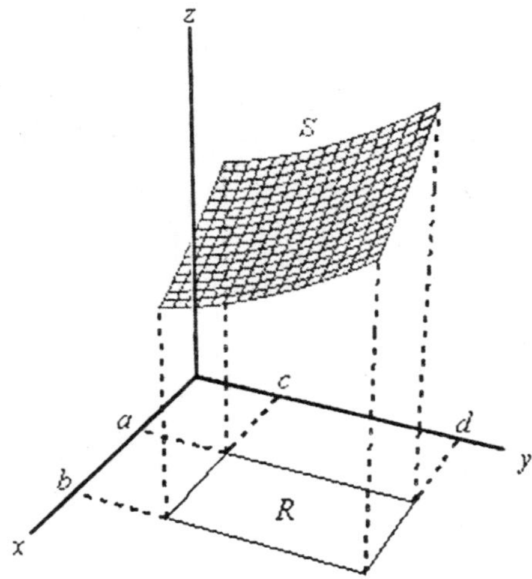

Now, just like with functions of one variable let's not worry about integrals quite yet. Let's first ask what the volume of the region under S (and above the xy-plane of course) is.

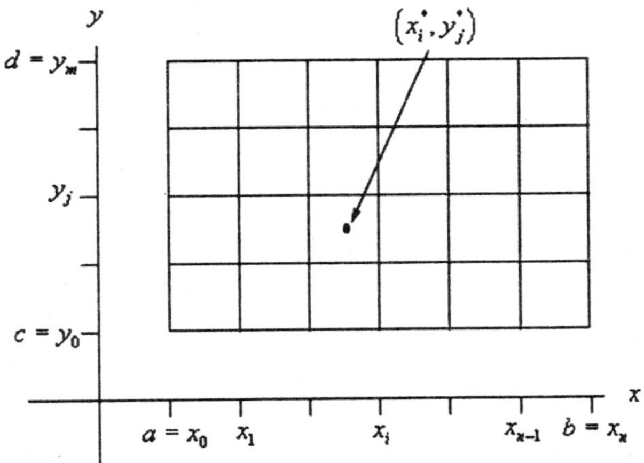

We will first approximate the volume much as we approximated the area above. We will first divide up $a \leq x \leq b$ into n subintervals and divide up $c \leq y \leq d$ into m subintervals. This will divide up R into a series of smaller rectangles and from each of these we will choose a point (xi, y_i). Here is a sketch of this set up.

Now, over each of these smaller rectangles we will construct a box whose height is given by $f\left(x_i^*, y_i^*\right)$. Here is a sketch of that.

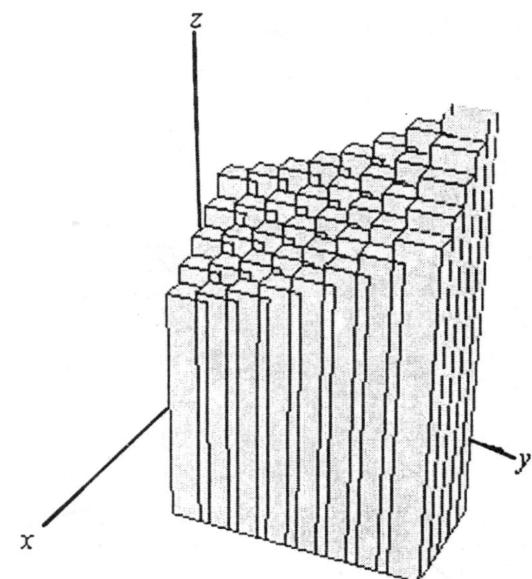

Each of the rectangles has a base area of ΔA and a height of $f\left(x_i^*, y_i^*\right)$ so the volume of each of these boxes is $f\left(x_i^*, y_i^*\right)\Delta A$. The volume under the surface S is then approximately,

$$V \approx \sum_{i=1}^{n}\sum_{j=1}^{m} f\left(x_i^*, y_i^*\right)\Delta A$$

We will have a *double* sum since we will need to add up volumes in both the x and y directions.

To get a better estimation of the volume we will take n and m larger and larger and to get the exact volume we will need to take the limit as both n and m go to infinity. In other words,

$$V = \lim_{n,m\to\infty}\sum_{i=1}^{n}\sum_{j=1}^{m} f\left(x_i^*, y_i^*\right)\Delta A$$

Now, this should look familiar. This looks a lot like the definition of the integral of a function of single variable. In fact this is also the definition of a double integral, or more exactly an integral of a function of two variables over a rectangle.

Here is the official definition of a double integral of a function of two variables over a rectangular region R as well as the notation that we'll use for it.

$$\iint_R f(x,y)\,dA = \lim_{n,m\to\infty} \sum_{i=1}^{n}\sum_{j=1}^{m} f\left(x_i^*, y_i^*\right)\Delta A$$

Note the similarities and differences in the notation to single integrals. We have two integrals to denote the fact that we are dealing with a two dimensional region and we have a differential here as well. Note that the differential is dA instead of the dx and dy that we're used to seeing. Note as well that we don't have limits on the integrals in this notation.

Instead we have the R written below the two integrals to denote the region that we are integrating over. Note that one interpretation of the double integral of $f(x,y)$ over the rectangle R is the volume under the function $f(x,y)$ (and above the xy-plane). Or,

$$\text{Volume} = \iint_R f(x,y)\,dA$$

We can use this double sum in the definition to estimate the value of a double integral if we need to. We can do this by choosing $\left(x_i^*, y_i^*\right)$ to be the midpoint of each rectangle. When we do this we usually denote the point as $\left(\overline{x}_i, \overline{y}_j\right)$. This leads to the Midpoint Rule,

$$\iint_R f(x,y)\,dA \approx \sum_{i=1}^{n}\sum_{j=1}^{m} f\left(\overline{x}_i, \overline{y}_j\right)\Delta A$$

In the next section we start looking at how to actually compute double integrals.

ITERATED INTEGRALS

In the previous section we gave the definition of the double integral. However, just like with the definition of a single integral the definition is very difficult to use in practice and so we need to start looking into how we actually compute double integrals. We will continue to assume that we are integrating over the rectangle

$$R = |a,b| \times |c,d|$$

We will look at more general regions in the next section.

FUBINI'S THEOREM

If $f(x,y)$ is continuous on $R = |a,b| \times |c,d|$ then,

$$\iint_R f(x,y)\,dA = \int_a^b \int_c^d f(x,y)\,dy\,dx = \int_c^d \int_a^b f(x,y)\,dx\,dy$$

These integrals are called iterated integrals.

Note that there are in fact two ways of computing a double integral and also notice that the inner differential matches up with the limits on the inner integral and similarly for the out differential and limits. In other words, if the inner differential is dy then the limits on the inner integral must be y limits of integration and if the outer differential is dy then the limits on the outer integral must be y limits of integration.

Now, on some level this is just notation and doesn't really tell us how to compute the double integral. Let's just take the first possibility above and change the notation a little.

$$\iint_R f(x,y)\,dA = \int_a^b\left[\int_c^d f(x,y)\,dy\right]dx$$

We will compute the double integral by first computing

$$\int_c^d f(x,y)\,dy$$

and we compute this by holding x constant and integrating with respect to y as if this were an single integral. This will give a function involving only x's which we can in turn integrate. We've done a similar process with partial derivatives. To take the derivative of a function with respect to y we treated the x's as constants and differentiated with respect to y as if it was a function of a single variable.

Double integrals work in the same manner. We think of all the x's as constants and integrate with respect to y or we think of all y's as constants and integrate with respect to x.

Let's take a look at some examples.

Example: Compute each of the following double integrals over the indicated rectangles.

1. $\iint_R 6xy^2\,dA, R=[2,4]\times[1,2]$
2. $\iint_R 2x-4y^3\,dA, R=[-5,4]\times[0,3]$
3. $\iint_R x^2y^2+\cos(\pi x)+\sin(\pi y)\,dA, R=[-2,-1]\times[0,1]$
4. $\iint_R \frac{1}{(2x+3y)^2}\,dA, R=[0,1]\times[1,2]$
5. $\iint_R xe^{xy}\,dA, R=[-1,2]\times[0,1]$

Solution:

1. $\iint_R 6xy^2\,dA, R=[2,4]\times[1,2]$

It doesn't matter which variable we integrate with respect to first, we will get the same answer regardless of the order of integration. To prove that let's work this one with each order to make sure that we do get the same answer.

Solution 1: In this case we will integrate with respect to y first. So, the iterated integral that we need to compute is,

$$\iint_R 6xy^2 dA = \int_1^4 \int_1^2 6xy^2 dy\, dx$$

When setting these up make sure the limits match up to the differentials. Since the dy is the inner differential (*i.e.* we are integrating with respect to y first) the inner integral needs to have y limits.

To compute this we will do the inner integral first and we typically keep the outer integral around as follows,

$$\iint_R 6xy^2 dA = \int_1^4 \left(2xy^2\right)\Big|_1^2 dx$$

$$= \int_2^4 16x - 2x\, dx$$

$$= \int_2^4 14x\, dx$$

Remember that we treat the x as a constant when doing the first integral and we don't do any integration with it yet. Now, we have a normal single integral so let's finish the integral by computing this.

$$\iint_R 6xy^2 dA = 7x^2\Big|_2^4 = 84$$

Solution 2: In this case we'll integrate with respect to x first and then y. Here is the work for this solution.

$$\iint_R 6xy^2 dA = \int_1^2 \int_2^4 6xy^2 dx\, dy$$

$$= \int_1^2 \left(3x^2y^2\right)\Big|_2^4 = dy$$

$$= \int_1^2 36y^2$$

$$= 12y^3\Big|_1^2$$

$$= 84$$

Sure enough the same answer as the first solution.

So, remember that we can do the integration in any order.

2. $\iint\limits_R 2x - 4y^3 dA = R = [-5,4]\times[0,3]$

For this integral we'll integrate with respect to y first.

$$\iint\limits_R 2x - 4y^3 dA = \int_{-5}^{4}\int_0^3 2x - 4y^3 dy\, dx$$
$$= \int_{-5}^{4}\left(2xy - y^4\right)\Big|_0^3 dx$$
$$= \int_{-5}^{4}\left(2xy - y^4\right)\Big|_0^3 dx$$
$$= \left(3x^2 - 81x\right)\Big|_{-5}^{4}$$
$$= -756$$

Remember that when integrating with respect to y all x's are treated as constants and so as far as the inner integral is concerned the $2x$ is a constant and we know that when we integrate constants with respect to y we just tack on a y and so we get $2xy$ from the first term.

3. $\iint\limits_R x^2y^2 + \cos(\pi x) + \sin(\pi y)\, dA, R = [-2,-1]\times[0,1]$

In this case we'll integrate with respect to x first.

$$\iint\limits_R x^2y^2 + \cos(\pi x) + \sin(\pi y) dA = \int_0^1\int_{-2}^{-1} x^2y^2 - \cos(\pi x) + dx\, dy$$
$$= \int_0^1\left(\frac{1}{3}x^2y^2 + \frac{1}{\pi}\sin(\pi x) + x\sin(\pi y)\right)\Bigg|_{-2}^{-1}$$
$$= \int_0^1 \frac{7}{3}y^2 + \sin(\pi y)\, dy$$
$$= \frac{7}{9}y^3 - \frac{1}{\pi}\cos(\pi y)\Big|^1$$
$$= \frac{7}{9} + \frac{2}{\pi}$$

4. $\iint\limits_R \frac{1}{(2x+3y)^2} dA, R = [0,1]\times[1,2]$

In this case because the limits for x are kind of nice (*i.e.* they are zero and one which are often nice for evaluation) let's integrate with respect to x first.

We'll also rewrite the integrand to help with the first integration.

$$\iint\limits_R (2x+3y)^{-2}\,dA = \int_1^2 \int_0^1 (2x+3y)^{-2}\,dcx\,dy$$

$$= \int_1^2 \left(-\frac{1}{2}(2x+3y)^{-1}\right)\Bigg|_0^1 dy$$

$$= -\frac{1}{2}\int_1^2 \frac{1}{2+3y} - \frac{1}{3y}\,dy$$

$$= -\frac{1}{2}\left(\frac{1}{3}\ln|2+3y| - \frac{1}{3}\ln|y|\right)\Bigg|_1^2$$

$$= -\frac{1}{6}(\ln 8 - \ln 2 - \ln 5)$$

5. $\iint\limits_R xe^{xy}\,dA, R = [-1,2]\times[0,1]$

Now, while we can technically integrate with respect to either variable first sometimes one way is significantly easier than the other way. In this case it will be significantly easier to integrate with respect to y first as we will see.

$$\iint\limits_R xe^{xy}\,dA = \int_{-1}^2 \int_0^1 xe^{xy}\,dy\,dx$$

The y integration can be done with the quick substitution,

$$u = xy$$
$$du = xfy$$

which gives

$$\iint\limits_R xe^{xy}\,dA = \int_{-1}^2 e^{xy}\Big|_0^1\,dx$$

$$= \int_{-1}^2 e^x - 1dx$$

$$= \left(e^x - x\right)\Big|_{-1}^2$$

$$= e^2 - 2 - \left(e^{-1}+1\right)$$

$$= e^2 - e^{-1} - 3$$

So, not too bad of an integral there provided you get the substitution. Now let's see what would happen if we had integrated with respect to x first.

$$\iint_R xe^{xy}\, dA = \int_0^1 \int_{-1}^2 xe^{xy}\, dx\, dy$$

In order to do this we would have to use integration by parts as follows,

$$u = x \qquad dv = e^{xy} dx$$

$$du = dx \qquad v = \frac{1}{y} e^{xy}$$

The integral is then,

$$\iint_R xe^{xy}\, dA = \int_0^1 \left(\frac{x}{y} e^{xy} - \int \frac{1}{y} e^{xy}\, dx \right)\Bigg|_{-1}^{2} dy$$

$$= \int_0^1 \left(\frac{x}{y} e^{xy} - \frac{1}{y^2} e^{xy} \right)\Bigg|_{-1}^{2} dy$$

$$= \int_0^1 \left(\frac{2}{y} e^{2y} - \frac{1}{y^2} e^{2y} \right) - \left(-\frac{1}{y} e^{-y} - \frac{1}{y^2} e^{-y} \right) dy$$

We're not even going to continue here as these are very difficult integrals to do.

As we saw in the previous set of examples we can do the integral in either direction. However, sometimes one direction of integration is significantly easier than the other so make sure that you think about which one you should do first before actually doing the integral.

The next topic of this section is a quick fact that can be used to make some iterated integrals somewhat easier to compute on occasion.

Fact

If $f(x, y) = g(x)h(y)$ and we are integrating over the rectangle $R = [a, b] \times [c, d]$ then,

$$\iint_R f(x, y)\, dA = \iint_R g(x)h(y)\, dA = \left(\int_a^b g(x)\, dx \right)\left(\int_c^d h(y)\, dy \right)$$

So, if we can break up the function into a function only of x times a function of y then we can do the two integrals individually and multiply them together.

Let's do a quick example using this integral.

Example: Evaluate, $\iint_R x\cos^2(y)\, dA, R = [-2, 3] \times \left[0, \frac{\pi}{2}\right]$

Solution: Since the integrand is a function of x times a function of y we can use the fact.

$$\iint_R x\cos^2(y)\,dA = \left(\int_{-2}^{3} x\,dx\right)\left(\int_0^{\frac{\pi}{2}} \cos^2(y)\,dy\right)$$

$$= \left(\frac{1}{2}x^2\right)\Bigg|_{-2}^{3}\left(\frac{1}{2}\int_0^{\frac{\pi}{2}} 1+\cos(2y)\,dy\right)$$

$$= \left(\frac{5}{2}\right)\left(\frac{1}{2}\left(y+\frac{1}{2}\sin(2y)\right)\Bigg|_0^{\frac{\pi}{2}}\right)$$

$$= \frac{5\pi}{8}$$

We have one more topic to discuss in this section. This topic really doesn't have anything to do with iterated integrals, but this is as good a place as any to put it and there are liable to be some questions about it at this point as well so this is as good a place as any. What we want to do is discuss single indefinite integrals of a function of two variables. In other words we want to look at integrals like the following.

$$\int x\sec^2(2y)+4xy\,dy$$

$$\int x^3 - e^{-\frac{x}{y}}\,dx$$

We know that these integrals are asking what function that we differentiated to get the integrand. However, in this case we need to pay attention to the differential (*dy* or *dx*) in the integral, because that will change things a little. In the case of the first integral we are asking what function we differentiated with respect to y to get the integrand while in the second integral we're asking what function differentiated with respect to x to get the integrand. For the most part answering these questions isn't that difficult. The important issue is how we deal with the constant of integration.

Here are the integrals.

$$\int x\sec^2(2y)+4xy\,dy = \frac{x}{2}\tan(2y)+2xy^2+g(x)$$

$$\int x^3 - e^{-\frac{x}{y}}\,dx = \frac{1}{4}x^4 + ye^{-\frac{x}{y}} + h(y)$$

Notice that the "constants" of integration are now functions of the opposite variable. In the first integral we are differentiating with respect to y and we know that any function involving only x's will differentiate to zero and so when integrating with respect to y we need to acknowledge that there may have been a function of only x's in the function and so the "constant" of integration is a function of x. Likewise, in the second integral, the "constant" of integration must be a function of y since we are integrating with respect to x. Again, remember if we differentiate the answer with respect to x then any function of only y's will differentiate to zero.

DOUBLE INTEGRALS OVER GENERAL REGIONS

In the previous section we looked at double integrals over rectangular regions. The problem with this is that most of the regions are not rectangular so we need to now look at the following double integral,

Case 1

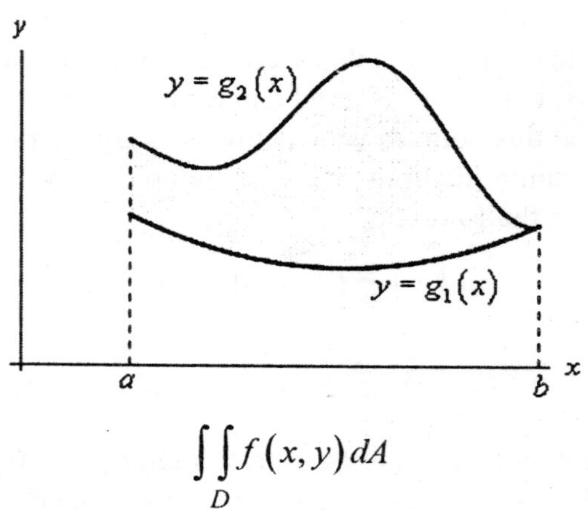

$$\iint\limits_D f(x, y)\, dA$$

where D is any region.

There are two types of regions that we need to look at. Here is a sketch of both of them.

Case 2

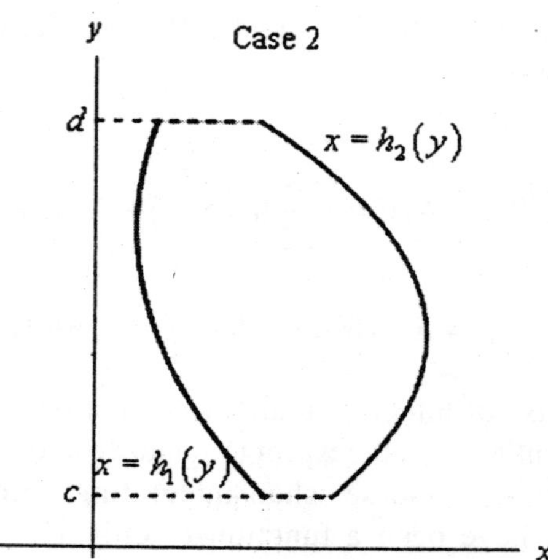

We will often use *set builder notation* to describe these regions. Here is the definition for the region in Case 1

$$D = \{(x,y) | a \le x \le b, g_1(x) \le y \le g_2(x)\}$$

and here is the definition for the region in Case 2.

$$D = \{(x,y) h_1(y) \le x \le h_2(y), c \le y \le d\}$$

This notation is really just a fancy way of saying we are going to use all the points,, in which both of the coordinates satisfy the two given inequalities. The double integral for both of these cases are defined in terms of iterated integrals as follows.

In Case 1 where $D = \{(x,y) | a \le x \le b, g_1(x) \le y \le g_2(x)\}$ the integral is defined to be,

$$\iint_D f(x,y)\,dA = \int_a^b \int_{g_1(x)}^{g_2(x)} f(x,y)\,dy\,dx$$

In Case 2 where $D = \{(x,y) | h_1(y) \le x \le h_2(y) c \le y \le d\}$ the integral is defined to be,

$$\iint_D f(x,y)\,dA = \int_c^d \int_{h_1(y)}^{h_2(y)} f(x,y)\,dx\,dy$$

Here are some properties of the double integral that we should go over before we actually do some examples. Note that all three of these properties are really just extensions of properties of single integrals that have been extended to double integrals.

Properties

1. $\iint_D f(x,y) + g(x,y)\,dA = \iint_D f(x,y)\,dA + \iint_D g(x,y)\,dA$
2. $\iint_D cf(x,y)\,dA = c\iint_D f(x,y)\,dA$, where c is any constant.
3. If the region D can be split into two separate regions D_1 and D_2 then the integral can be written as

$$\iint_D f(x,y)\,dA = \iint_D f(x,y)\,dA + \iint_{D_2} f(x,y)\,dA$$

Let's take a look at some examples of double integrals over general regions.

Example: Evaluate each of the following integrals over the given region D.

1. $\iint_D e^{\frac{x}{y}}\,dA, D = \{(x,y) | 1 \le y \le 2, y \le x \le y^3\}$

2. $\iint_D 4xy - y^3 dA$, D is the region bounded by $y = \sqrt{x}$ and $y = x^3$.

3. $\iint_D 6x^2 - 40y dA$, D is the triangle with vertices, $(0,3),(1,1)$ and $(5,3)$.

Solution:

1. $\iint_D e^{\frac{x}{y}} dA, D = \left\{(x,y) \middle| 1 \le y \le 2, y \le x \le y^3\right\}$

This first one is set up to just use the formula above so let's do that.

$$\iint_D e^{\frac{x}{y}} dA = \int_1^2 \int_y^{y^3} e^{\frac{x}{y}} dx\, dy = \int_1^2 ye^{\frac{x}{y}} \Bigg|_y^{y^3}$$

$$= \int_1^2 ye^{y^2} - ye^1 dy$$

$$= \left(\frac{1}{2}e^{y^2} - \frac{1}{2}y^2e^1\right)\Bigg|_1^2 = \frac{1}{2}e^4 - 2e^1$$

2. $\iint_D 4xy - y^3 dA$, D is the region bounded by $y = \sqrt{x}$ and $y = x^3$.

In this case we need to determine the two inequalities for *x* and *y* that we need to do the integral. The best way to do this is the graph the two curves. Here is a sketch.

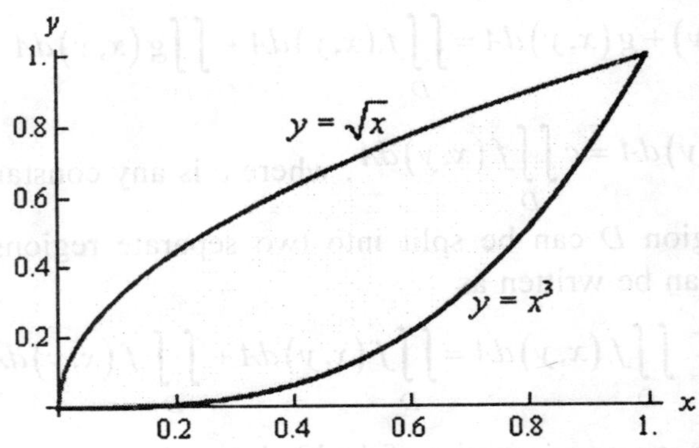

So, from the sketch we can see that that two inequalities are,

$$0 \le x \le 1 \quad x^3 \le y \le \sqrt{x}$$

We can now do the integral,

$$\iint_D 4xy - y^3\, dA = \int_0^1 \int_{x^3}^{\sqrt{x}} 4xy - y^3\, dy\, dx$$

$$= \int_0^1 \left(2xy^2 - \frac{1}{4}y^4 \right)\Bigg|_{x^3}^{\sqrt{x}} dx$$

$$= \int_0^1 \frac{7}{4}x^2 - 2x^7 + \frac{1}{4}x^{12}\, dx$$

$$= \left(\frac{7}{12}x^3 - \frac{1}{4}x^8 + \frac{1}{52}x^{13} \right)\Bigg|_0^1 = \frac{55}{156}$$

3. $\iint_D 6x^2 - 40y\, dA$, D is the triangle with vertices, $(0,3)$, $(1,1)$ and $(5,3)$

We got even less information about the region this time. Let's start this off by sketching the triangle.

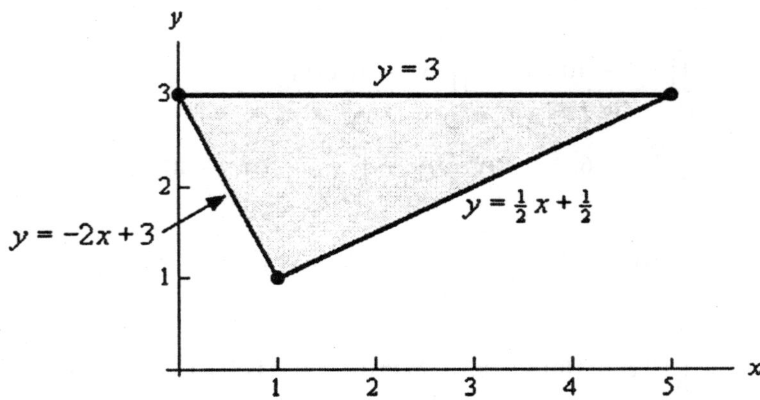

Since we have two points on each edge it is easy to get the equations for each edge and so we'll leave it to you to verify the equations. Now, there are two ways to describe this region. If we use functions of x, as shown in the image we will have to break the region up into two different pieces since the lower function is different depending upon the value of x. In this case the region would be given by $D = D_1 \cup D_2$ where,

$$D_1 = \{(x,y) \mid 0 \le x \le 1,\ -2x+3 \le y \le 3\}$$

$$D_2 = \left\{(x,y) \mid 1 \le x \le 5,\ \frac{1}{2}x + \frac{1}{2} \le y \le 3\right\}$$

Note the is the "union" symbol and just means that D is the region we get by combing the two regions. If we do this then we'll need to do two separate integrals, one for each of the regions. To avoid this we could turn things around and solve the two equations for x to get,

$$y=-2x+3 \quad \Rightarrow \quad x=-\frac{1}{2}y+\frac{3}{2}$$

$$y=\frac{1}{2}x+\frac{1}{2} \quad \Rightarrow \quad x=2y-1$$

If we do this we can notice that the same function is always on the right and the same function is always on the left and so the region is,

$$D=\left\{(x,y)\left|-\frac{1}{2}y+\frac{1}{3}\le x\le 2y-1, 1\le y\le 3\right.\right\}$$

Writing the region in this form means doing a single integral instead of the two integrals we'd have to do otherwise.

Either way should give the same answer and so we can get an example in the notes of splitting a region up let's do both integrals.

Solution 1:

$$\iint_D 6x^{21}-40y\,dA=\iint_{D_1} 6x^2-40y\,dA+\iint_{D_2} 6x^2-40y\,dA$$

$$=\int_0^1\int_{-2x+3}^3 6x^2-40y\,dy\,dx+\int_1^5\int_{\frac{1}{2}x+\frac{1}{2}}^3 6x^2-40y\,dy\,dx$$

$$=\int_0^1 (6x^2y-20y^2)\Big|_{-2x=3}^3 dx+\int_1^5 (6x^2y-20y^2)\Big|_{\frac{1}{2}x+\frac{1}{2}}^3 dx$$

$$=\int_0^1 12x^3-180+20(3-2x)^2\,dx+\int_1^5 -3x^3+15x^2-180+20\left(\frac{1}{2}x+\frac{1}{2}\right)^2 dx$$

$$=\left(3x^4-180x-\frac{10}{3}(3-2x)^3\right)\Bigg|_0^1+\left(-\frac{3}{4}x^4+5x^3-180x+\frac{40}{3}\left(\frac{1}{2}x+\frac{1}{2}\right)^3\right)\Bigg|_1^5$$

$$=-\frac{935}{3}$$

That was a lot of work. Notice however, that after we did the first substitution that we didn't multiply everything out. The two quadratic terms can be easily integrated with a basic substitution and so we didn't bother to multiply them out. We'll do that on occasion to make some of these integrals a little easier.

Solution 2: This solution will be a lot less work since we are only going to do a single integral.

$$\iint\limits_D 6x^2 - 40y\,dA = \int_1^3 \int_{-\frac{1}{2}y+\frac{3}{2}}^{2y-1} 6x^2 - 40y\,dx\,dy$$

$$= \int_1^3 \left(2x^2 - 40xy\right)\Big|_{-\frac{1}{2}y+\frac{3}{2}}^{2y-1} dy$$

$$= \int_1^3 100y - 100y^2 + 2(2y-1)^3 - 2\left(-\frac{1}{2}y+\frac{3}{2}\right)^3$$

$$= \left(50y^2 - \frac{100}{3}y^3 + \frac{1}{4}(2y-1)\right)^4 + \left(-\frac{1}{2}y+\frac{3}{2}\right)^4\Bigg|_1^3$$

$$= -\frac{935}{3}$$

So, the numbers were a little messier, but other than that there was much less work for the same result. As the last part of the previous example has shown us we can integrate these integrals in either order (*i.e.* x followed by y or y followed by x), although often one order will be easier than the other.

In fact there will be times when it will not even be possible to do the integral in one order while it will be possible to do the integral in the other order.

Let's see a couple of examples of these kinds of integrals.

Example: Evaluate the following integrals by first reversing the order of integration.

1. $\int_0^3 \int_{x^2}^9 x^3 e^{y^3}\,dy\,dx$
2. $\int_0^8 \int_{\sqrt[3]{y}}^2 \sqrt{x^4+1}\,dx\,dy$

Solution:

1. $\int_0^3 \int_{x^2}^2 x^3 e^{y^3}\,dy\,dx$

First, notice that if we try to integrate with respect to y we can't do the integral because we would need a y^2 in front of the exponential in order to do the y integration. We are going to hope that if we reverse the order of integration we will get an integral that we can do. Now, when we say that we're going to reverse the order of integration this means that we want to integrate with respect to x first and then y. Note as well that we can't just interchange the integrals, keeping the original limits, and be done with it.

This would not fix our original problem and in order to integrate with respect to x we can't have x's in the limits of the integrals. Even if we ignored that the answer would not be a constant as it should be. So, let's see how we reverse the order of integration. The best way to reverse the order of integration is to first sketch the region given by the original limits of integration. From the integral we see that the inequalities that define this region are,

$$0 \le x \le 3$$

$$x^2 \le y \le 9$$

These inequalities tell us that we want the region with $y = x^2$ on the lower boundary and $y = 9$ on the upper boundary that lies between $x = 0$ and $x = 3$. Here is a sketch of that region.

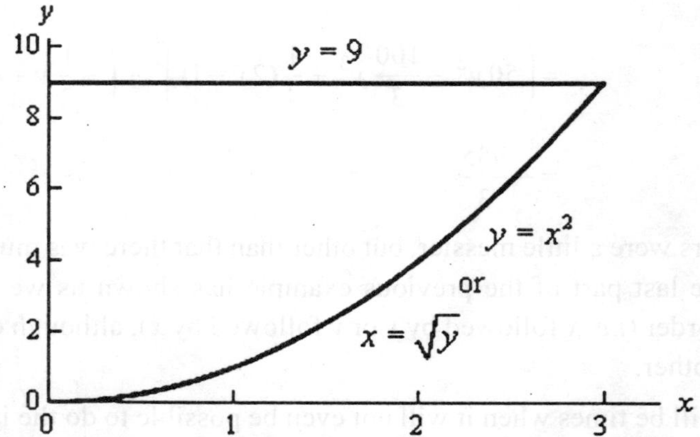

Since we want to integrate with respect to x first we will need to determine limits of x (probably in terms of y) and then get the limits on the y's. Here they are for this region.

$$0 \le x \le \sqrt{y}$$

$$0 \le y \le 9$$

Any horizontal line drawn in this region will start at $x = 0$ and end at $x = \sqrt{y}$ and so these are the limits on the x's and the range of y's for the regions is 0 to 9.

The integral, with the order reversed, is now,

$$\int_0^3 \int_{x^2}^9 x^3 e^{y^3}\, dy\, dx = \int_0^9 \int_0^{\sqrt{y}} x^3 e^{y^3}\, dx\, dy$$

and notice that we can do the first integration with this order. We'll also hope that this will give us a second integral that we can do. Here is the work for this integral.

$$\int_0^3 \int_{x^2}^9 x^3 e^{y^3}\, dy\, dx = \int_0^9 \int_0^{\sqrt{y}} x^3 e^{y^3}\, dx\, dy$$

$$= \int_0^9 \frac{1}{4} x^4 e^{y^3} \Big|_0^{\sqrt{y}} dy$$

$$= \int_0^9 \frac{1}{4} y^e e^{y^3}\, dy$$

$$= \frac{1}{12} e^{y^3} \Big|_0^9$$

$$= \frac{1}{12}(e^{729} - 1)$$

2. $\int_0^8 \int_{\sqrt[3]{y}}^2 \sqrt{x^4 + 1}\, dx\, dy$

As with the first integral we cannot do this integral by integrating with respect to x first so we'll hope that by reversing the order of integration we will get something that we can integrate. Here are the limits for the variables that we get from this integral.

$$\sqrt[3]{y} \le x \le 2$$
$$0 \le y \le 8$$

and here is a sketch of this region.

So, if we reverse the order of integration we get the following limits.

$$0 \le x \le 2$$
$$0 \le y \le x^3$$

The integral is then,

$$\int_0^8 \int_{\sqrt[3]{y}}^2 \sqrt{x^4 = 1}\, dx\, dy = \int_0^2 \int_0^{x^3} \sqrt{x^4 + 1}\, dy\, dx$$

$$= \int_0^2 y\sqrt{x^4 + 1} \Big|_0^{x^3} dx$$

$$= \int_0^2 x^3 \sqrt{x^4 + 1}\, dx = \frac{1}{6}\left(17^{\frac{3}{2}} - 1\right)$$

The final topic of this section is two geometric interpretations of a double integral. The first interpretation is an extension of the idea that we used to develop the idea of a double integral in the first section of this chapter.

We did this by looking at the volume of the solid that was below the surface of the function $z = f(x, y)$ and over the rectangle R in the xy-plane. This idea can be extended to more general regions.

The volume of the solid that lies below the surface given by $z = f(x, y)$ and above the region D in the xy-plane is given by,

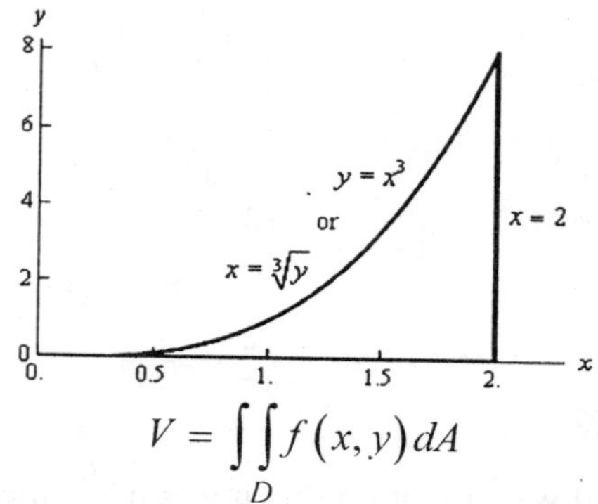

$$V = \iint_D f(x, y)\, dA$$

Example: Find the volume of the solid that lies below the surface given by $z = 16xy + 200$ and lies above the region in the *xy*-plane bounded by $y = x^2$ and $y = 8 - x^2$.

Solution: Here is the graph of the surface and we've tried to show the region in the *xy*-plane below the surface.

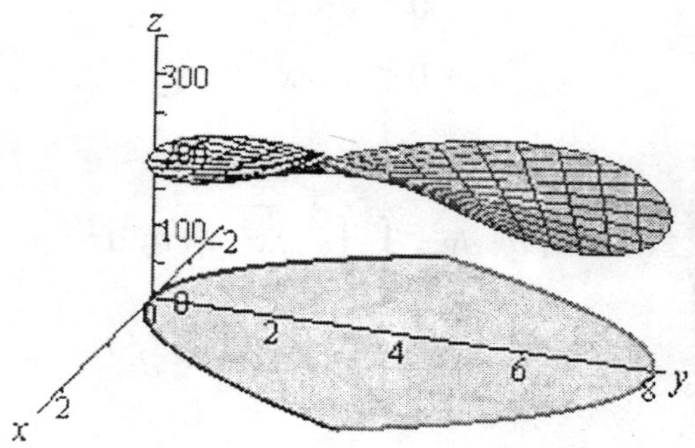

Here is a sketch of the region in the *xy*-plane by itself.

By setting the two bounding equations equal we can see that they will intersect at $x = 2$ and $x = -2$. So, the inequalities that will define the region *D* in the *xy*-plane are,

$$-2 \le x \le 2$$
$$x^2 \le y \le 8 - x^2$$

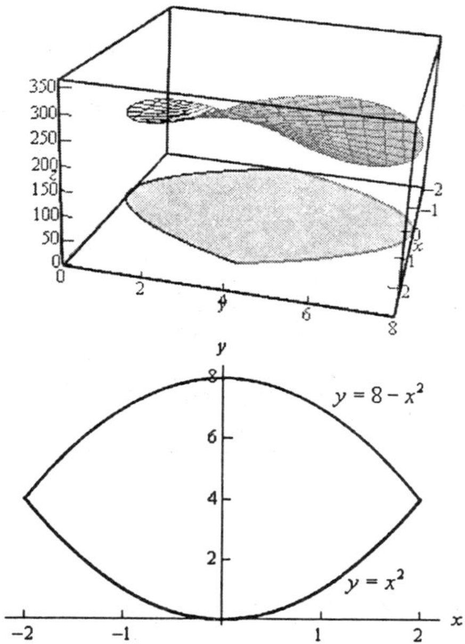

The volume is then given by,

$$V = \iint\limits_D 16xy + 200\,dA$$

$$= \int_{-2}^{2}\int_{x^2}^{8-x^2} 16xy + 200\,dy\,dx$$

$$= \int_{-2}^{2} \left.(8xy^2 + 200y)\right|_{x^2}^{8-x^2} dx$$

$$= \int_{-2}^{2} -128x^3 - 400x^2 + 512x + 1600\,dx$$

$$= \left.\left(32x^4 - \frac{400}{3}x^3 + 256x^2 + 1600x\right)\right|_{-2}^{2} = \frac{1280}{3}$$

Example: Find the volume of the solid enclosed by the planes $4x+2y+z=10$, $y=3x, z=0\, x=0$.

Solution: This example is a little different from the previous one. Here the region D is not explicitly given so we're going to have to find it. First, notice that the last two planes are really telling us that we won't go past the *xy*-plane and the *yz*-plane when we reach them.

The first plane, $4x+2y+z=10$, is the top of the volume and so we are really looking for the volume under $z=10-4x-2y$, and above the region D in the xy-plane.

The second plane, $y=3x$ (yes that is a plane), gives one of the sides of the volume as shown below.

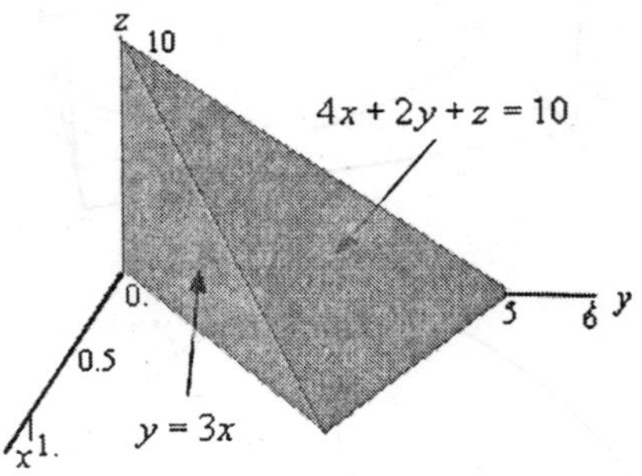

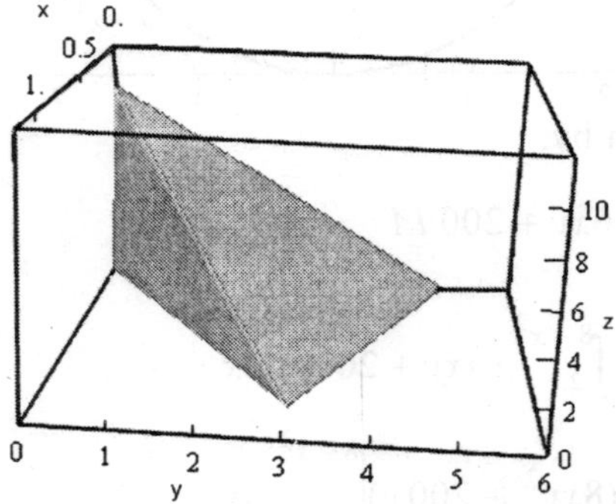

The region D will be the region in the xy-plane ($i.e. z=0$) that is bounded by, $y = 3x$, $x = 0$ and the line where $z + 4x + 2y = 10$ intersects the xy-plane. We can determine where $z + 4x + 2y = 10$ intersects the xy-plane by plugging $z = 0$ into it.

$$0+4x+2y=10 \quad \Rightarrow \quad 2y+y=5 \quad y=-2x+5$$

So, here is a sketch the region D.

The region D is really where this solid will sit on the xy-plane and here are the inequalities that define the region.

$$0 \le x \le 1$$

$$3x \le y \le -2x+5$$

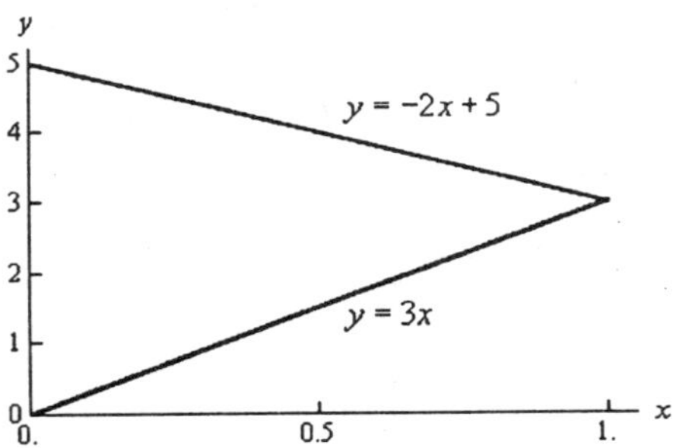

Here is the volume of this solid.

$$V = \iint_D 10 - 4x - 2y\,dA$$
$$= \int_0^1 \int_{3x}^{-2x+5} 10 - 4x - 2y\,dy\,dx$$
$$= \int_0^1 \left(10y - 4xy - y^2\right)\Big|_{3x}^{-2x+5} dx$$
$$= \int_0^1 25x^2 - 50x + 25\,dx$$
$$= \left(\frac{25}{3}x^3 - 25x^2 + 25x\right)\Big|_0^1 = \frac{25}{3}$$

The second geometric interpretation of a double integral is the following.

$$\text{Area of } D = \iint_D dA$$

This is easy to see why this is true in general. Let's suppose that we want to find the area of the region shown below.

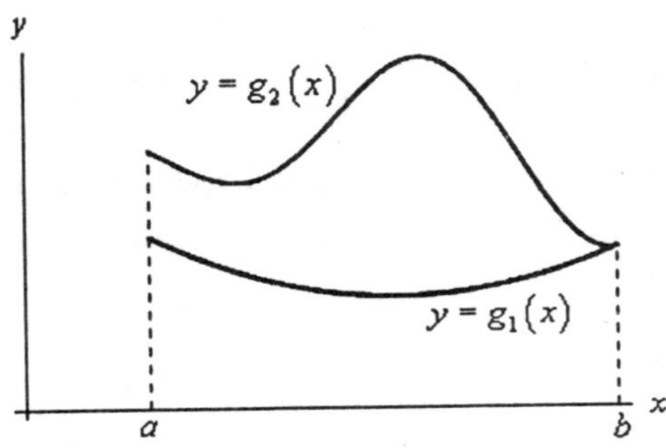

We know that this area can be found by the integral,

$$A = \int_a^b g_2(x) - g_1(x)\,dx$$

Or in terms of a double integral we have,

$$\text{Area of } D = \iint_D dA$$

$$= \int_a^b \int_{g_1}^{g_2} dy\,dx$$

$$= \int_a^b y\Big|_{g_1(x)}^{g_2(x)} dx = \int_a^b g_2(x) - g_1(x)\,dx$$

DOUBLE INTEGRALS IN POLAR COORDINATES

To this point we've seen quite a few double integrals. However, in every case in region D could be easily described in terms of simple functions in Cartesian coordinates. In this section we want to look at some regions that are much easier to describe in terms of polar coordinates. For instance, we might have a region that is a disk, ring, or a portion of a disk or ring.

In these cases using Cartesian coordinates could be somewhat cumbersome. For instance let's suppose we wanted to do the following integral,

$$\iint_D f(x,y)\,dA,\ D \text{ is the disk of radius 2}$$

To this we would have to determine a set of inequalities for x and y that describe this region. These would be,

$$-2 \le x \le 2$$

$$-4\sqrt{4-x^2} \le y \le \sqrt{4-x^2}$$

With these limits the integral would become,

$$\iint_D f(x,y)\,dA = \int_{-2}^{2} \int_{\sqrt{4-x^2}}^{\sqrt{4-x^2}} f(x,y)\,dy\,dx$$

Due to the limits on the inner integral this is liable to be an unpleasant integral to compute.

However, a disk of radius 2 can be defined in polar coordinates by the following inequalities,

$$0 \le \theta \le 2\pi$$

$$0 \le r \le 2$$

These are very simple limits and, in fact, are constant limits of integration which almost always makes integrals somewhat easier.

So, if we could convert our double integral formula into one involving polar coordinates we would be in pretty good shape. The problem is that we can't just convert the *dx* and the *dy* into a *dr* and a $d\theta$. In computing double integrals to this point we have been using the fact that $dA = dx\,dy$ and this really does require Cartesian coordinates to use. Once we've moved into polar coordinates $dA \neq dr\,d\theta$ and so we're going to need to determine just what *dA* is under polar coordinates.

So, let's step back a little bit and start off with a general region in terms of polar coordinates and see what we can do with that. Here is a sketch of some region using polar coordinates.

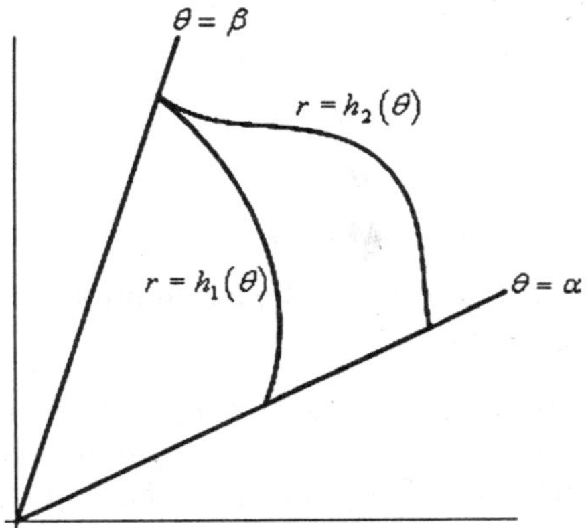

So, our general region will be defined by inequalities,

$$\alpha \leq \theta \leq \beta$$
$$h_1(\theta) \leq r \leq h_2(\theta)$$

Now, to find *dA* let's redo the figure above as follows,

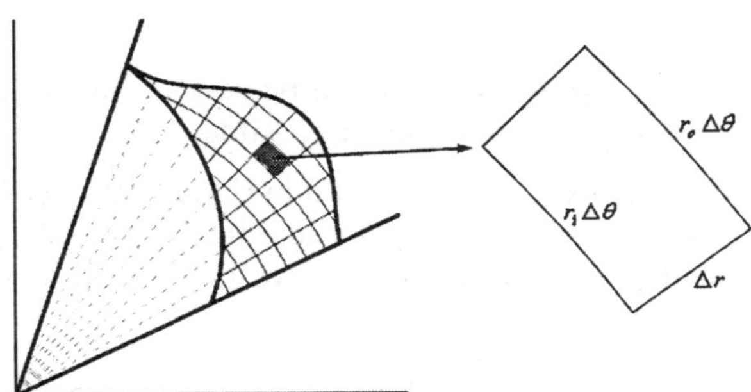

As shown, we'll break up the region into a mesh of radial lines and arcs. Now, if we pull one of the pieces of the mesh out as shown we have something that is almost, but not quite a rectangle. The area of this piece is ΔA .

The two sides of this piece both have length $\Delta r = r_o - r_i$ where is the radius of the outer arc and r_i is the radius of the inner arc. Basic geometry then tells us that the length of the inner edge is $r_i\Delta\theta$ while the length of the out edge is $r_o\Delta\theta$ where $\Delta\theta$ is the angle between the two radial lines that form the sides of this piece.

Now, let's assume that we've taken the mesh so small that we can assume that $r_i \approx r_o = r$ and with this assumption we can also assume that our piece is close enough to a rectangle that we can also then assume that,

$$\Delta A \approx r\Delta\theta\Delta r$$

Also, if we assume that the mesh is small enough then we can also assume that,

$$dA \approx \Delta A$$

$$d\theta \approx \Delta\theta$$

$$dr \approx \Delta r$$

With these assumptions we then get $dA \approx rdr\,d\theta$.

In order to arrive at this we had to make the assumption that the mesh was very small. This is not an unreasonable assumption.

Recall that the definition of a double integral is in terms of two limits and as limits go to infinity the mesh size of the region will get smaller and smaller. In fact, as the mesh size gets smaller and smaller the formula above becomes more and more accurate and so we can say that,

$$dA \approx rdr\,d\theta$$

We'll see another way of deriving this once we reach the Change of Variables section later in this chapter. This second way will not involve any assumptions either and so it maybe a little better way of deriving this. Before moving on it is again important to note that $dA \approx dr\,d\theta$.

The actual formula for dA has an r in it. It will be easy to forget this r on occasion, but as you'll see without it some integrals will not be possible to do.

Now, if we're going to be converting an integral in Cartesian coordinates into an integral in polar coordinates we are going to have to make sure that we've also converted all the x's and y's into polar coordinates as well. To do this we'll need to remember the following conversion formulas,

$$x = r\cos\theta$$

$$y = r\sin\theta$$

$$r^2 = x^2 + y^2$$

We are now ready to write down a formula for the double integral in terms of polar coordinates.

$$\iint_D f(x,y)\,dA = \int_\alpha^\beta \int_{h_2(\theta)}^{h_2(\theta)} f(r\cos\theta, r\sin\theta)\, r\,dr\,d\theta$$

It is important to not forget the added *r* and don't forget to convert the Cartesian coordinates in the function over to polar coordinates.

Let's look at a couple of examples of these kinds of integrals.

Example: Evaluate the following integrals by converting them into polar coordinates.

1. $\iint_D 2xy\,dA$, *D* is the portion of the region between the circles of radius 2and radius 5 centered at the origin that lies in the first quadrant.

2. $\iint_D e^{x^2+y^2}\,dA$, *D* is the unit circle centered at the origin.

Solution:

1. $\iint_D 2xy\,dA$, *D* is the portion of the region between the circles of radius 2 and radius 5 centered at the origin that lies in the first quadrant.

First let's get *D* in terms of polar coordinates. The circle of radius 2 is given by $r = 2$ and the circle of radius 5 is given by $r = 5$. We want the region between them so we will have the following inequality for *r*.

$$2 \le r \le 5$$

Also, since we only want the portion that is in the first quadrant we get the following range of θ's.

$$0 \le \theta \le \frac{\pi}{2}$$

Now that we've got these we can do the integral.

$$\iint_D 2xy\,dA = \int_0^{\frac{\pi}{2}} \int_2^5 2(r\cos\theta)(r\sin\theta)\,r\,dr\,d\theta$$

Don't forget to do the conversions and to add in the extra *r*. Now, let's simplify and make use of the double angle formula for sine to make the integral a little easier.

$$\iint_D 2xy\,dA = \int_0^{\frac{\pi}{2}}\int_2^5 r^2 \sin(2\theta)\,dr\,d\theta$$

$$= \int_0^{\frac{\pi}{2}} \frac{1}{4}r^4 \sin(2\theta)\Big|_2^5 \,d\theta$$

$$= \int_0^{\frac{\pi}{2}} \frac{609}{4}\sin(2\theta)\,d\theta$$

$$= -\frac{609}{8}\cos(2\theta)\Big|_0^{\frac{\pi}{2}}$$

$$= \frac{609}{4}$$

2. $\iint_D e^{x^2+y^2}\,dA$, D is the unit circle centered at the origin.

In this case we can't do this integral in terms of Cartesian coordinates. We will however be able to do it in polar coordinates. First, the region D is defined by,

$$0 \le \theta \le 2\pi$$

$$0 \le r \le 1$$

In terms of polar coordinates the integral is then,

$$\iint_D e^{x^2+y^2}\,dA = \int_0^{2\pi}\int_0^1 re^{r^2}\,dr\,d\theta$$

Notice that the addition of the r gives us an integral that we can now do. Here is the work for this integral.

$$\iint_D e^{x^2+y^2}\,dA = \int_0^{2\pi}\int_0^1 re^{r^2}\,dr\,d\theta$$

$$= \int_0^{2\pi} \frac{1}{2}e^{r^2}\Big|_0^1 \,d\theta$$

$$= \int_0^{2\pi} \frac{1}{2}(e-1)\,d\theta$$

$$= \pi(e-1\_$$

Let's not forget that we still have the two geometric interpretations for these integrals as well.

Example: Determine the area of the region that lies inside $r = 3 + 2\sin\theta$ and outside $r = 2$.

Solution: Here is a sketch of the region, D, that we want to determine the area of.

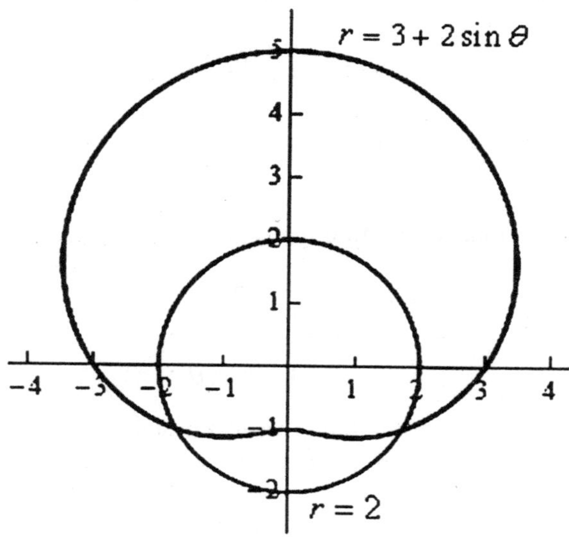

To determine this area we'll need to know that value of $\theta$for which the two curves intersect. We can determine these points by setting the two equations and solving.

$$3 + 2\sin\theta = 2$$

$$\sin\theta = -\frac{1}{2} \quad \Rightarrow \quad \theta = \frac{7\pi}{6}, \frac{11\pi}{6}$$

Here is a sketch of the figure with these angles added.

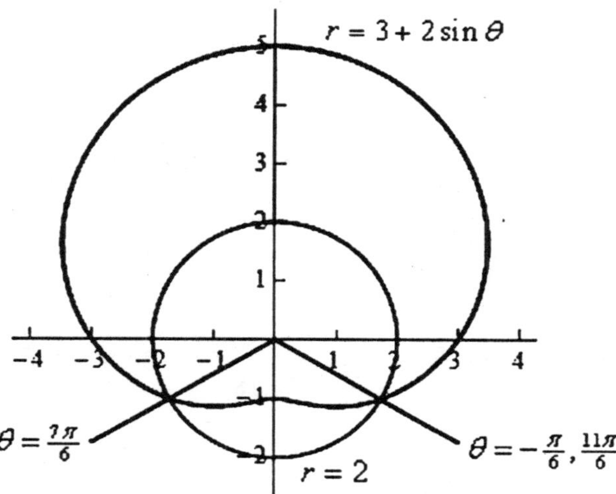

Note as well that we've acknowledged that $-\frac{\pi}{6}$ is another representation for the angle

$\frac{11\pi}{6}$. This is important since we need the range of θto actually enclose the regions as we increase from the lower limit to the upper limit. If we'd chosen to use $\frac{11\pi}{6}$ then as we increase from $\frac{7\pi}{6}$ to $\frac{11\pi}{6}$ we would be tracing out the lower portion of the circle and that is not the region that we are after.

So, here are the ranges that will define the region.

$$-\frac{\pi}{6} \le \theta \le \frac{7\pi}{6}$$

$$2 \le r \le 3 + 2\sin\theta$$

To get the ranges for r the function that is closest to the origin is the lower bound and the function that is farthest from the origin is the upper bound.

The area of the region D is then,

$$A = \iint_D dA$$

$$= \int_{-\pi/6}^{7\pi/6} \int_2^{3+2\sin\theta} r\, dr\, d\theta$$

$$= \int_{-\pi/6}^{7\pi/6} \frac{1}{2} r^2 \bigg|_2^{3+2\sin\theta} d\theta$$

$$= \int_{-\pi/6}^{7\pi/6} \frac{5}{2} + 6\sin\theta + 2\sin^2\theta\, d\theta$$

$$= \int_{-\pi/6}^{7\pi/6} \frac{7}{2} + 6\sin\theta - \cos(2\theta)\, d\theta$$

$$= \left(\frac{7}{2}\theta - 6\cos\theta - \frac{1}{2}\sin(2\theta)\right)\bigg|_{-\frac{\pi}{6}}^{\frac{7\pi}{6}}$$

$$= \frac{11\sqrt{3}}{2} + \frac{14\pi}{3} = 24.187$$

Example: Determine the volume of the region that lies under the sphere $x^2 + y^2 + z^2 = 9$, above the plane $z = 0$ and inside the cylinder $x^2 + y^2 = 5$.

Solution: We know that the formula for finding the volume of a region is,

$$V = \iint_D f(x,y)\,dA$$

In order to make use of this formula we're going to need to determine the function that we should be integrating and the region D that we're going to be integrating over.

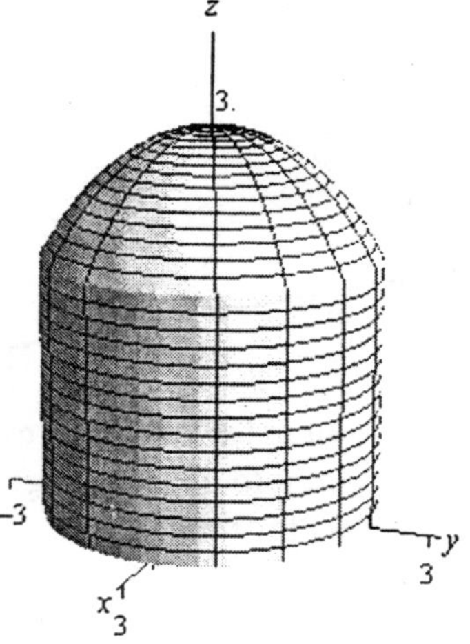

The function isn't too bad. It's just the sphere, however, we do need it to be in the form $z = f(x,y)$. We are looking at the region that lies under the sphere and above the plane $z = 0$(just the xy-plane right?) and so all we need to do is solve the equation for z and when taking the square root we'll take the positive one since we are wanting the region above the xy-plane. Here is the function.

$$z = \sqrt{9 - x^2 - y^2}$$

The region D isn't too bad in this case either. As we take points, (x, y), from the region we need to completely graph the portion of the sphere that we are working with. Since we only want the portion of the sphere that actually lies inside the cylinder given by $x^2 + y^2 = 5$ this is also the region D. The region D is the disk $x^2 + y^2 \le 5$ in the xy-plane. For reference purposes here is a sketch of the region that we are trying to find the volume of.

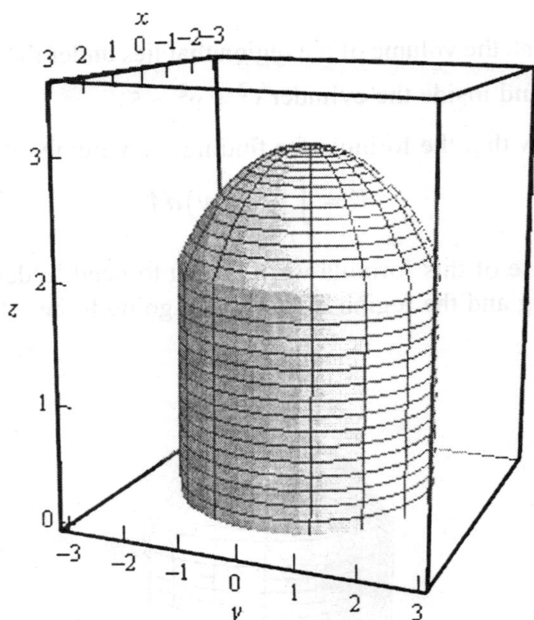

So, the region that we want the volume for is really a cylinder with a cap that comes from the sphere. We are definitely going to want to do this integral in terms of polar coordinates so here are the limits (in polar coordinates) for the region,

$$0 \le \theta \le 2\pi$$

$$0 \le r \le \sqrt{5}$$

and we'll need to convert the function to polar coordinates as well.

$$z \Rightarrow \sqrt{9-\left(x^2+y^2\right)} = \sqrt{9-r^2}$$

The volume is then,

$$V = \iint\limits_D \sqrt{9-x^2-y^2}\, dA$$

$$= \int_0^{2\pi} \int_0^{\sqrt{5}} r\sqrt{9-r^2}\, dr\, d\theta$$

$$= \int_0^{2\pi} -\frac{1}{3}\left(9 - r^2\right)^{\frac{3}{2}}\Bigg|_0^{\sqrt{5}} d\theta$$

$$= \int_0^{2\pi} \frac{19}{3} d\theta$$

$$= \frac{38\pi}{3}$$

Example: Find the volume of the region that lies inside $z = x^2 + y^2$ and below the plane $z = 16$.

Solution: Let's start this example off with a quick sketch of the region.

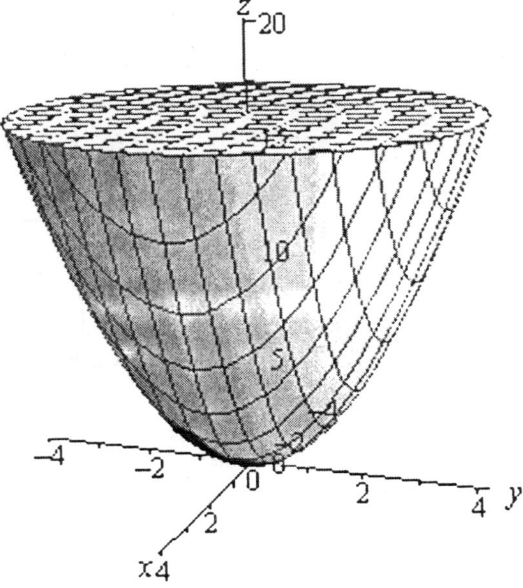

Now, in this case the standard formula is not going to work. The formula

$$V = \iint_D f(x, y)\, dA$$

finds the volume under the function $f(x, y)$ and we're actually after the area that is above a function. This isn't the problem that it might appear to be however. First, notice that

$$V = \iint_D 16 dA$$

will be the volume under $z = 16$ (of course we'll need to determine D eventually) while

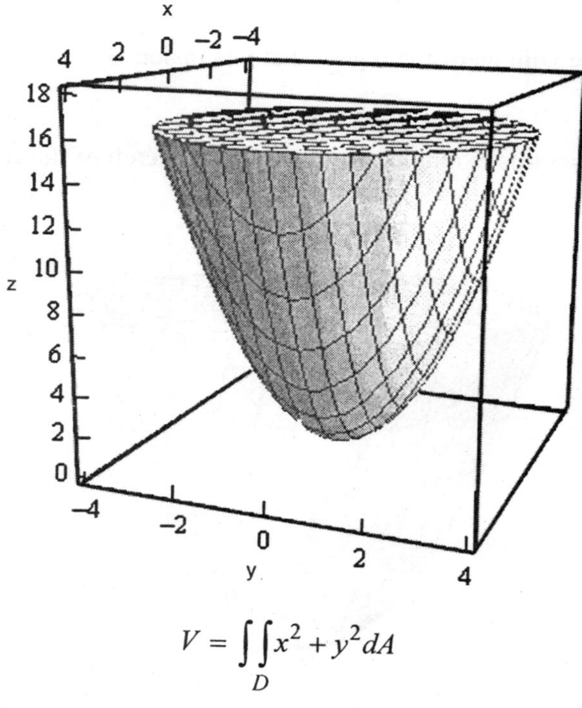

$$V = \iint_D x^2 + y^2 dA$$

is the volume under $z = x^2 + y^2$, using the same D.

The volume that we're after is really the difference between these two or,

$$V = \iint_D 16 dA - \iint_D x^2 + y^2 dAz = \iint_D 16 - \left(x^2 + y^2\right) dA$$

Now all that we need to do is to determine the region D and then convert everything over to polar coordinates. Determining the region D in this case is not too bad.

If we were to look straight down the z-axis onto the region we would see a circle of radius 4 centered at the origin. This is because the top of the region, where the elliptic paraboloid intersects the plane, is the widest part of the region.

We know the z coordinate at the intersection so, setting $z = 16$ in the equation of the paraboloid gives,

$$16 = x^2 + y^2$$

which is the equation of a circle of radius 4 centered at the origin. Here are the inequalities for the region and the function we'll be integrating in terms of polar coordinates.

$$0 \le \theta \le 2\pi \quad 0 \le r \le 4 \quad z = 16 - r^2$$

The volume is then,

$$\begin{aligned} V &= \iint_D 16 - (x^2 + y^2)\,dA \\ &= \int_0^{2\pi} \int_0^4 r(16 - r^2)\,dr\,d\theta \\ &= \int_0^{2\pi} \left(8r^2 - \frac{1}{4}r^4 \right)\Bigg|_0^4 d\theta \\ &= \int_0^{2\pi} 64\,d\theta \\ &= 128\pi \end{aligned}$$

In both of the previous volume problems we would have not been able to easily compute the volume without first converting to polar coordinates so, as these examples show, it is a good idea to always remember polar coordinates. There is one more type of example that we need to look at before moving on to the next section. Sometimes we are given an iterated integral that is already in terms of x and y and we need to convert this over to polar so that we can actually do the integral. We need to see an example of how to do this kind of conversion.

Example: Evaluate the following integral by first converting to polar coordinates.

$$\int_0^1 \int_0^{\sqrt{1-y^2}} \cos\left(x^2 + y^2\right) dx\,dy$$

Solution: First, notice that we cannot do this integral in Cartesian coordinates and so converting to polar coordinates may be the only option we have for actually doing the integral. Notice that the function will convert to polar coordinates nicely and so shouldn't be a problem. Let's first determine the region that we're integrating over and see if it's a region that can be easily converted into polar coordinates. Here are the inequalities that define the region in terms of Cartesian coordinates.

$$0 \le y \le 1$$

$$0 \le x \le \sqrt{1 - y^2}$$

Now, the upper limit for the x's is,

$$x = \sqrt{1-y^2}$$

and this looks like the right side of the circle of radius 1 centered at the origin. Since the lower limit for the x's is $x = 0$ it looks like we are going to have a portion (or all) of the right side of the disk of radius 1 centered at the origin. The range for the y's however, tells us that we are only going to have positive y's. This means that we are only going to have the portion of the disk of radius 1 centered at the origin that is in the first quadrant.

So, we know that the inequalities that will define this region in terms of polar coordinates are then,

$$0 \le \theta \le \frac{\pi}{2}$$
$$0 \le r \le 1$$

Finally, we just need to remember that,

$$dx\,dy = dA = r dr\,d\theta$$

and so the integral becomes,

$$\int_0^1 \int_0^{\sqrt{1-y^2}} \cos\left(x^2+y^2\right) dx\,dy = \int_0^{\frac{\pi}{2}} \int_0^1 \cos\left(r^2\right) dr\,d\theta$$

Note that this is an integral that we can do. So, here is the rest of the work for this integral.

$$\int_0^1 \int_0^{\sqrt{1-y^2}} \cos(x^2+y^2) dx\,dy = \int_0^{\frac{\pi}{2}} \frac{1}{2}\sin(r^2)\Big|_0^1 \,d\theta$$
$$= \int_0^{2\pi} \frac{1}{2}\sin(1)\,d\theta$$
$$= \frac{\pi}{4}\sin(1)$$

TRIPLE INTEGRALS

Now that we know how to integrate over a two-dimensional region we need to move on to integrating over a three-dimensional region. We used a double integral to integrate over a two-dimensional region and so it shouldn't be too surprising that we'll use a triple integral to integrate over a three dimensional region. The notation for the general triple integrals is,

$$\iiint_E f(x,y,z)\,dV$$

Let's start simple by integrating over the box,

$$B = |a,b| \times |c,d| \times |r,s|$$

Note that when using this notation we list the x's first, the y's second and the z's third.

The triple integral in this case is,

$$\iiint_E f(x,y,z)\,dA = \int_r^s \int_c^d \int_a^b f(x,y,z)\,dx\,dy\,dz$$

Note that we integrated with respect to x first, then y, and finally z here, but in fact there is no reason to the integrals in this order. There are 6 different possible orders to do the integral in and which order you do the integral in will depend upon the function and the order that you feel will be the easiest. We will get the same answer regardless of the order however.

Let's do a quick example of this type of triple integral.

Example: Evaluate the following integral.

$$\iiint_B 8xyz\,dV, B = [2,3]\times[1,2]\times[0,1]$$

Solution: Just to make the point that order doesn't matter let's use a different order from that listed above. We'll do the integral in the following order.

$$\iiint_B 8xyz\,dV = \int_1^2 \int_2^3 \int_0^1 8xyz\,dz\,dx\,dy$$

$$= \int_1^2 \int_2^3 4xyz^2\Big|_0^1\,dx\,dy$$

$$= \int_1^2 \int_2^3 4xy\,dx\,dy$$

$$= \int_1^2 2x^2y\Big|_2^3\,dy$$

$$= \int_1^2 10y\,dy = 15$$

Before moving on to more general regions let's get a nice geometric interpretation about the triple integral out of the way so we can use it in some of the examples to follow.

Fact

The volume of the three-dimensional region E is given by the integral,

$$V = \iiint_E dV$$

Let's now move on the more general three-dimensional regions. We have three different possibilities for a general region. Here is a sketch of the first possibility.

In this case we define the region E as follows,

$$E = \{(x, y, z) | (x, y) \in D, u_1(x, y) \le z \le u_2(x, y)\}$$

where $(x, y) \in D$ is the notation that means that the point (x, y) lies in the region D from the xy-plane.

In this case we will evaluate the triple integral as follows,

$$\iiint_E f(x, y, z)\, dV = \iint_D \left[\int_{u_1(x,y)}^{1/2(x,y)} f(x, y, z)\, dz \right] dA$$

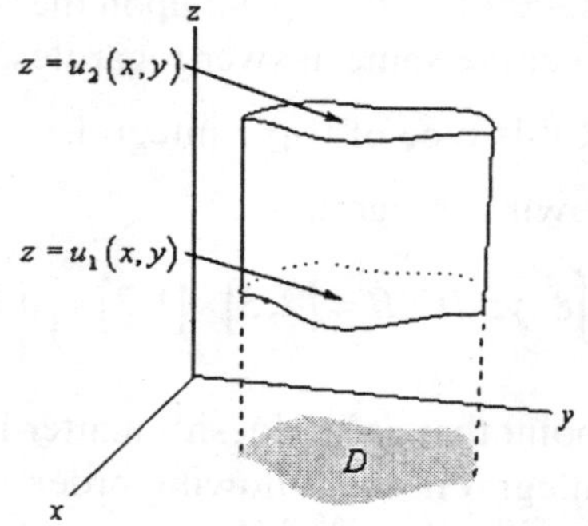

where the double integral can be evaluated in any of the methods that we saw in the previous couple of sections. In other words, we can integrate first with respect to x, we can integrate first with respect to y, or we can use polar coordinates as needed.

Example: Evaluate $\iiint_E 2x dV$ where E is the region under the plane $2x + 3y + z = 6$ $2x + 3y + z = 62x + 3y + z = 6$ that lies in the first octant.

Solution: We should first define *octant*. Just as the two-dimensional coordinates system can be divided into four quadrants the three-dimensional coordinate system can be divided into eight octants. The first octant is the octant in which all three of the coordinates are positive.

Here is a sketch of the plane in the first octant.

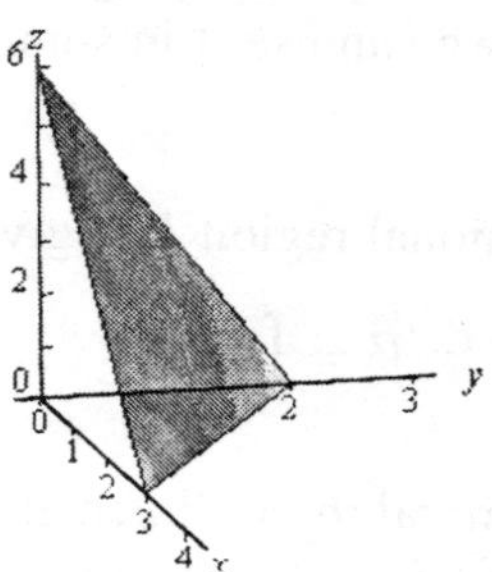

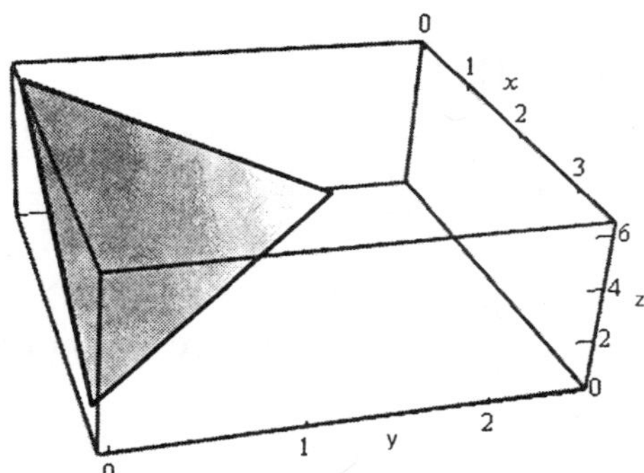

We now need to determine the region D in the xy-plane. We can get a visualization of the region by pretending to look straight down on the object from above. What we see will be the region D in the xy-plane. So D will be the triangle with vertices at, $(0,0)$, $(3,0)$ and $(0,2)$. Here is a sketch of D.

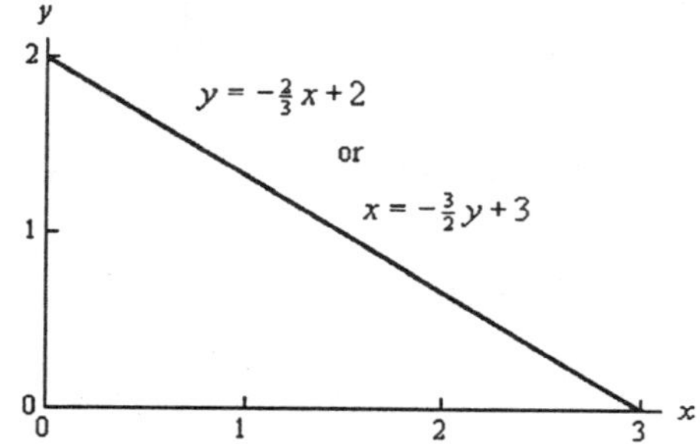

Now we need the limits of integration. Since we are under the plane and in the first octant (so we're above the plane $z=0$) we have the following limits for z.

$$0 \le z \le 6-2x-3y$$

We can integrate the double integral over D using either of the following two sets of inequalities.

$$0 \le x \le 3 \qquad\qquad 0 \le x \le -\frac{3}{2}y+3$$

$$o \le y \le -\frac{2}{3}x+2 \qquad\qquad 0 \le y \le 2$$

Since neither really holds an advantage over the other we'll use the first one. The integral is then,

$$\iiint_E 2x\,dV = \iint_D \left[\int_0^{6-2x-3y} 2x\,dz\right] dA$$

$$= \iint_D 2xz\Big|_0^{6-2x-3y}\,dA$$

$$= \int_0^3 \int_0^{-\frac{2}{3}x+2} 2x(6-2x-3y)\,dy\,dx$$

$$= \int_0^3 (12xy - 4x^2y - 3xy^2)\Big|_0^{-\frac{2}{3}x+2}\,dx$$

$$= \int_0^3 \frac{4}{3}x^3 - 8x^2 + 12x\,dx$$

$$= \left(\frac{1}{3}x^4 - \frac{8}{3}x^3 + 6x^2\right)\Big|_0^3$$

$$= 9$$

Let's now move onto the second possible three-dimensional region we may run into for triple integrals. Here is a sketch of this region.

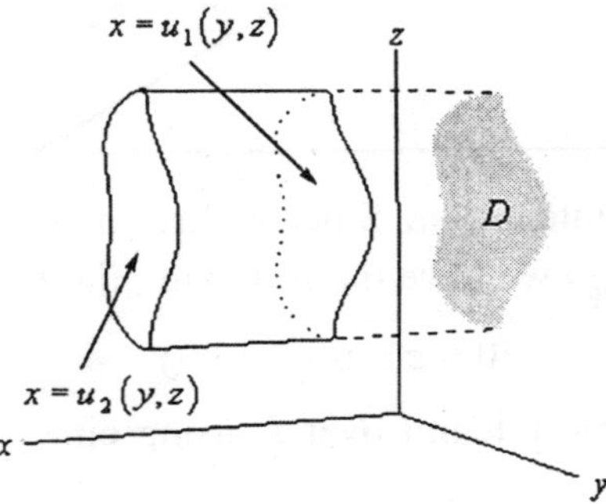

For this possibility we define the region E as follows,

$$E = \{(x,y,z) \,|\, (y,z) \in D, u_1(y,z) \le x \le u_2(y,z)\}$$

So, the region D will be a region in the yz-plane. Here is how we will evaluate these integrals.

$$\iiint_E f(x,y,z)\,dV = \iint_D \left[\int_{u_1(y,z)}^{u_2(y,z)} f(x,y,z)\,dx\right] dA$$

As with the first possibility we will have two options for doing the double integral in the yz-plane as well as the option of using polar coordinates if needed.

Example: Determine the volume of the region that lies behind the plane $x+y+z=8$ and in front of the region in the yz-plane that is bounded by $z=\frac{3}{2}\sqrt{y}$ and $z=\frac{3}{4}y$.

Solution: In this case we've been given D and so we won't have to really work to find that. Here is a sketch of the region D as well as a quick sketch of the plane and the curves defining D projected out past the plane so we can get an idea of what the region we're dealing with looks like.

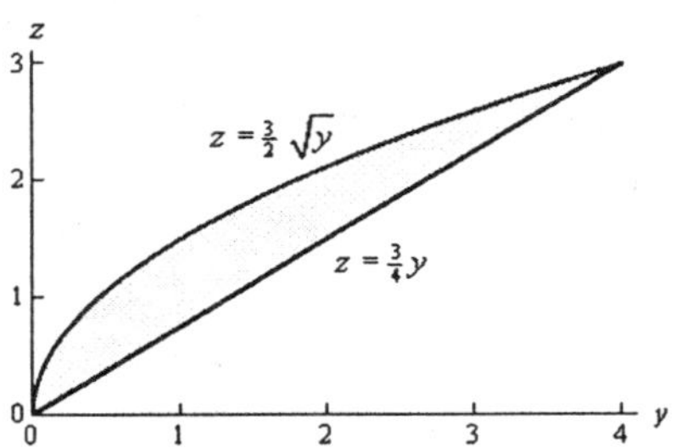

Now, the graph of the region above is all okay, but it doesn't really show us what the region is. So, here is a sketch of the region itself.

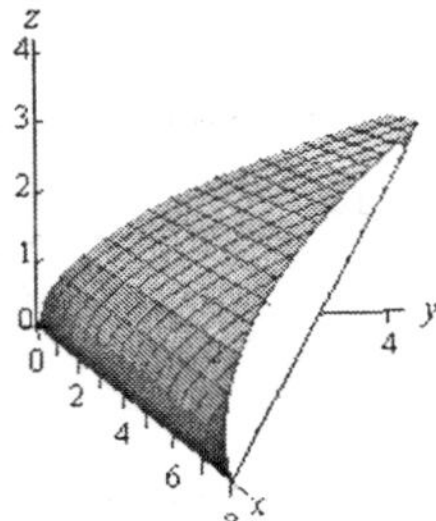

Here are the limits for each of the variables.

$$0 \le y \le 4$$

$$\frac{3}{4}y \le z \le \frac{3}{2}\sqrt{y}$$

$$0 \le x \le 8-y-z$$

The volume is then,

$$V = \iiint\limits_D dV = \iint\limits_D \left[\int_0^{8-y-z} dx\right] dA$$

$$= \int_0^4 \int_{3y/4}^{3\sqrt{y}/2} 8-y-z\,dz\,dy$$

$$= \int_0^4 \left(8z - yz - \frac{1}{2}z^2\right)\Bigg|_{\frac{3y}{4}}^{\frac{3\sqrt{y}}{2}} dy$$

$$= \int_0^4 12y^{\frac{1}{2}} - \frac{57}{8}y - \frac{3}{2}y^{\frac{3}{2}} + \frac{33}{32}y^2\,dy$$

$$= \left(8y^{\frac{3}{2}} - \frac{57}{16}y^2 - \frac{3}{5}y^{\frac{5}{2}} + \frac{11}{32}y^3\right)\Bigg|_0^4 = \frac{49}{5}$$

We now need to look at the third (and final) possible three-dimensional region we may run into for triple integrals. Here

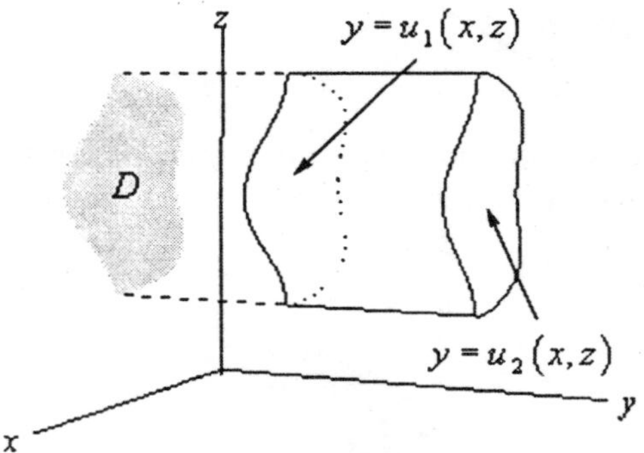

is a sketch of this region. In this final case E is defined as,

$$E = \{(x,y,z) | (x,y) \in D, u_1(x,y) \le y \le u_2(x,z)\}$$

and here the region D will be a region in the xz-plane. Here is how we will evaluate these integrals.

$$\iiint_E f(x,y,z)\,dV = \iint_D \left[\int_{u_1(x,z)}^{u_2(x,z)} f(x,y,z)\,dy\right] dA$$

where we will can use either of the two possible orders for integrating D in the xz-plane or we can use polar coordinates if needed.

Example: Evaluate $\iiint_E \sqrt{3x^2 + 3z^2}\,dV$ where E is the solid bounded by $y = 2x^2 + 2z^2$ and the plane.

Solution: Here is a sketch of the solid E.

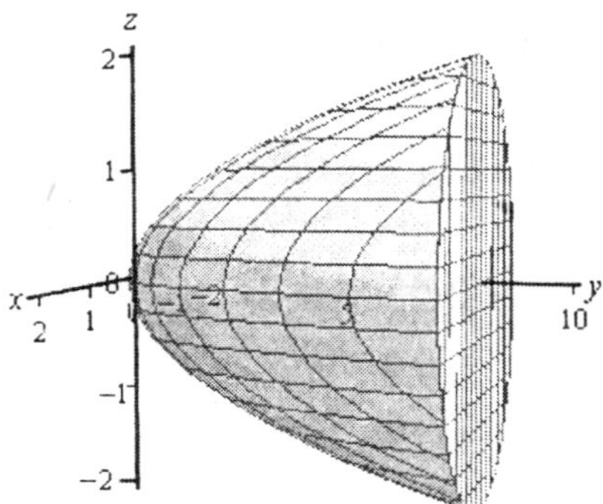

The region D in the xz-plane can be found by "standing" in front of this solid and we can see that D will be a disk in the xz-plane.

This disk will come from the front of the solid and we can determine the equation of the disk by setting the elliptic paraboloid and the plane equal.

$$2x^2 + 2z^2 = 8 \quad \Rightarrow \quad x^2 + z^2 = 4$$

This region, as well as the integrand, both seems to suggest that we should use something like polar coordinates. However we are in the xz-plane and we've only seen polar coordinates in the xy-plane. This is not a problem. We can always "translate" them over to the xz-plane with the following definition.

$$x = r\cos\theta$$

$$z = r\sin\theta$$

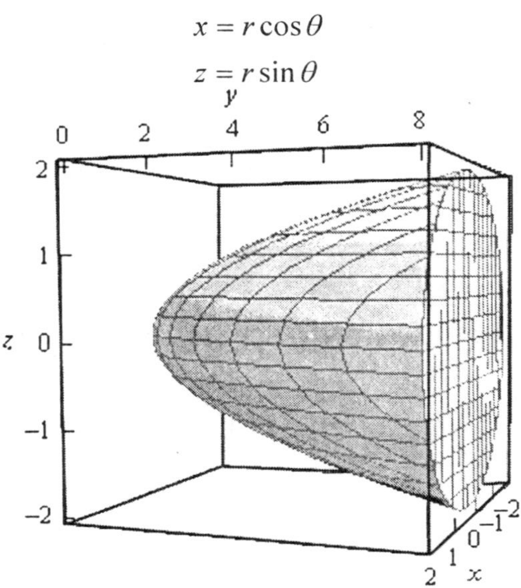

Since the region doesn't have y's we will let z take the place of y in all the formulas. Note that these definitions also lead to the formula,

$$x^2 + z^2 = r^2$$

$$x^2 + z^2 = r^2 x^2 + z^2 = r^2$$

With this in hand we can arrive at the limits of the variables that we'll need for this integral.

$$2x^2 + 2z^2 \le y \le 8$$
$$0 \le r \le 2$$
$$0 \le \theta \le 2\pi$$

The integral is then,

$$\iiint_E \sqrt{3x^2 + 3z^2}\, dV = \iint_D \left[\int_{2x^2 \div 2z^2}^{8} \sqrt{3x^2 + 3z^2}\, dy \right] dA$$

$$= \iint_D \left(y\sqrt{3x^2 + 3z^2} \right) \Bigg|_{2x^2 + 2z^2}^{8} dA$$

$$= \iint_D \sqrt{3(x^2 + z^2)} \left(8 - (2x^2 + 2z^2)\right) dA$$

Now, since we are going to do the double integral in polar coordinates let's get everything converted over to polar coordinates. The integrand is,

$$\sqrt{3}\left(x^2 + z^2\right)\left(8 - \left(2x^2 + 2z^2\right)\right) = \sqrt{3r^2}\left(8 - 2r^2\right)$$

$$= \sqrt{3}r\left(8 - 2r^2\right)$$

$$= \sqrt{3}\left(8r - 2r^3\right)$$

The integral is then,

$$\iiint_E \sqrt{3x^2 + 3z^2}\, dV = \iint_D \sqrt{3}\,(8r - 2r^3)\, dA$$

$$= \sqrt{3} \int_0^{2\pi} \int_0^{2} (8r - 2r^3)\, r dr\, d\theta$$

$$= \sqrt{3} \int_0^{2\pi} \left(\frac{8}{3} r^3 - \frac{2}{5} r^5 \right) \Bigg|_0^2 d\theta$$

$$= \sqrt{3} \int_0^{2\pi} \frac{128}{15}\, d\theta$$

$$= \frac{156\sqrt{3}\,\pi}{15}$$

TRIPLE INTEGRALS IN CYLINDRICAL COORDINATES

In this section we want do take a look at triple integrals done completely in Cylindrical Coordinates. Recall that cylindrical coordinates are really nothing more than an extension of polar coordinates into three dimensions. The following are the conversion formulas for cylindrical coordinates.

$$x = r\cos\theta$$

$$y = r\sin\theta$$

$$z = z$$

In order to do the integral in cylindrical coordinates we will need to know what dV will become in terms of cylindrical coordinates. We will be able to show in the Change of Variables section of this chapter that,

$$dV = r\,dz\,dr\,d\theta$$

The region, E, over which we are integrating becomes,

$$E = \{(x,y,z)|(x,y) \in D, u_1(x,y) \le z \le u_2(x,y)\}$$

$$= \{(r,\theta,z)|\alpha \le \beta, h_1(\theta) \le r \le h_2(\theta), u_1(r\cos\theta, r\sin\theta) \le z \le u_2(r\cos\theta, r\sin\theta)\}$$

Note that we've only given this for E's in which D is in the xy-plane. We can modify this accordingly if D is in the yz-plane or the xz-plane as needed.

In terms of cylindrical coordinates a triple integral is,

$$\iiint\limits_R f(x,y,z)\,dV = \int_\alpha^\beta \int_{h_1(\theta)}^{h_2(\theta)} \int_{u_1(r\cos\theta,\, r\sin\theta)}^{u_2(r\cos\theta,\, r\sin\theta)} rf(r\cos\theta, r\sin\theta, z)\,dz\,dr\,d\theta$$

Don't forget to add in the r and make sure that all the x's and y's also get converted over into cylindrical coordinates. Let's see an example.

Example: Evaluate $\iiint\limits_E y\,dV$ where E is the region that lies below the plane $z = x + 2$ above the xy-plane and between the cylinders $x^2 + y^2 = 1$ and $x^2 + y^2 = 4$.

Solution: There really isn't too much to do with this one other than do the conversions and then evaluate the integral.

We'll start out by getting the range for z in terms of cylindrical coordinates.

$$0 \le z \le x + 2 \quad \Rightarrow \quad 0 \le z \le r\cos\theta + 2$$

Remember that we are above the xy-plane and so we are above the plane $z = 0$

Next, the region D is the region between the two circles $x^2 + y^2 = 1$ and $x^2 + y^2 = 4$ in the xy-plane and so the ranges for it are, $0 \le \theta \le 2\pi \quad 1 \le r \le 2$. Here is the integral.

$$\iiint_E y\,dV = \int_0^{2\pi}\int_0^{r\cos\theta+2}(r\sin\theta)\,r\,dz\,dr\,d\theta$$

$$= \int_0^{2\pi}\int_1^{2} r^2(r\cos\theta+2)\,dr\,d\theta$$

$$= \int_0^{2\pi}\int_1^{2}\frac{1}{2}r^3\sin(2\theta)+2r^2\sin\theta\,dr\,d\theta$$

$$= \int_0^{2\pi}\left(\frac{1}{8}r^4\sin(2\theta)+\frac{2}{3}r^3\sin\theta\right)\Bigg|_1^2 d\theta$$

$$= \int_0^{2\pi}\frac{15}{8}\sin(2\theta)+\frac{14}{3}\sin\theta\,d\theta$$

$$= -\left(-\frac{15}{16}\cos(2\theta)-\frac{14}{3}\cos\theta\right)\Bigg|_0^{2\pi}$$

$$= 0$$

Just as we did with double integral involving polar coordinates we can start with an iterated integral in terms of x, y, and z and convert it to cylindrical coordinates.

Example: Convert $\int_{-1}^{1}\int_0^{\sqrt{1-y^2}}\int_{x^2+y^2}^{\sqrt{x^2+y^2}} xyz\,dz\,dx\,dy$ into an integral in cylindrical coordinates.

Solution: Here are the ranges of the variables from this iterated integral.

$$-1 \le y \le 1$$

$$0 \le x \le \sqrt{1-y^2}$$

$$x^2+y^2 \le z \le \sqrt{x^2+y^2}$$

The first two inequalities define the region D and since the upper and lower bounds for the x's are $x=\sqrt{1-y^2}$ and we know that we've got at least part of the right half a circle of radius 1 centered at the origin. Since the range of y's is $-1 \le y \le 1$ we know that we have the complete right half of the disk of radius 1 centered at the origin. So, the ranges for D in cylindrical coordinates are,

$$-\frac{\pi}{2} \le \theta \le \frac{\pi}{2}$$

$$0 \le r \le 1$$

All that's left to do now is to convert the limits of the z range, but that's not too bad.

$$r^2 \le z \le r$$

On a side note notice that the lower bound here is an elliptic paraboloid and the upper bound is a cone. Therefore E is a portion of the region between these two surfaces.

The integral is,

$$\int_{-1}^{1}\int_{0}^{\sqrt{1-y^2}}\int_{x^2+y^2}^{\sqrt{x^2+y^2}} xyz\,dz\,dx\,dy = \int_{-\pi/2}^{\pi/2}\int_{0}^{1}\int_{r^2}^{y} r\left(r\cos\theta\right)\left(r\sin\theta\right)z\,dz\,dr\,d\theta$$

$$= \int_{-\pi/2}^{\pi/2}\int_{0}^{1}\int_{r^2}^{y} zr^3\cos\theta\sin\theta\,dz\,dr\,d\theta$$

TRIPLE INTEGRALS IN SPHERICAL COORDINATES

In the previous section we looked at doing integrals in terms of cylindrical coordinates and we now need to take a quick look at doing integrals in terms of spherical coordinates. First, we need to recall just how spherical coordinates are defined.

The following sketch shows the relationship between the Cartesian and spherical coordinate systems.

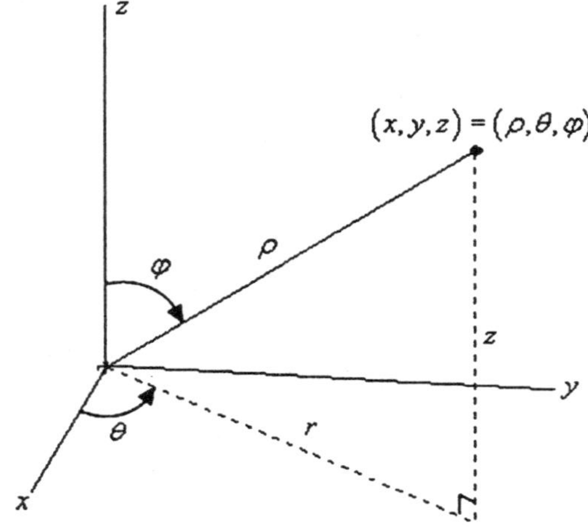

Here are the conversion formulas for spherical coordinates.

$$x = \rho\sin\varphi\cos\theta$$
$$y = \rho\sin\varphi\sin\theta$$
$$z = \rho\cos\varphi$$
$$x^2 + y^2 + z^2 = \rho^2$$

We also have the following restrictions on the coordinates.

$$\rho \geq 0 \quad 0 \leq \varphi \leq \pi$$

For our integrals we are going to restrict E down to a spherical wedge. This will mean that we are going to take ranges for the variables as follows,

$$a \le \rho \le b$$

$$\alpha \le \theta \le \beta$$

$$\delta \le \varphi \le \gamma$$

Here is a quick sketch of a spherical wedge in which the lower limit for both $\rho$and $\varphi$are zero for reference purposes. Most of the wedges we'll be working with will fit into this pattern.

From this sketch we can see that E is really nothing more than the intersection of a sphere and a cone. In the next section we will show that

$$dV = \rho^2 \sin\varphi d\,\rho d\theta d\varphi$$

Therefore the integral will become,

$$\iiint_E f(x, y, z)\,dV = \int_a^b \int_a^\beta \int_\delta^\gamma \rho^2 \sin\varphi f(\rho\sin\varphi\cos\theta, \rho\sin\varphi\sin\theta, \rho\cos\varphi)\,d\rho\,d\theta\,d\varphi$$

This looks bad, but given that the limits are all constants the integrals here tend to not be too bad.

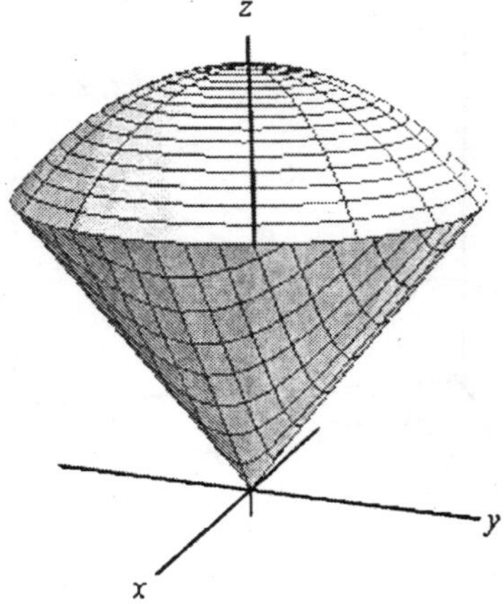

Example: Evaluate $\iiint_E 16z\,dV$ where E is the upper half of the sphere $x^2 + y^2 + z^2 = 1$.

Solution: Since we are taking the upper half of the sphere the limits for the variables are,

$$0 \le \theta \le 1$$

$$0 \le \theta \le 2\pi$$

$$0 \le \varphi \le \frac{\pi}{2}$$

The integral is then,

$$\iiint_E 16z\, dV = \int_0^{\frac{\pi}{2}} \int_0^{2\pi} \int_2^{1} \rho^2 \sin\varphi(16\rho\cos\varphi)\, d\rho\, d\theta\, d\varphi$$

$$= \int_0^{\frac{\pi}{2}} \int_0^{2\pi} \int_0^{1} 8\rho^3 \sin(2\varphi)\, d\rho\, d\theta\, d\varphi$$

$$= \int_0^{\frac{\pi}{4}} \int_0^{2\pi} 2\sin(2\varphi)\, d\theta\, d\varphi$$

$$= \int_0^{\frac{\pi}{2}} 4\pi \sin(2\varphi)\, d\varphi$$

$$= -2\pi\cos(2\varphi)\Big|_0^{\frac{\pi}{2}}$$

$$= 4\pi$$

Example:

$$\int_0^3 \int_0^{\sqrt{9-y^2}} \int_{\sqrt{x^2+y^2}}^{\sqrt{18-x^2-y^2}} x^2+y^2+z^2\, dz\, dx\, dy = \int_0^{\pi/4} \int_0^{\pi/2} \int_0^{3\sqrt{2}} \rho^4 \sin\varphi d\rho d\theta d\varphi$$

Convert into spherical coordinates.

Solution: Let's first write down the limits for the variables.

$$0 \le y \le 3$$

$$0 \le x \le \sqrt{18 - x^2 - y^2}$$

$$\sqrt{x^2+y^2} \le z \le \sqrt{18-x^2-y^2}$$

The range for x tells us that we have a portion of the right half of a disk of radius 3 centered at the origin. Since we are restricting y's to positive values it looks like we will have the quarter disk in the first quadrant.

Therefore since D is in the first quadrant the region, E, must be in the first octant and this in turn tells us that we have the following range for θ(since this is the angle around the z-axis).

$$0 \le \theta \le \frac{\pi}{2}$$

Now, let's see what the range for z tells us. The lower bound, $z = \sqrt{x^2+y^2}$, is the upper half of a cone. At this point we don't need this quite yet, but we will later. The upper bound, $z = \sqrt{18-x^2-y^2}$, is the upper half of the sphere,

$$x^2 + y^2 + z^2 = 18$$

and so from this we now have the following range for ρ

$$0 \le \rho \le \sqrt{18} = 3\sqrt{2}$$

Now all that we need is the range for φ. There are two ways to get this. One is from where the cone and the sphere intersect. Plugging in the equation for the cone into the sphere gives,

$$\left(\sqrt{x^2+y^2}\right)^2 + z^2 = 18$$

$$z^2 + z^2 = 18$$

$$z^2 = 9$$

$$z = 3$$

Note that we can assume z is positive here since we know that we have the upper half of the cone and/or sphere. Finally, plug this into the conversion for z and take advantage of the fact that we know that $\rho = 3\sqrt{2}$ since we are intersecting on the sphere. This gives,

$$\rho\cos\varphi = 3$$

$$3\sqrt{2}\cos\varphi = 3$$

$$\cos\varphi = \frac{1}{\sqrt{2}} = \frac{\sqrt{2}}{2} \quad \Rightarrow \quad \varphi = \frac{\pi}{4}$$

So, it looks like we have the following range,

$$0 \le \varphi \le \frac{\pi}{4}$$

The other way to get this range is from the cone by itself. By first converting the equation into cylindrical coordinates and then into spherical coordinates we get the following,

$$z = r$$

$$\rho\cos\varphi = \rho\sin\varphi$$

$$1 = \tan\varphi \quad \Rightarrow \quad \varphi = \frac{\pi}{4}$$

So, recalling that $\rho^2 = x^2 + y^2 + z^2$, the integral is then,

$$\int_0^3 \int_0^{\sqrt{9-y^2}} \int_{\sqrt{x^2+y^2}}^{\sqrt{18-x^2-y^2}} x^2 + y^2 + z^2 dz\, dx\, dy$$

$$= \int_0^{\pi/4} \int_0^{\pi/2} \int_0^{3\sqrt{2}} p^4 \sin\varphi d\rho d\theta\varphi$$

CHANGE OF VARIABLES

We had the substitution rule that told us that,

$$\int_a^b f(g(x))g'(x) = \int_c^d f(u)\,du \quad \text{where } u = g(x)$$

In essence this is taking an integral in terms of x's and changing it into terms of u's. We want to do something similar for double and triple integrals. In fact we've already done this to a certain extent when we converted double integrals to polar coordinates and when we converted triple integrals to cylindrical or spherical coordinates.

The main difference is that we didn't actually go through the details of where the formulas came from. If you recall, in each of those cases we commented that we would justify the formulas for dA and dV eventually. Now is the time to do that justification.

While often the reason for changing variables is to get us an integral that we can do with the new variables another reason for changing variables is to convert the region into a nicer region to work with. When we were converting the polar, cylindrical or spherical coordinates we didn't worry about this change since it was easy enough to determine the new limits based on the given region. That is not always the case however. So, before we move into changing variables with multiple integrals we first need to see how the region may change with a change of variables. First we need a little notation out of the way. We call the equations that define the change of variables a transformation. Also we will typically start out with a region, R, in xy-coordinates and transform it into a region in uv-coordinates.

Example: Determine the new region that we get by applying the given transformation to the region R.

1. R is the ellipse $x^2 + \frac{y^2}{36} = 1$ and the transformation is $x = \frac{u}{2}, y = 3v$,

2. R is the region bounded by $y = -x + 4, y = x + 1$ and $y = \frac{x}{3} - \frac{4}{3}$ and the transformation is $y = \frac{1}{2}(u + v), y = \frac{1}{2}(u - v)$

Solution:

1. R is the ellipse $x^2 + \frac{y^2}{36} = 1$ and the transformation is, $x = \frac{u}{2}, y = 3v$.

There really isn't too much to do with this one other than to plug the transformation into the equation for the ellipse and see what we get.

$$\left(\frac{u}{2}\right)^2 + \frac{(3v)^2}{36} = 1$$

$$\frac{u^2}{4}+\frac{9v^2}{36}=1$$

$$u^2+v^2=4$$

So, we started out with an ellipse and after the transformation we had a disk of radius 2.

2. R is the region bounded by, $y=-x+4, y=x+1$ and $y=\frac{x}{3}-\frac{4}{3}$ and the transformation is, $y=\frac{1}{2}(u+v), y=\frac{1}{2}(u-v)$.

As with the first part we'll need to plug the transformation into the equation, however, in this case we will need to do it three times, once for each equation. Before we do that let's sketch the graph of the region and see what we've got.

So, we have a triangle. Now, let's go through the transformation. We will apply the transformation to each edge of the triangle and see where we get.

Let's do $y=-x+4$ first. Plugging in the transformation gives,

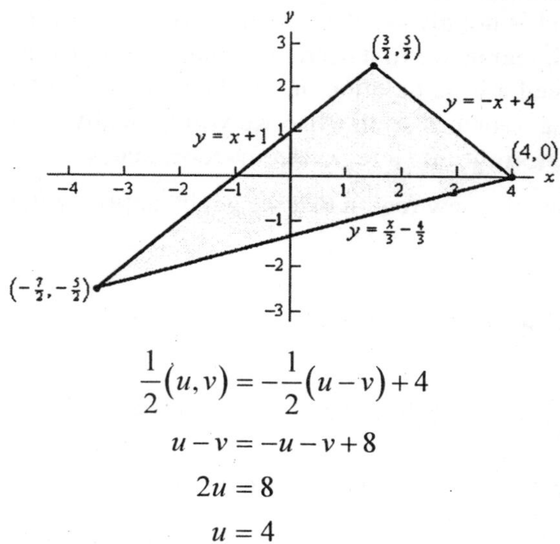

$$\frac{1}{2}(u,v)=-\frac{1}{2}(u-v)+4$$

$$u-v=-u-v+8$$

$$2u=8$$

$$u=4$$

The first boundary transforms very nicely into a much simpler equation.

Now let's take a look at $y=x+1$,

$$\frac{1}{2}(u-v)=-\frac{1}{2}(u+v)+1$$

$$u-v=u+v=2$$

$$-2v=2$$

$$v=-1$$

Again, a much nicer equation that what we started with.

Finally, let's transform $y = \frac{x}{3} - \frac{4}{3}$.

$$\frac{1}{2}(u - v) = \frac{1}{3}\left(\frac{1}{2}(u + v)\right) - \frac{4}{3}$$

$$3u - 3v = u + v - 8$$

$$4v = 2u + 8$$

$$v = \frac{u}{2} + 2$$

So, again, we got a somewhat simpler equation, although not quite as nice as the first two.

Let's take a look at the new region that we get under the transformation.

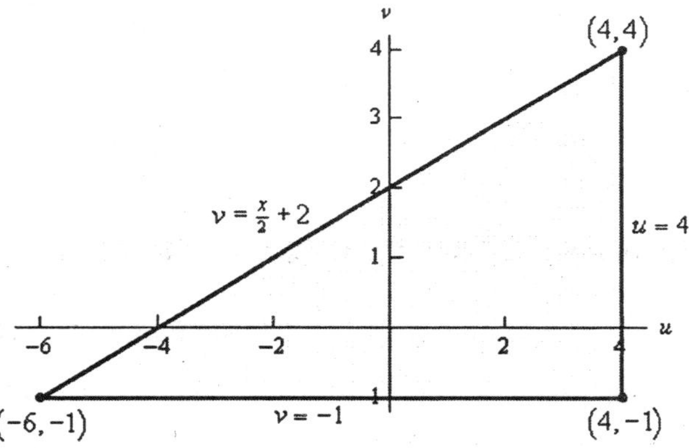

We still get a triangle, but a much nicer one.

Note that we can't always expect to transform a specific type of region (a triangle for example) into the same kind of region. It is completely possible to have a triangle transform into a region in which each of the edges are curved and in no way resembles a triangle. Notice that in each of the above examples we took a two dimensional region that would have been somewhat difficult to integrate over and converted it into a region that would be much nicer in integrate over.

As we noted at the start of this set of examples, that is often one of the points behind the transformation. In addition to converting the integrand into something simpler it will often also transform the region into one that is much easier to deal with. Now that we've seen a couple of examples of transforming regions we need to now talk about how we actually do change of variables in the integral.

We will start with double integrals. In order to change variables in a double integral we will need the Jacobian of the transformation. Here is the definition of the Jacobian.

Definition: The Jacobian of the transformation, $x = g(u,v), y = h(u,v)$ is

$$\frac{\partial(x,y)}{\partial(u,v)} = \begin{vmatrix} \frac{\partial x}{\partial u} & \frac{\partial x}{\partial v} \\ \frac{\partial y}{\partial u} & \frac{\partial y}{\partial v} \end{vmatrix}$$

The Jacobian is defined as a determinant of a 2x2 matrix, if you are unfamiliar. Here is how to compute the determinant.

$$\begin{vmatrix} a & b \\ c & d \end{vmatrix} = ad - bc$$

Therefore, another formula for the determinant is,

$$\frac{\partial(x,y)}{\partial(u,v)} = \begin{vmatrix} \frac{\partial x}{\partial u} & \frac{\partial x}{\partial v} \\ \frac{\partial y}{\partial u} & \frac{\partial y}{\partial v} \end{vmatrix} = \frac{\partial x}{\partial u}\frac{\partial y}{\partial v} - \frac{\partial x}{\partial v}\frac{\partial y}{\partial u}$$

Now that we have the Jacobian out of the way we can give the formula for change of variables for a double integral.

CHANGE OF VARIABLES FOR A DOUBLE INTEGRAL

Suppose that we want to integrate $f(x,y)$ over the region *R*. Under the transformation $x = g(u,v), y = h(u,v)$ the region becomes *S* and the integral becomes,

$$\iint_D f(x,y)\,dA = \iint_S f(g(u,v),h(u,v))\left|\frac{\partial(x,y)}{\partial(u,v)}\right| du\,dv$$

Note that we used *du dv* instead of *dA* in the integral to make it clear that we are now integrating with respect to *u* and *v*. Also note that we are taking the absolute value of the Jacobian.

If we look just at the differentials in the above formula we can also say that

$$dA = \left|\frac{\partial(x,y)}{\partial(u,v)}\right| du\,dv$$

Example: Show that when changing to polar coordinates we have $dA = r\,dr\,d\theta$

Solution: So, what we are doing here is justifying the formula that we used back when we were integrating with respect to polar coordinates. All that we need to do is use the formula above for dA.

The transformation here is the standard conversion formulas,

$$x = r\cos\theta$$
$$y = r\sin\theta$$

The Jacobian for this transformation is,

$$\begin{aligned}\frac{\partial(x,y)}{\partial(r,\theta)} &= \begin{vmatrix}\frac{\partial x}{\partial r} & \frac{\partial x}{\delta\theta}\\ \frac{\partial y}{\partial r} & \frac{\partial y}{\partial\theta}\end{vmatrix}\\ &= \begin{vmatrix}\cos\theta & -r\sin\theta\\ \sin\theta & r\cos\theta\end{vmatrix}\\ &= r\cos^2\theta\left(r\sin^2\theta\right)\\ &= r\left(\cos^2\theta+\sin^2\theta\right)\\ &= r\end{aligned}$$

We then get,

$$dA = \left|\frac{\partial(x,y)}{\partial(r,\theta)}\right| dr\,d\theta = |r|\,dr\ d\theta = rdr\ d\theta$$

So, the formula we used in the section on polar integrals was correct.

Now, let's do a couple of integrals.

Example: Evaluate $\iint\limits_R x+y\ dA$ where R is the trapezoidal region with vertices given by $(0.0),(5,0),\left(\frac{5}{2},\frac{5}{2}\right)$ and $\left(\frac{5}{2},-\frac{5}{2}\right)$ using the transformation $x = 2u+3v$ and $y = 2u-3v$.

Solution: First, let's sketch the region R and determine equations for each of the sides.

Each of the equations was found by using the fact that we know two points on each line (*i.e.* the two vertices that form the edge). While we could do this integral in terms of x and y it would involve two integrals and so would be some work. Let's use the transformation and see what we get. We'll do this by plugging the transformation into each of the equations above.

Let's start the process off with $y = x$ .7

$$2u-3v = 2u+3v$$
$$6v = 0$$
$$v = 0$$

Transforming $y = -x$ is similar.

$$2u - 3v = -(2u + 3v)$$
$$4u = 0$$
$$u = 0$$

Next we'll transform $y = -x + 5$.

$$2u - 3v = -(2u + 3v) + 5$$
$$4u = 5$$
$$u = \frac{5}{4}$$

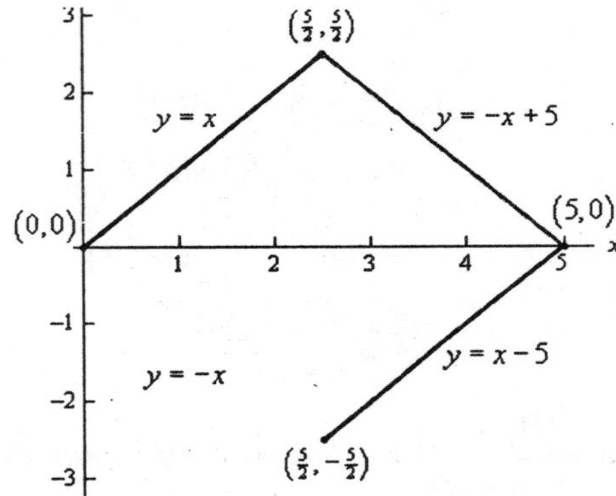

Finally, let's transform $y = x - 5$.

$$2u - 3v = -2u + 3v - 5$$
$$-6v = -5$$
$$v = \frac{5}{6}$$

The region S is then a rectangle whose sides are given by $u = 0, v = 0, u = \frac{5}{4}$ and $v = \frac{5}{6}$ and so the ranges of u and v are,

$$0 \le u \le \frac{5}{4} \quad 0 \le v \le \frac{5}{6}$$

Next, we need the Jacobian.

$$\frac{\partial(x, y)}{\partial(u, v)} = \begin{vmatrix} 2 & 3 \\ 2 & -3 \end{vmatrix} = -6 - 6 = -12$$

The integral is then,

$$\iint_R x - y\,dA = \int_0^{\frac{5}{6}} \int_0^{\frac{5}{4}} (2u+3v)+(2u-3v)\left|-12\right| du\,dv$$

$$= \int_0^{\frac{5}{6}} \int_0^{\frac{5}{4}} 48u\,du\,dv$$

$$= \int_0^{\frac{5}{6}} 24u^2 \Big|_0^{\frac{5}{4}} dv$$

$$= \int_0^{\frac{5}{4}} \frac{75}{2} dv$$

$$= \frac{75}{2} v \Big|_0^{\frac{5}{6}}$$

$$= \frac{125}{4}$$

Example: Evaluate $\iint_R x^2 - xy + y^2 dA$ where R is the ellipse given by $x^2 - xy + y^2 = 2$ and using the transformation $x = \sqrt{2}u - \sqrt{\frac{2}{3}}v, y = \sqrt{2}u + \sqrt{\frac{2}{3}}v$.

Solution: The first thing to do is to plug the transformation into the equation for the ellipse to see what the region transforms into.

$$2 = x^2 - xy + y^2$$

$$= \left(\sqrt{2}u - \sqrt{\frac{2}{3}}v\right)^2 - \left(\sqrt{2}u - \sqrt{\frac{2}{3}}v\right)\left(\sqrt{2}u + \sqrt{\frac{2}{3}}v\right) + \left(\sqrt{2}u - \sqrt{\frac{2}{3}}v\right)^2$$

$$= 2u^2 - \frac{4}{\sqrt{3}}uv + \frac{2}{3}v^2 - \left(2u^2 - \frac{2}{3}v^2\right) + 2u^2 + \frac{4}{\sqrt{3}}uv + \frac{2}{3}v^2$$

$$= 2u^2 + 2v^2$$

Or, upon dividing by 2 we see that the equation describing R transforms into

$$u^2 + v^2 = 1$$

or the unit circle. Again, this will be much easier to integrate over than the original region.

Note as well that we've shown that the function that we're integrating is

$$x^2 - xy + y^2 = 2\left(u^2 + v^2\right)$$

in terms of u and v so we won't have to redo that work when the time to do the integral comes around.

Finally, we need to find the Jacobian.

$$\frac{\partial(x,y)}{\partial(x,y)} = \begin{vmatrix} \sqrt{2} & -\sqrt{\frac{2}{3}} \\ \sqrt{2} & \sqrt{\frac{2}{3}} \end{vmatrix} = \frac{2}{\sqrt{3}} + \frac{2}{\sqrt{3}} = \frac{4}{\sqrt{3}}$$

The integral is then,

$$\iint_R x^2 - xy + y^2 \, dA = \iint_S 2\left(u^2 + v^2\right)\left|\frac{4}{\sqrt{3}}\right| du\, dv$$

Before proceeding a word of caution is in order. Do not make the mistake of substituting $x^2 - xy + y^2 = 2$ or $u^2 + v^2 = 1$ in for the integrands.

These equations are only valid on the boundary of the region and we are looking at all the point interior to the boundary as well and for those points neither of these equations will be true!

At this point we'll note that this integral will be much easier in terms of polar coordinates and so to finish the integral out will convert to polar coordinates.

$$\begin{aligned} \iint_R x^2 - xy + y^2 dA &= \iint_S 2\left(u^2 + v^2\right)\left|\frac{4}{\sqrt{3}}\right| du\, dv \\ &= \frac{8}{\sqrt{3}} \int_0^{2\pi} \int_0^1 \left(r^2\right) r \, dr\, d\theta \\ &= \frac{8}{\sqrt{3}} \int_0^{2\pi} \frac{1}{4} r^4 \Big|_0^1 \, d\theta \\ &= \frac{8}{\sqrt{3}} \int_0^{2\pi} \frac{1}{4} d\theta \\ &= \frac{4\pi}{\sqrt{3}} \end{aligned}$$

Let's now briefly look at triple integrals. In this case we will again start with a region R

and use the transformation $x = g(u,v,w)$, $y = h(u,v,w)$, and $z = k(u,v,w)$ to transform the region into the new region *S*.

To do the integral we will need a Jacobian, just as we did with double integrals. Here is the definition of the Jacobian for this kind of transformation.

$$\frac{\partial(x,y,z)}{\partial(u,v,w)} = \begin{vmatrix} \frac{\partial x}{\partial u} & \frac{\partial x}{\partial v} & \frac{\partial x}{\partial w} \\ \frac{\partial y}{\partial u} & \frac{\partial y}{\partial y} & \frac{\partial y}{\partial w} \\ \frac{\partial z}{\partial u} & \frac{\partial z}{\partial v} & \frac{\partial z}{\partial w} \end{vmatrix}$$

In this case the Jacobian is defined in terms of the determinant of a 3x3 matrix. We saw how to evaluate these when we looked at cross products back.

If you need a refresher on how to compute them you should go back and review that section. The integral under this transformation is,

$$\iiint_R f(x,y,z)\,dV = \iiint_S f\big(g(u,v,w),h(u,v,w),k(u,v,w)\big)\left|\frac{\partial(x,y,z)}{\partial(u,v,w)}\right| du\ dv\ dw$$

As with double integrals we can look at just the differentials and note that we must have

$$dV = \left|\frac{\partial(x,y,z)}{\partial(u,v,w)}\right| du\,dv\,dw$$

We're not going to do any integrals here, but let's verify the formula for dV for spherical coordinates.

Example: Verify that $dV = \rho^2 \sin\varphi d\,\rho d\theta\,d\varphi$ when using spherical coordinates.

Solution: Here the transformation is just the standard conversion formulas.

$$x = \rho\sin\varphi\cos\theta$$
$$y = \rho\sin\varphi\sin\theta$$
$$z = \rho\cos\varphi$$

The Jacobian is,

$$\frac{\partial(z,y,z)}{\partial(\rho,\theta,\varphi)} = \begin{vmatrix} \sin\varphi\cos\theta & -\rho\sin\varphi\sin\theta & \rho\cos\varphi\cos\theta \\ \sin\varphi\cos\theta & \rho\sin\varphi\sin\theta & \rho\cos\varphi\sin\theta \\ \cos\varphi & 0 & -\rho\sin\varphi \end{vmatrix}$$

$$= -\rho^2\sin^3\varphi\cos^2\theta - \rho^2\sin\varphi\cos^2\varphi\sin^2\theta + 0$$

$$-\rho^2\sin^3\varphi - \rho^2\sin\varphi\cos^2\varphi$$

$$= -\rho^2\sin^3\varphi - \rho^2\sin\varphi\cos^2\varphi$$

$$= -\rho^2\sin\varphi\left(\sin^2\varphi + \cos^2\varphi\right) = -\rho^2\sin\varphi$$

Finally, dV becomes,

$$dV = \left|-\rho^2\sin\varphi\right| d\rho\, d\theta\, d\varphi = \rho^2\sin\varphi\, d\rho\, d\theta\, d\varphi$$

Recall that we restricted $\varphi$to the range $0 \le \varphi \le \pi$ for spherical coordinates and so we know that $\sin\varphi \le 0$ and so we don't need the absolute value bars on the sine.

SURFACE AREA

In this section we will look at the lone application (aside from the area and volume interpretations) of multiple integrals in this material. This is not the first time that we've looked at surface area , however, in that setting we were looking at the surface area of a solid of revolution. In other words we were looking at the surface area of a solid obtained by rotating a function about the x or y axis.

Here we want to find the surface area of the surface given by $z = f(x,y)$ where (x,y) is a point from the region D in the xy-plane. In this case the surface area is given by,

$$S = \iint_D \sqrt{[f_x]^2 + \left[f_y\right]^2 + 1dA}$$

Let's take a look at a couple of examples.

Example: Find the surface area of the part of the plane $3x + 2y + z = 6$ that lies in the first octant.

Solution: Remember that the first octant is the portion of the xyz-axis system in which all three variables are positive.

Let's first get a sketch of the part of the plane that we are interested in.

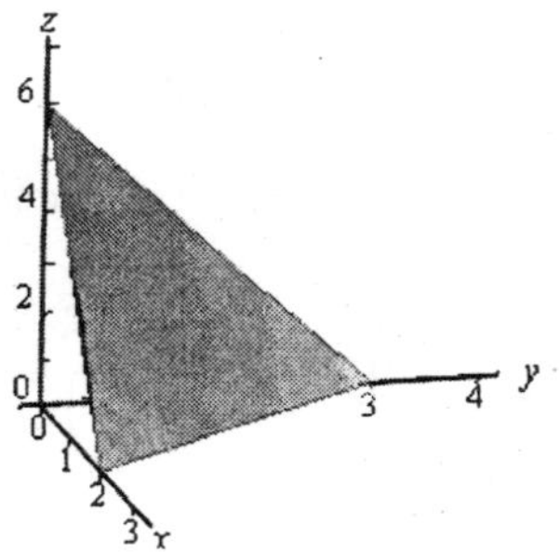

We'll also need a sketch of the region *D*..

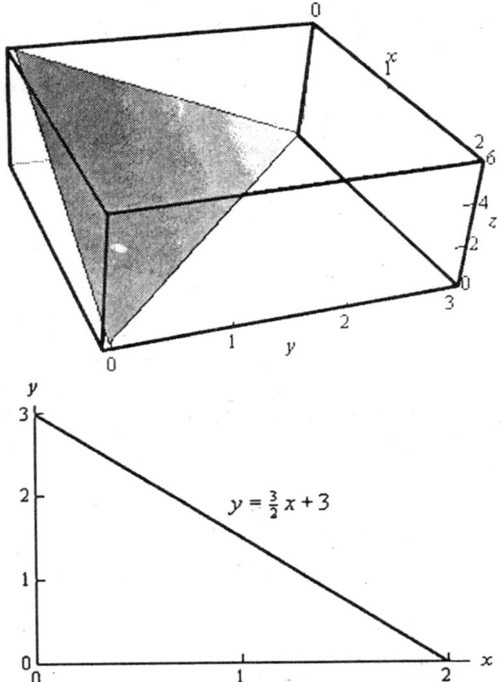

Remember that to get the region *D* we can pretend that we are standing directly over the plane and what we see is the region *D*.

We can get the equation for the hypotenuse of the triangle by realizing that this is nothing more than the line where the plane intersects the *xy*-plane and we also know that $z = 0$ on the *xy*-plane. Plugging $z = 0$ into the equation of the plane will give us the equation for the hypotenuse. Notice that in order to use the surface area formula we need to have the function in the form $z = f(x, y)$ and so solving for *z* and taking the partial derivatives gives,

$$z = 6 - 3x - 2y$$
$$f_x = -3$$
$$f_y = -2$$

The limits defining D are,

$$0 \le x \le 2$$
$$0 \le y \le -\frac{3}{2}x + 3$$

The surface area is then,

$$S = \iint_D \sqrt{[-3]^2 + [-2]^2 + 1}\,dA$$
$$= \int_0^2 \int_0^{-\frac{3}{2}x+3} \sqrt{14}\,dy\,d\theta$$
$$= \sqrt{14}\int_0^2 -\frac{3}{2}x + 3\,dx$$
$$= \sqrt{14}\left(-\frac{3}{2}x + 3x\right)\Big|_0^2$$
$$= 3\sqrt{14}$$

Example: Determine the surface area of the part of $z = xy$ that lies in the cylinder given by $x^2 + y^2 = 1$.

Solution: In this case we are looking for the surface area of the part of $z = xy$ where (x, y) comes from the disk of radius 1 centered at the origin since that is the region that will lie inside the given cylinder.

Here are the partial derivatives,

$$f_x = y$$
$$f_y = x$$

The integral for the surface area is,

$$S = \iint_D \sqrt{x^2 + y^2 + 1}\,dA$$

Given that D is a disk it makes sense to do this integral in polar coordinates.

$$
\begin{aligned}
S &= \iint\limits_D \sqrt{x^2 + y^2 + 1dA} \\
&= \int_0^{2\pi} \int_0^1 r\sqrt{1+r^2}\,dr\,d\theta \\
&= \int_0^{2\pi} \frac{1}{2}\left(\frac{2}{3}\right)\left(1+r^2\right)^{\frac{3}{2}}\Big|_0^1 \,d\theta \\
&= \int_0^{2\pi} \frac{1}{3}\left(2^{\frac{3}{2}} - 1\right) d\theta \\
&= \frac{2\pi}{3}\left(2^{\frac{3}{2}} - 1\right)
\end{aligned}
$$

AREA AND VOLUME REVISITED

This section is here only so we can summarize the geometric interpretations of the double and triple integrals that we saw in this chapter.

Since the purpose of this section is to summarize these formulas we aren't going to be doing any examples in this section.

We'll first look at the area of a region. The area of the region D is given by,

$$\text{Area of } D = \iint\limits_D dA$$

Now let's give the two volume formulas. First the volume of the region E is given by,

$$\text{Volume of } E = \iiint\limits_E dA$$

Finally, if the region E can be defined as the region under the function $z = f(x, y)$ and above the region D in xy-plane then,

$$\text{Volume of } E = \iint\limits_D f(x, y)\,dA$$

Note as well that there are similar formulas for the other planes. For instance, the volume of the region behind the function $y = f(x, y)$ and in front of the region D in the xz-plane is given by,

$$\text{Volume of } E = \iint_D f(x, z)\, dA$$

Likewise, the the volume of the region behind the function $x = f(x, y)$ and in front of the region D in the yz-plane is given by,

$$\text{Volume of } E = \iint_D f(x, z)\, dA$$

$$= \int_0^{2\pi} \frac{1}{2}\left(\frac{2}{3}\right)(1 + r^2)^{\frac{3}{2}}\Bigg|_0^1 \, d\theta$$

$$= \int_0^{2\pi} \frac{1}{3}\left(2^{\frac{3}{2}} - 1\right) d\theta$$

$$= \frac{2\pi}{2}\left(2^{\frac{3}{2}} - 1\right)$$

AREA AND VOLUME REVISITED

This section is here only so we can summarize the geometric interpretations of the double and triple integrals that we saw in this chapter. Since the purpose of this section is to summarize these formulas we aren't going to be doing any examples in this section.

We'll first look at the area of a region. The area of the region D is given by,

$$\text{Area of } D = \iint_D dA$$

Now let's give the two volume formulas. First the volume of the region E is given by,

$$\text{Volume of } E = \iiint_E dA$$

Finally, if the region E can be defined as the region under the function $z = f(x, y)$ and above the region D in xy-plane then,

$$\text{Volume of } E = \iint_D f(x, y)\, dA$$

Note as well that there are similar formulas for the other planes. For instance, the volume of the region behind the function $y = f(x, y)$ and in front of the region D in the xz-plane is given by,

$$\text{Volume of } E = \iint_D f(x, z)\, dA$$

Likewise, the the volume of the region behind the function $x = f(x, y)$ and in front of the region D in the yz-plane is given by,

$$\text{Volume of } E = \iint_D f(x, z)\, dA$$

Chapter 5

Integrals Over Regions

The relationship between double integrals and iterated integrals (Fubini's Theorem) over more general regions is detailed. The distinction between a double integral and an iterated integral in two variables is explained. Indeed, the double integral is often evaluated by converting it to an equivalent iterated integral, which is usually easier to compute; but nonetheless double integrals and iterated integration are distinct concepts. Essentially, an iterated integral is like the inverse of mixed partial differentiation and a double integral is a direct extension of the Riemann integral (Riemann sums) of one variable to functions of two independent variables.

A type I, or vertically simple region R_1 is a region of the plane that can be described by the inequalities

$$D_1 : a \leq x \leq b, \quad g_1(x) \leq y \leq g_2(x),$$

where $g_1(x)$ and $g_2(x)$ are continuous functions of x on $[a, b]$. Similarly, a type II, or horizontally simple region R_1, in the plane is a region that can be described by the inequalities

$$D_2 : c \leq y \leq d, \quad h_1(x) \leq y \leq h_2(y),$$

where $h_1(x)$ and $h_2(x)$ are continuous functions of y on $[c, d]$.

Definition: *(Double Integral over a Region)* Let $f(x, y)$ be a function that is continuous on the region D that can be contained in a rectangle R Define the function $F(x, y)$ on R as if (x, y) is in D and 0 otherwise. If F is integrable over R, we say that *f is integrable over D,* and the *double integral* of R over D is defined as

$$\iint_D f(x, y)\, dA = \iint_R f(x, y)\, dA$$

The function $F(x, y)$ may have discontinuities on the boundary of D, but if $f(x, y)$ is continuous on D and the boundary of is fairly "well behaved", then it can be shown that

$\iint_R F(x, y)\,dA$ exists and hence that $\iint_D f(x, y)\,dA$ exists. This procedure is valid for the type I and type II simple regions.

Proposition: *(Double Integral over a Region)* If D_1 is a type I region, then

$$\iint_{D2} f(x, y)\,dA = \int_c^d \int_{h1(y)}^{h2(y)} f(x, y)dl\,x\,dly$$

whenever both integrals exist. Similarly, for a type II region

whenever both integrals exist.

Example: *(Double Integral over a Region)* Compute the double integral over the nonrectangular region.

1. Compute the double integral $\int_{-2}^{1} \int_{y^2+4y}^{3y+2} dl\,x\,dl\,y$

Solution: The region is horizontally simple or a type II region as $-2 \le y \le 1$ and $y^2 + 4\,y \le x \le 3y + 2$. We compute

$$\int_{-2}^{1} \int_{y^2+4y}^{3y+2} dl\,x\,dl\,y = \int_{-2}^{1} (-y^2 - y + 2)dl\,y\,dl\,x = \frac{9}{2}$$

This region could also be considered as a vertically simple or a type I region as $-4 \le x \le 5\,\frac{x-2}{3} \le y \le -2 + \sqrt{x+4}$. and We compute

$$\int_{-4}^{5} \int_{(x-2)/3}^{-2+\sqrt{x+4}} dl\,y\,dl\,x = \int_{-4}^{5} \left(-2 + \sqrt{x+4}) - \left(\frac{x-2}{3} \right) \right) dlx = \frac{9}{2}$$

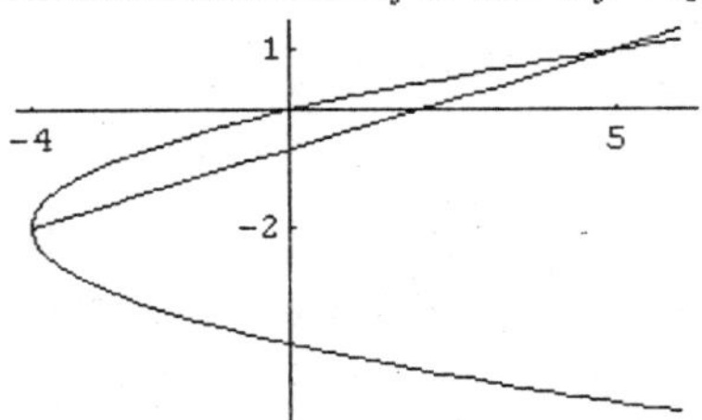

2. Compute the double integral $\int_0^{\pi/2} \int_0^{\sin x} e^y \cos x dl\,y\,dl\,x$

Solution: The region is vertically simple or a type I region as $0 \le x \le \frac{\pi}{2}$ and $0 \le y \le \sin x$

We compute

$$\int_0^{\pi/2}\int_0^{\sin x} e^y \cos x dl\, y dl\, y dl\, x = \int_0^{\pi/2}(e^{x \sin x} \cos x - \cos x) dl\, x = e - 2$$

The region is also horizontally simple or a type II region as $0 \leq y \leq 1$ and $\text{Arcsin}(y) \leq x \leq \frac{\pi}{2}$. We compute

$$\int_0^1 \int_{\text{Arcsin}(y)}^{\pi/2} e^y \cos x\, dl\, x\, dl\, y = e - 2$$

3. Evaluate the integral $\int_0^1 \int_x^1 \sin(y^2)\, dl\, y\, dl\, x$

Solution. If we try to evaluate the integral as it stands, we are faced with the task of first evaluating $\int \sin(y^2) dl\, y$. But this is impossible to do so in finite terms since $\int \sin(y^2) dl\, y$. is not an elementary function. So we must change the order of integration. This is accomplished by first expressing the given iterated integral as a double integral. We have

$$\int_0^1 \int_x^1 \sin(y^2) dl\, y\, dl\, x = \iint_D \sin(y^2) d\, A$$

where $D = \{(x, y) \mid 0 \leq x \leq 1, x \leq y \leq 1\}$. This region has an alternate description: $D = \{(x, y) \mid 0 \leq y \leq 1, x \leq y \leq 1\}$

Thus we can express the double integral as an iterated integral in the reverse order:

$$\int_0^1 \int_x^1 \sin(y^2) dl\, y\, dl\, x = \iint_D \sin(y^2) d\, A$$

$$= \int_0^1 \int_0^y \sin(y^2) dl\, x\, dl\, y = \int_0^1 [x \sin(y^2)]\Big|_{x=0}^{x=y} dl\, y = \frac{1}{2}(1 - \cos 1)$$

4. Evaluate $\iint_D x\, y\, d\, A$ where D is the region bounded by the line $y = x - 1$ and the parabola $y^2 = 2x + 6$

Solution: The region D is both a type I and a type II region, but the description of D as a type I is more complicated because the lower boundary consists of two parts. Therefore we express D as a type II region:

$$D = \left\{(x, y) \mid -2 \leq y \leq 4, \frac{y^2}{2} - 3 \leq x \leq y + 1\right\}.$$

Then the double integral becomes

$$\iint_D x\,y\,d\,A = \int_{-2}^{4}\int_{\frac{y^2}{2}-3}^{y+1} x\,y\,dl\,x\,dl\,y = \frac{1}{2}\int_{-2}^{4}\left(-\frac{y^2}{4}+4y^2+2y^2-8y\right)dl\,y = 36.$$

If we had expressed D as a type I region, then we would have obtained

$$\iint_D x\,y\,d\,A = \int_{-3}^{-1}\int_{-\sqrt{2x+6}}^{\sqrt{2x+6}} xy\,dl\,y\,dl\,x + \int_{x-1}^{\sqrt{2x+6}} xy\,dl\,y\,dl\,x$$

but this would have involved more work than the first part.

he intersection of $y = x - 1$ and $y^2 = 2x + 6$

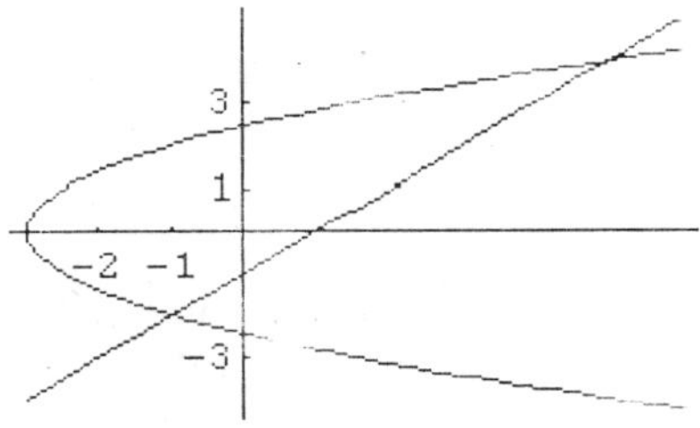

$$\left\{(x, y)\,|-2 \le y \le 4, \frac{y^2}{2} - 3 \le x \le y + 1\right\}$$

DOUBLE INTEGRALS

In calculus of a single variable the definite integral

$$\int_a^b f(x)$$

for $f(x)^{3}0$ is the area under the curve f(x) from x=a to x=b. For general f(x) the definite integral is equal to the area above the x-axis minus the area below the x-axis.

The definite integral can be extended to functions of more than one variable. Consider a function of 2 variables z=f(x,y). The definite integral is denoted by

$$\iint_R f(x, y)\,dA$$

where R is the région of integration in the xy-plane. For positive f(x,y), the definite integral is equal to the volume under the surface z=f(x,y) and above xy-plane for x and y in the region R. This is shown in the figure below.

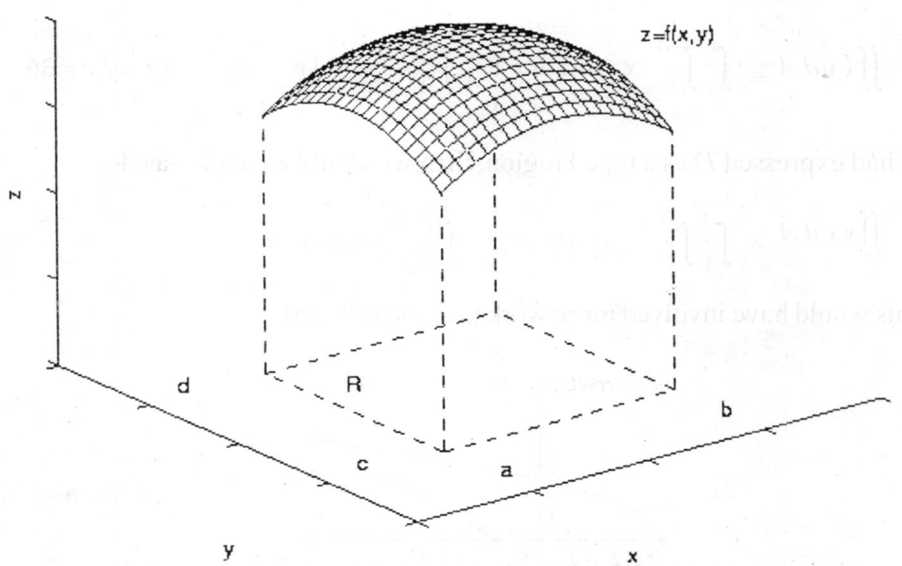

·For general f(x,y), the definite integral is equal to the volume above the xy-plane minus the volume below the xy-plane.

APPLICATIONS

Double integrals arise in a number of areas of science and engineering, including computations of

- Area of a 2D region
- Volume
- Mass of 2D plates
- Force on a 2D plate
- Average of a function
- Center of Mass and Moment of Inertia
- Surface Area

BRIEF DISCUSSION OF RIEMANN SUMS

As in the case of an integral of a function of one variable, a double integral is defined as a limit of a Riemann sum. Suppose we subdivide the region R into subrectangles as in the figure below (say there are M rectangles in the x direction and N rectangles in the y direction).

Label the rectangles R_{ij} where 1£ i£ M and 1 £ j £ N. Think of the definite integral as representing volume. The volume under the surface above rectangle R_{ij} is approximately $f(x_i,y_j)A_{ij}$, where A_{ij} is area of the rectangle and $f(x_i,y_j)$ is the approximate height of the surface in the rectangle. Here (x_i, y_j) is some point in the rectangle R_{ij}. If we sum over all rectangles we have

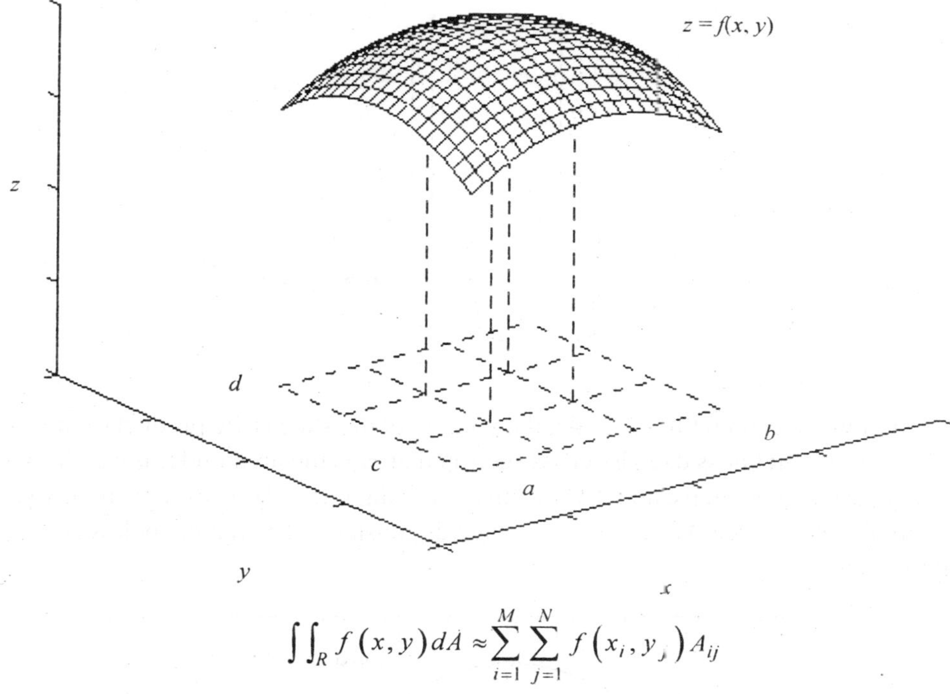

$$\iint_R f(x,y)\,dA \approx \sum_{i=1}^{M} \sum_{j=1}^{N} f(x_i, y_j) A_{ij}$$

In the limit as the size of the rectangles goes to 0, the sum on the right converges to a value which is the definite integral.

The quantity f(x,y)dA in the definite integral represents the volume in some infinitesimal region around the point (x,y).

The region is so small that the f(x,y) only varies infinitesimally in the region. The double integral sign says: add up volumes in all the small regions in R.

DOUBLE INTEGRALS OVER A RECTANGULAR REGION

Suppose that f(x,y) is continuous on a rectangular region in the xy plane as shown above. The double integral

$$\iint_R f(x,y)\,dA \approx$$

represents the volume under the surface. We can compute the volume by slicing the three-dimensional region like a loaf of bread. Suppose the slices are parallel to the y-axis. An example of slice between x and x+dx is shown in the figure.

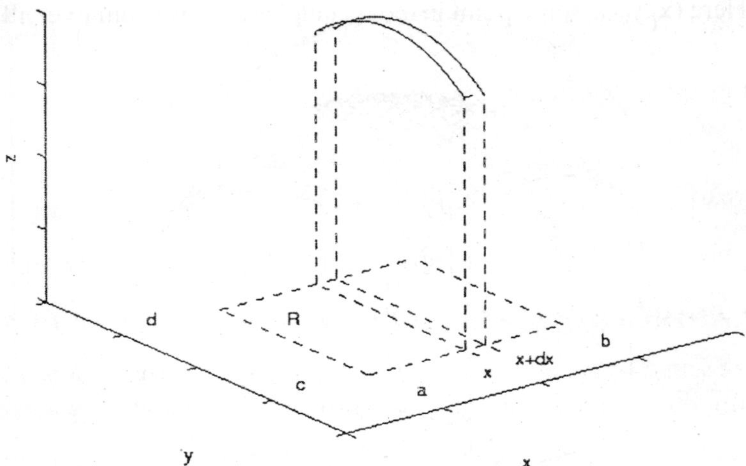

In the limit of infinitesimal thickness dx, the volume of the slice is the product of the cross-sectional area and the thickness dx. The cross sectional area is the area under the curve f(x,y) for fixed x and y varying between c and d. (Note that if the thickness dx is infinitesimal, x varies only infinitesimally on the slice. We can assume that x is constant.) The picture below shows the cross-sectional area.

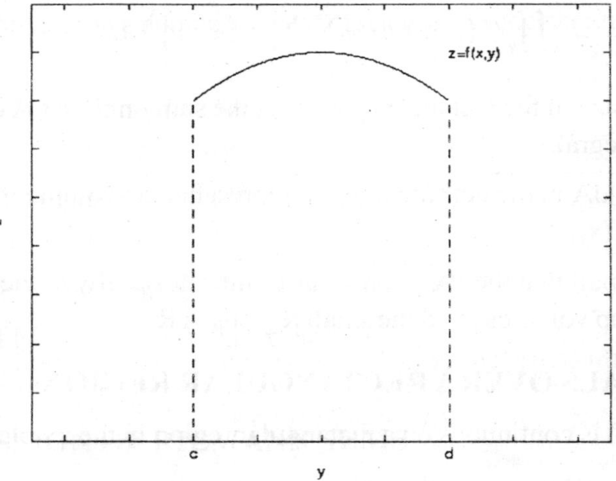

The area is given by the integral

$$C(x) = \int_c^d f(x,y)\,dy$$

The variable of integation is y and x is a CONSTANT. The cross-sectional area depends on x and this is why we write C=C(x). The volume of the slice between x and x+dx is C(x)dx. The total volume is the sum of the volumes of all the slices between x=a and x=b:

If substitute for C(x), we obtain:

$$V = \int_a^d \left[\int_c^d f(x,y)\, dy \right] dx = \int_a^b \int_c^d f(x,y)\, dydx$$

This is an example of an *iterated* integral. One integrates with respect to y first, then x. The integrals with respect to y and x are called the inner and outer integrals, respectively. Alternatively, one can make slices that are parallel to the x-axis. In this case the volume is given by

$$V = \int_c^d \left[\int_a^d f(x,y)\, dy \right] dy = \int_c^b \int_a^d f(x,y)\, dydx$$

The inner integral corresponds to the cross-sectional area of a slice between y and $y + dy$.

The quantities $f(x,y)$dydx and $f(x, y)$dxdy represent the value of the double integral in the infinitesimal rectangle between x and $x + dx$ and y and $y + dy$. The length and width of the rectangle are dx and dy, respectively. Hence dydx (or dxdy) is the area of the rectangle. We can make the connection $dA = dydx$ (or $dA = dxdy$).

Example: Consider the double integral:

$$V = \iint_R \left(x^2 + xy^3\right) dA,$$

where R is the rectangle $0 \leq x \leq 1$, $1 \leq y \leq 2$. Suppose we integrate with respect to y first. Then

$$V = \int_0^1 \int_1^2 \left(x^2 + xy^3\right) dydx,$$

The inner integral is

$$V = \int_1^2 \left(x^2 + xy^3\right) dy = \left(x^2 y + x\frac{y^4}{4}\right) \Big|_{y=1}^{y=2}$$

Note that we treat x as a constant as we integrate with respect to y. The integral is equal to

$$x^2(2) + x\left(2^4/4\right) - x^2 - x/4 = x^2 + \frac{15}{4}x$$

We are now left with the integral

$$\int_0^1 \left(x^2 + \frac{15}{4}x\right) dx = \left(\frac{x^3}{3} + \frac{15}{8}x^2\right) \Big|_0^1 = \frac{1}{3} + \frac{15}{8}$$

Alternatively, we can integrate with respect to x first and then y. We have

$$V = \int_1^2 \int_0^1 \left(x^2 + xy^3\right) dx dy,$$

The inner integral is

$$V = \int_0^1 \left(x^2 + xy^3\right) dx = \left(\frac{x^3}{3} + y^3 \frac{x^3}{2}\right) \Big|_{x=0}^{x=1} = \frac{1}{3} + \frac{1}{2} y^3$$

Note that we treat y as a constant as we integrate with respect to x. The outer integral is

$$\int_1^2 \left(\frac{1}{3} + \frac{1}{2} y^3\right) dy = \left(\frac{y}{3} + \frac{y^4}{8}\right) \Big|_1^2 = \frac{1}{3} + \frac{15}{8}.$$

The two approaches yield the same answer.

In applications, integrals like the one above arise in computations of

1. Area: if f(x,y)=1, then the double integrals gives the area of region R.
2. Volume: the integral is equal to volume under the surface $z = f(x,y)$ above the region R.
3. *Mass:* if R is a plate and $f(x, y)$ is density per unit area of the plate, then the integral is equal to the mass of the plate.
4. *Force:* if $f(x, y)$ is the force per unit area on the plate in the downward direction, then integral is the total force on the plate.
5. *Average:* the integral divided by the area of the region R is the the average of the function $f(x, y)$ on R.

DOUBLE INTEGRALS OVER GENERAL REGIONS

In this section we consider double integrals over more general regions. Suppose that the region R is defined by $G_1(x) \leq y \leq G_2(x)$ with $a \leq x \leq b$.

This is called a vertically simple region. The double integral is given by

$$\iint_R f(x,y)\, dA = \int_a^b \int_{G_1(x)}^{G_2(x)} f(x,y)\, dy dx$$

To derive this formula we slice the three-dimensional region into slices parallel to the y-axis. The figure below shows a top view of slice between x and $x + dx$.

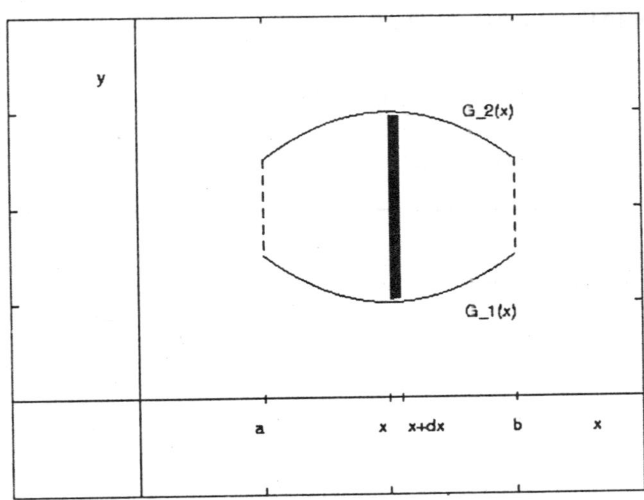

The inner integral

$$C(x) = \int_{G_1(x)}^{G_2(x)} f(x,y)\,dy$$

is the cross-sectional area of the slice between x and x+dx. The volume of the slice between x and x+dx is C(x)dx. The total volume is the sum of the volumes of all the slices:

$$\int_a^b C(x)\,dx.$$

If R is the horizontally simple region - that is, if the region is defined by $c \le y \le d$ and $H_1(y) \le x \le H_2(y)$, then

$$\iint_R f(x,y)\,dA = \int_c^b \int_{H_1(y)}^{H_2(y)} f(x,y)\,dx\,dy\cdot$$

In this case the slices are parallel to the x-axis. The inner integral

$$C(y) = \int_{H_1(y)}^{H_2(y)} f(x,y)\,dx$$

is the cross-sectional area of the slice between y and y+dy. The volume of the slice between y and y+dy is C(y)dy. The total volume is

$$\int_c^d C(y)\,dy\cdot$$

In some cases a region may be neither vertically nor horizontally simple. However, in general, a region can be split up into vertically and horizontally simple regions.

Example: Compute the double integral

$$\iint_R 30xy\,dA,$$

where R is bounded by y=x and y=x^2.

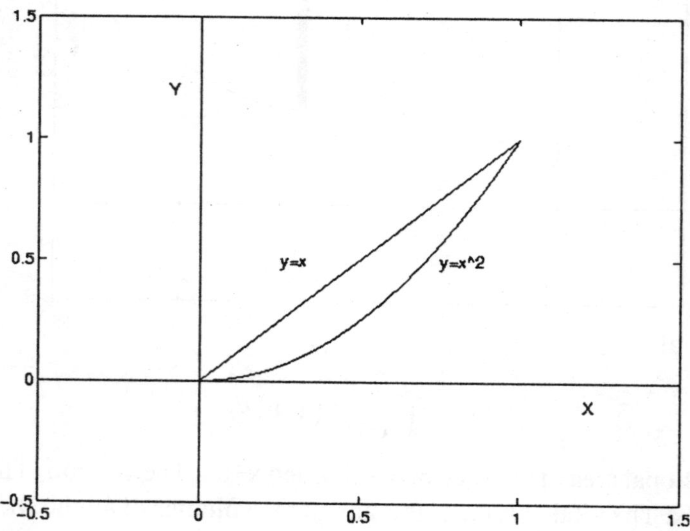

We can treat the region R as a vertically simple region. In this case the integral is given by

$$\iint_R 30xy\,dA = \int_0^1 \int_{x^2}^{x} 30xy\,dy\,dx$$

The inner integral is (remember x is a constant)

$$\int_{x^2}^{x} 30xy\,dy = 15xy^2 \Big|_{y=x^2}^{y=x} = 15x\left(x^2 - x^4\right)$$

The outer integral is

$$\int_0^1 15x\left(x^2 - x^4\right)dx = \frac{5}{4}.$$

One can also treat the region as a horizontally simple region. The left hand function $y = x$ can be written $x = H\text{-}1\ (y) = y$. The right hand function $y = x^2$ can be written $x =$ sqrt(y). The iterated integral is

$$\iint_R 30xy\,dA = \int_0^1 \int_y^{\sqrt{y}} 30xy\,dx\,dy$$

You should check that the answer is 5/4.

DOUBLE INTEGRALS IN POLAR COORDINATES

For some regions R, it convenient to convert to polar coordinates in order to evaluate the double integral

$$\iint_R f(x, y)\, dA$$

Consider the sector $a \le r \le b$, $c \le \theta \le d$ shown in the figure below.

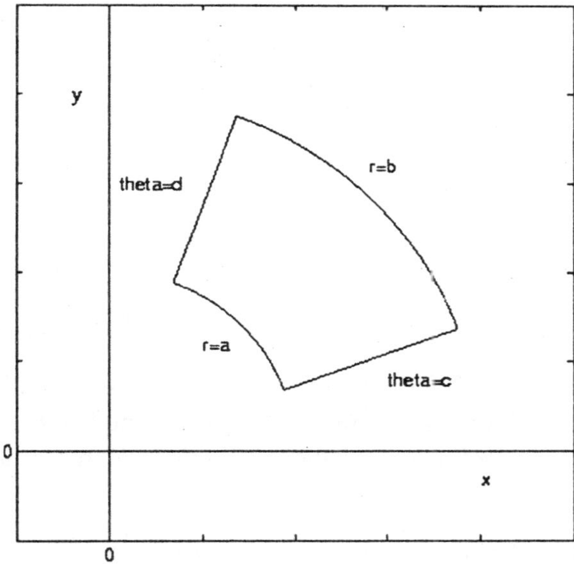

Recall that $x = r\cos\theta$ and $y = r\sin\theta$. The double integral is given by:

$$\iint_R f(x, y)\, dA = \int_a^b \int_c^d f(\tau\cos\theta, \tau\sin\theta)\, \tau\, d\theta\, d\tau$$

In the above formula one integrates with respect to θ first, then r. Alternatively, one could integrate with respect to r first, then θ.

DISCUSSION OF THE ITERATED INTEGRAL IN POLAR COORDINATES

In the case of double integral in polar coordinates we made the connection $dA = dxdy$. dxdy is the area of an infinitesimal rectangle between x and $x + dx$ and y and $y + dy$. In polar coordinates, $dA = rd\theta\, dr$ is the area of an infinitesimal sector between r and r+dr and θ and θ+d θ. See the figure below.

The area of the region is the product of the length of the region in θ direction and the width in the r direction.

The width is dr. The length is d θ r, the arclength of a part of a circle of angle d θ. (The radius is essentially constant in the region since dr is infinitesimal.)

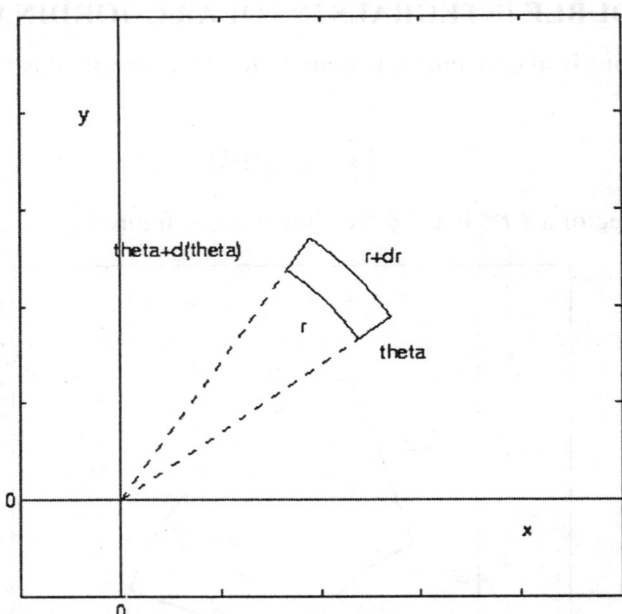

Example: Consider the integral with f(x,y) = $2x + 3y^2$ where *R* is the region between the circles $x^2 + y^2 = 1$ and $x^2 + y^2 = 4$. In this case $1 \le r \le 2$ and $0 \le \theta \le 2\pi$. We can convert the function $f(x, y)$ into polar coordinates with the substitutions $x = r\cos\theta$ and $y = r\sin\theta$. The iterated integral is

$$\iint_R f(x,y)\,dA = \int_1^2 \int_o^{2\pi} \left(2\tau\cos\theta + 3\tau^2\sin^2\theta\right)\tau\, d\theta\, d\tau$$

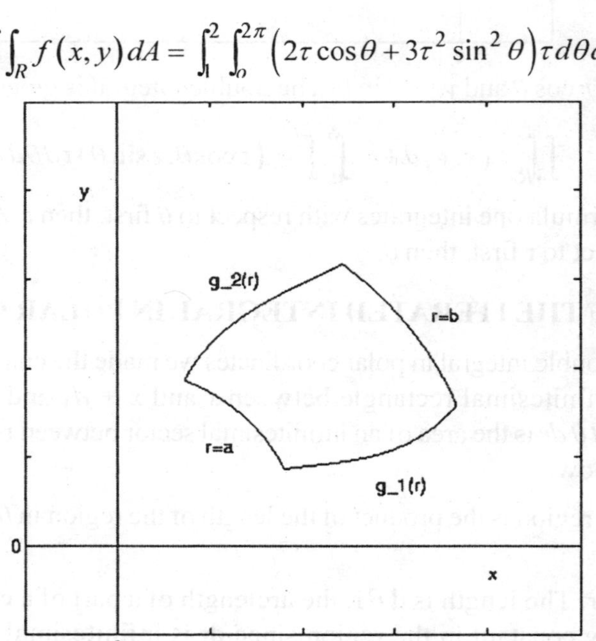

We integrate with respect to θ first, then r. Alternatively, we have

$$\iint_R f(x,y)\,dA = \int_0^{2\pi}\int_1^2 \left(2\tau\cos\theta + 3\tau^2\sin^2\theta\right)\tau\,d\tau\,d\theta$$

Examples of how to evaluate iterated integrals are given in the double integrals page.

GENERAL REGIONS

If the region R is of the form $g_1(r) \le \theta \le g_2(r)$ with $a \le r \le b$, as shown in the figure below, then the double integral is given by the iterated integral

$$\iint_R f(x,y)\,dA = \int_a^b \int_{g_1(\tau)}^{g_2(\tau)} f(\tau\cos\theta, \tau\sin\theta)\,\tau\,d\theta\,d\tau$$

If the region R is of the form $h_1\theta \le r \le h_2\,\theta$ with $c\ \theta \le d$, as shown in the figure below,

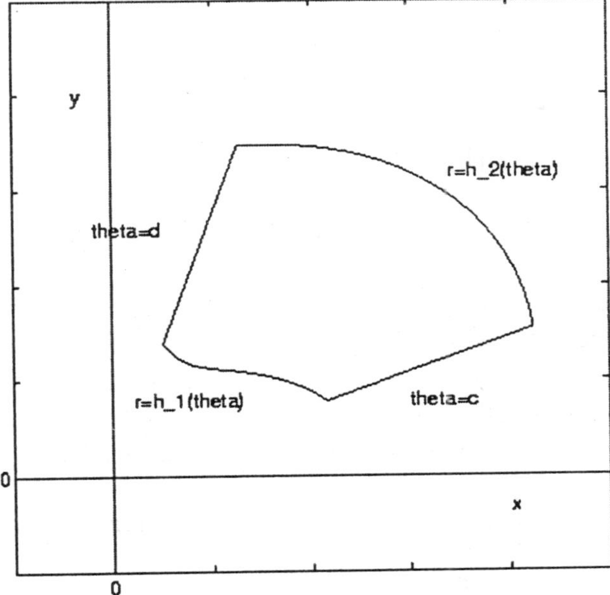

then the double integral is given by the iterated integral

$$\iint_R f(x,y)\,dA = \int_c^d \int_{\eta_1(\theta)}^{\eta_1(\tau)} f(\tau\cos\theta, \tau\sin\theta)\,\tau\,d\tau\,d\theta$$

TRIPLE INTEGRALS

Integration of a function of three variables, w=f(x,y,z), over a three-dimensional region R in xyz-space is called a triple integral and is denoted

$$\iiint_R f(x,y,z)\,dV$$

TRIPLE INTEGRALS IN BOX-LIKE REGIONS

Suppose that R is the box with $a \le x \le b$, $c \le y \le d$, and $r \le z \le s$.

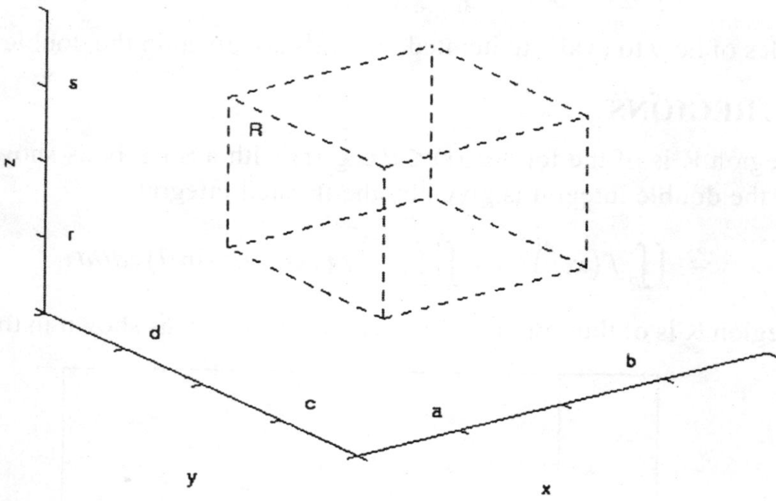

The triple integral is given by

$$\iiint_R f(x,y,z)\,dV = \int_a^b \left[\int_c^d \left[\int_r^s f(x,y,z)\,dz \right] \right] dx = \int_a^b \int_c^d \int_r^s f(x,y,z)\,dzdydx$$

To compute the iterated integral on the left, one integrates with respect to z first, then y, then x. When one integrates with respect to one variable, *all other variables are assumed to be constant*. For a box-like region, the integral is independent of the order of integration, assuming f(x,y,z) is continuous. Hence, there are total of 6 ways to order the integrations. For example we can integrate with respect to x, then z, then y. In this case we have

$$\iiint_R f(x,y,z)\,dV = \int_c^d \int_r^s \int_a^b f(x,y,z)\,dxdzdy$$

Consider the following example:

$$\int_0^1 \int_1^2 \int_1^2 (x^2 + yz)\,dzdydx$$

The inner integral is

$$\int_1^2 \left(x^2 + yz\right) dz.$$

Integrating with respect to z, treating x and y as constants, we obtain

$$\int_1^2 \left(x^2 + yz\right) dz = x^2 z + yz^2/2 \Big|_{z=1}^{z=2} = x^2 + 3y/2$$

Note that z has completely disappeared from the expression on the right. The middle integral is with respect to y and x is treated as a constant. We have

$$\int_1^2 \left(x^2 + 3y/2\right)dy = x^2 y + 3y^2/4\Big|_{z=1}^{z=2} = x^2 + 9/2$$

Note that y has disappeared from the expression on the right. The outer integral is with respect to x. We have

$$\int_0^1 \left(x^2 + 9/4\right)dx = 31/4$$

You should verify that the same answer is obtained if the order of integration is changed.

DISCUSSION AND APPLICATIONS

To get a better understanding of triple integrals let us consider the following example where the triple integral arises in the computation of mass. Suppose that that the region R in xyz-space corresponds to an object and f(x,y,z) is the density per unit volume at the point (x,y,z). If the density is constant, then the mass of the object is the product of the density and the volume of R. If the density varies with position, then we cannot apply this general formula.

We can compute the mass by slicing the R into a bunch of infinitesimal boxes. Consider the box between x and x+dx, y and y+dy, and z and z+dz. Here dx, dy, and dz are infinitesimals. In this small box the density is essentially constant and is equal to f(x,y,z). The mass of the small box is the product of density and volume. The volume of the box is dxdydz. Hence the mass of the the small box is f(x,y,z)dxdydz. The triple integral gives the total mass of the object and is equal to the sum of the masses of all the infinitesimal boxes in R.

Triple integrals also arise in computation of

1. Volume (if $f(x,y,z) = 1$, then the triple integral equals the volume of R)
2. Force on a 3D object
3. Average of a Function over a 3D region
4. Center of Mass and Moment of Inertia

TRIPLE INTEGRALS IN GENERAL REGIONS

We would like to be able to integrate triple integrals for more general regions. General regions are classifed into three types. Suppose that the region R is such that $g_1(x, y) \le z \le g_2(x, y)$, where (x, y) lies in region D in the xy plane, as shown in the figure:

The region D is the projection of R onto the xy plane. The triple integral is given by

$$\iiint_R f(x,y,z)\,dV = \iint_D \left[\int_{g_1(x,y)}^{g_2(x,y)} f(x,y,z)\,dz \right] dA$$

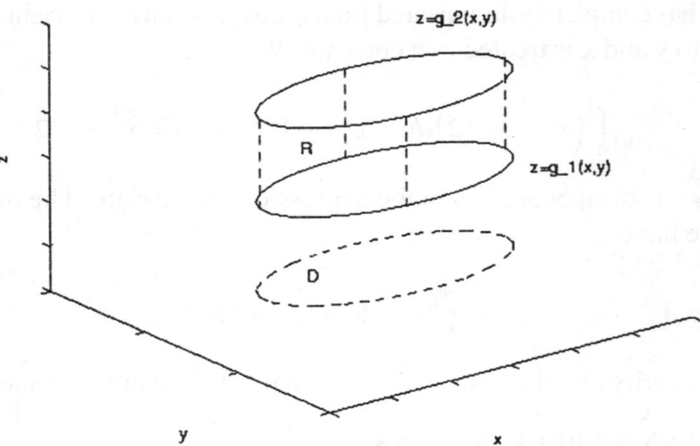

The inner integral

$$\int_{g_1(x,y)}^{g_2(x,y)} f(x,y,z)dz = \eta,(x,y)$$

is with respect to z. The result is a function of x and y. The remainder of the calculation

$$\iint_D \eta(x,y)dz]dA$$

is a double integral over the region D in the xy plane.

If the region R is defined such that $g_1(x,z) \leq y \leq g_2(x,z)$, (x,z) lies in region D in the xz plane, then

$$\iiint_R f(x,y,z)dV = \iint_D \left[\int_{g_1(x,z)}^{g_2(x,z)} f(x,y,z)dy \right] dA$$

The inner integral is with respect to y. The remainder of the calculation involves a double integral over the region D in the xz plane.

Finally suppose that R is such that $g_1(y,z) \leq x \leq g_2(y,z)$, where (y,z) lies in a region D in the yz plane, then

$$\iiint_R f(x,y,z)dV = \iint_D \left[\int_{g_1(y,z)}^{g_2(y,z)} f(x,y,z)dx \right] dA$$

The inner integral is with respect x. The remainder of the calculation involves a double integral over the region D in the yz plane.

Example: Consider the triple integral

$$\iiint_R (x+2y)dV$$

where R is the tetrahedral region bounded by the planes x=0, y=0, z=0 and x+y+z=2.

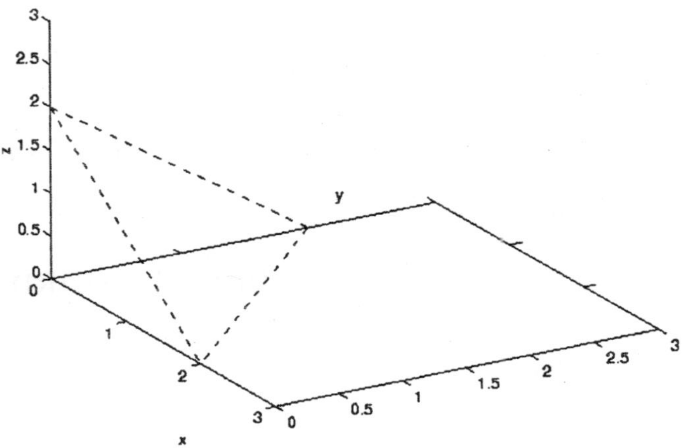

There are several ways to compute the integral. We can rewrite the equation of the plane x+y+z=2 as z=2-x-y. Note that 0<=z=<2-x-y. Hence, we have

$$\iiint_R (x+2y)\,dV = \iint_D \left[\int_0^{2-x-y} (x+2y)\,dz\right] dA$$

The inner integral is (remember x and y are constants in this integration)

$$\int_0^{2-x-y} (x+2y)\,dz = (x+2y)z\Big|_{z=1}^{z=2-z-y} = (x+2y)(2-x-y)$$

The projection of the region R onto the xy-plane is the triangle D shown in the figure below:

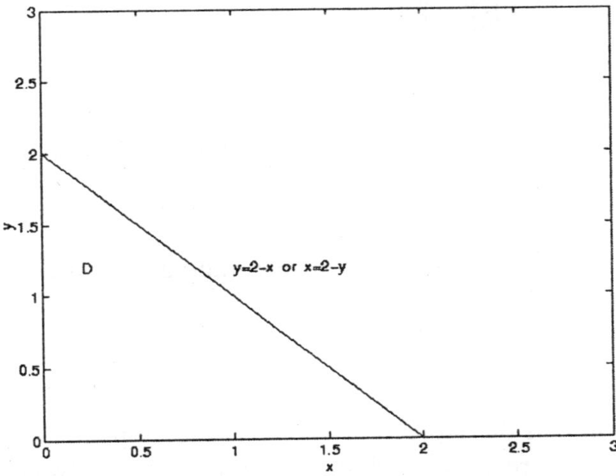

Hence, we are left with the double integral

$$\iint_D (x+2y)(2-x-y)\,dA = \int_0^2 \int_0^{2-x} (x+2y)(2-x-y)\,dydx$$

We can also evaluate the double integral by integrating with respect to x first, then y. In this case

$$\iint_D (x+2y)(2-x-y)\,dA = \int_0^2 \int_0^{2-x} (x+2y)(2-x-y)\,dxdy$$

It can be shown that the double integral equals 2/3.

TRIPLE INTEGRALS IN CYLINDRICAL COORDINATES

Cylindrical coordinates are obtained from Cartesian coordinates by replacing the x and y coordinates with polar coordinates r and θ and leaving the z coordinate unchanged.

It is simplest to get the ideas across with an example. Consider an object which is bounded above by the inverted paraboloid $z=16-x^2-y^2$ and below by the xy-plane. Suppose that the density of the object is given by f(x,y,z)=8+x+y. What is the mass of the object?

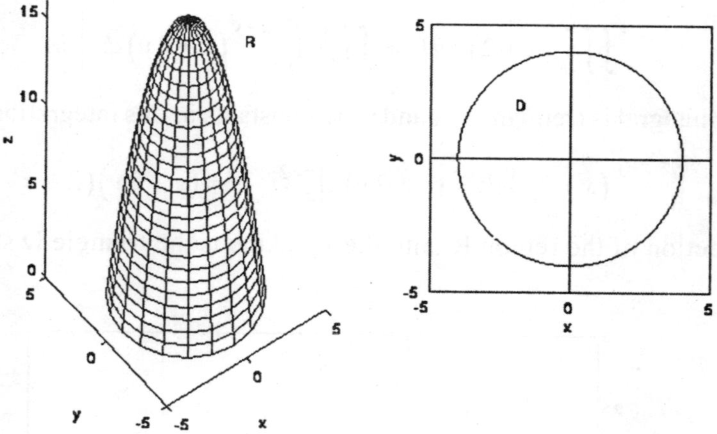

The object is shown above. The mass is given by the triple integral:

$$\iiint_R f(x,y,z)\,dV$$

Since z satisfies 0 £ z £ $16-x^2-y^2$, the triple integral becomes

$$\iint_D \left[\int_0^{18-x^{2-y^2}} (8+x+y)\,dz \right] dA$$

where the region D is the projection of R onto the xy-plane. It can be shown that D is the disk of radius 4 centered at the origin. (The circle $x^2+y^2=16$ is the intersection of the paraboloid and the plane z=0.)

Because of the circular symmetry of the object in the xy-plane it is convenient to convert to polar coordinates. We make the substitutions

$$x = \tau\cos(\theta) \quad y = \tau\sin(\theta)$$

With these substitutions, the paraboloid becomes $z=16-r^2$ and the region D is given by $0 \leq r \leq 4$ and $0 \leq \theta \leq 2\pi$. Hence, the triple integral is given by

$$\iint_D \left[\int_0^{16-x^2-y^2} (8+x+y)\,dz \right] dA = \int_0^{2\pi} \int_0^4 \int_0^{18-\tau^2} (8+\tau\cos\theta+\tau\sin\theta)\,\tau\, dz\, d\tau\, d\theta$$

Note that we can change the order of integration of r and θ so the integral can also be expressed

$$\int_0^4 \int_0^{2\pi} \int_0^{18-\tau^2} (8+\tau\cos\theta+\tau\sin\theta)\,\tau\, dz\, d\theta\, d\tau$$

Evaluating the iterated integral, we have find that the mass of the object is 1024p.

Discussion

In rectangular coordinates the volume element dV is given by dV=dxdydz, and corresponds to the volume of an infinitesimal region between x and x+dx, y and y+dy, and z and z+dz. In cylindrical coordinates, we have dV=rdzdrd θ, which is the volume of an infinitesimal sector between z and z+dz, r and r+dr, and θ and θ +d θ. As shown in the picture, the sector is nearly cube-like in shape. The length in the r and z directions is dr and dz, respectively. The length in the θ direction is r*d θ, and this yields the result for the volume. This result can also derived via the Jacobian.

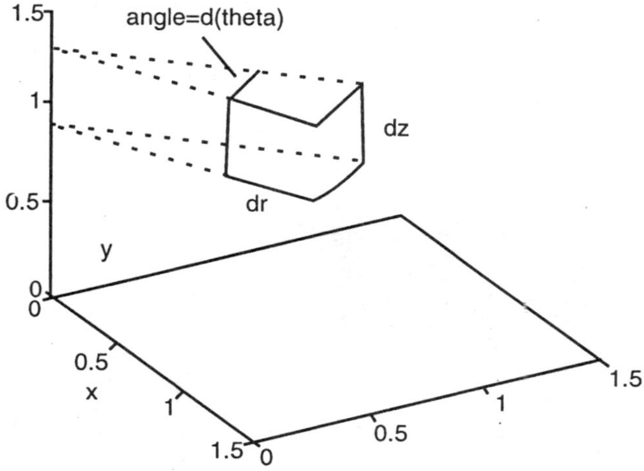

For some problems one must integrate with respect to r or θ first. For example, if $g_1(\theta,z)\ \pi r \leq g_2(\theta,z)$, then

$$\iiint_R f(x,y,z)\,dV = \iint_D \left[\int_{g_1(\theta,z)}^{g_2(\theta,z)} f(\tau\cos\theta, \tau\sin\theta, z)\,\tau\, d\tau \right] dA,$$

where D is the projection of R onto the θ-z plane. If $g_1(r,z) \le \theta \le g_2(r,z)$,

$$\iiint_R f(x,y,z)\,dV = \iint_D \left[\int_{g_1(\tau,z)}^{g_2(\tau,z)} f(\tau\cos\theta, \tau\sin\theta, z)\,\tau\, d\theta \right] dA$$

where D is the projection of R onto the rz plane.

TRIPLE INTEGRALS IN SPHERICAL COORDINATES

Recall that in spherical coordinates a point in xyz space characterized by the three coordinates ρ, θ, and ϕ. These are related to x,y, and z by the equations

$$x = \rho\sin\phi\cos\theta \quad y = \rho\sin\phi\cos\theta \quad z = \rho\cos\theta$$

where

$$\rho \ge 0 \quad 0 \le \theta \le 2\pi \quad 0 \le \phi \le \pi$$

CORRECTION The above formulas for y and z are incorrect:

$y = \rho \sin\phi \sin\theta$

$z = \rho \cos \phi$.

Recall that

$$\rho = \sqrt{x^2 + y^2 + z^2}$$

Consider the following example: a solid lies between a sphere or radius 2 and a sphere or radius 3 in the region y ³ 0 and z ³ 0. Find its mass if the density f(x,y,z) is equal to the distance to the origin.

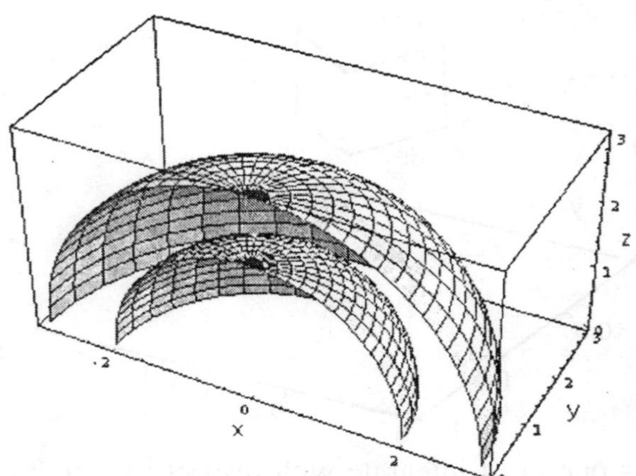

The mass is given by

$$\iiint_R \sqrt{x^2 + y^2 + x^2}\, dV$$

where R is the region in the xyz space occupied by the solid.

In spherical coordinates the solid occupies the region with

$$2 \le \rho \le 3 \quad 0 \le \theta \le \pi \quad 0 \le \phi \le \pi/2$$

The integrand in spherical coordinates becomes r. Finally, the volume element is given by

$$dV = \rho^2 \sin\phi\, d\rho\, d\phi\, d\theta$$

We will not derive this result here. It can be derived via the Jacobian. See a textbook for a geometric derivation. Putting everything together, we get the iterated integral

$$\text{Mass} = \int_0^{\pi} \int_0^{\pi/2} \int_2^3 \rho\rho^2 \sin\phi\, d\rho\, d\phi\, d\theta$$

In this example, since the limits of integration are constants, the order of integration can be changed. Integrating with respect to ρ, ϕ, and θ, we find that the integral equals $65\pi/4$.

In general integrals in spherical coordinates will have limits that depend on 1 or 2 of the variables. In these cases the order of integration does matter. We will not go over the details here.

Summary

To convert an integral from Cartesian coordinates to cylindrical or spherical coordinates:

- Express the limits in the appropriate form
- Express the integrand in terms of the appropriate variables
- Multiply by the correct volume element

  Cylindrical Coordinates: $dV = \tau d\tau dz d\theta$

  Spherical Coordinates: $dV = \rho^2 \sin\phi\, d\rho\, d\phi\, d\theta$
- Evaluate the integral

SURFACE AREA

Let $f(x,y)$ be a differentiable function. As we have seen, $z = f(x, y)$ defines a surface in xyz-space. In some applications, it necessary to know the surface area of the surface above some region R in the xy-plane. See the figure.

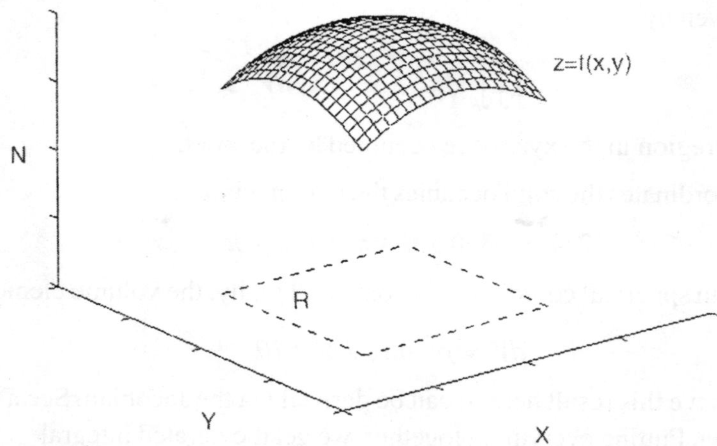

The formula for the surface area is

$$\iint_R \sqrt{1+\left[f_x(x,y)\right]^2+\left[f_y(x,y)\right]^2}\,dA$$

This is a double integral.

Example: What is the surface area of the plane $z = 2x + 3y$ above the rectangle with $-1 \le x \le 2$ and $0 \le y \le 2$? In this case $f_x = 2$ and $f_y = 3$. Applying the above formula, the surface area S is given by

$$S = \int_{-1}^{2}\int_{o}^{2}\sqrt{14}dydx = \int_{0}^{2}\int_{-1}^{2}\sqrt{14}dxdy.$$

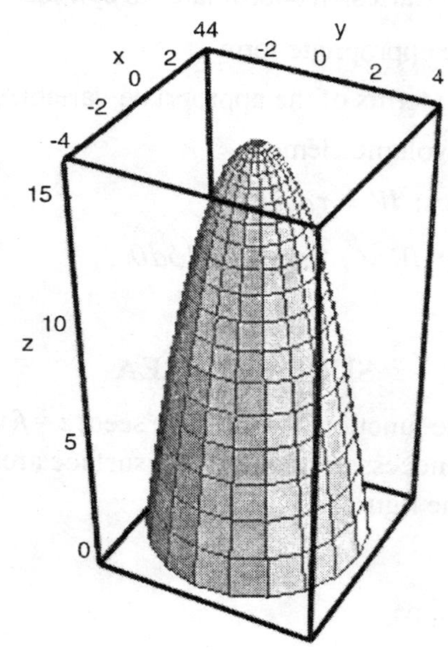

Since, the region of integration R is a rectangle and the integrand is continuous, the value of the integral is independent of the order of integration. It can be shown that $S = 6*$ sqrt(14).

Example: Find the surface area of the part of the paraboloid $z=16-x^2-y^2$ that lies above the *xy* plane.

The region R in the xy-plane is the disk $0 \le x^2+y^2 \le 16$ (disk or radius 4 centered at the origin).

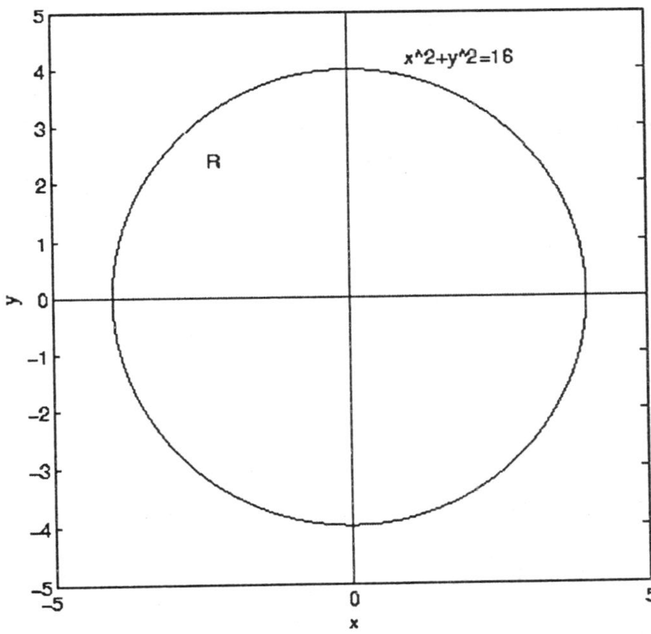

For this problem, f_x=-2x and f_y=-2y. Hence, the surface area S is given by

$$S = \iint_R \left(\sqrt{1+4x^2+4y^2}\right) dA$$

Since R is a disk, it convenient to convert the above integral into polar coordinates. The disk R satisfies $0 \le r \le 4$ and $0 \le \theta \le 2\pi$. In addition,

$$4x^2+4y^2 = 4\left(x^2+y^2\right) = 4\left(\tau^2\cos^2\theta + \tau^2\sin^2\theta\right) = 4\tau^2$$

The surface area is given by the integral

$$S = \int_0^{2\pi}\int_0^{4}\left(\sqrt{1+4\tau^2}\right)\tau d\tau d\theta S = \int_0^{4}\int_0^{2\pi}\left(\sqrt{1+4\tau^2}\right)\tau d\theta d\tau$$

Both iterated integrals above can be computed in a straightforward manner. The final answer is

$$S = \frac{\pi}{6}\left(65^{3/2} - 1\right)$$

DERIVATION OF THE SURFACE AREA FORMULA

It is instructive to derive the surface area formula. We start by assuming that the surface is the plane:

$$z - z_0 = a\left(x - x_0\right) + b\left(y - y_0\right)$$

Consider a part of the plane above a rectangle in the xy-plane with x_0 £ x £ x_0+dx and y_0 £ y £ y_0+dy, as shown in the figure below.

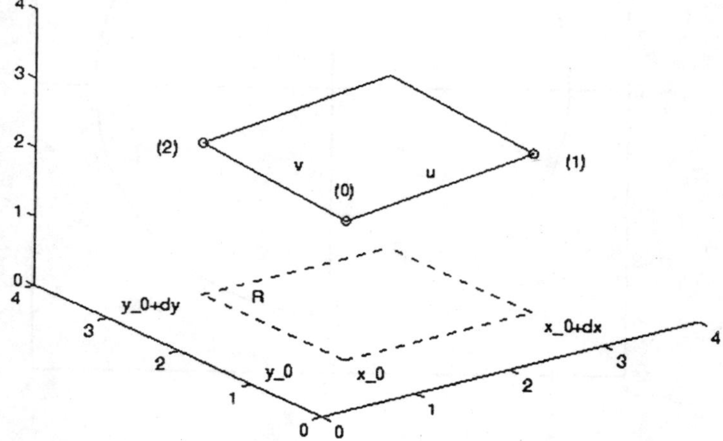

Let u be the vector from point 0 to point 1 and v be the vector from point 0 to point 2. The area of the plane above the rectangle R is

$$\text{Area} = \left|u \times v\right|$$

Given the formula for the plane, point 0 is (x_0,y_0,z_0), point 1 is (x_0+dx,y_0,z_0+adx), point 2 is (x_0,y_0+dy,z_0+bdy). Hence,

$$u =< dx, 0, adx > \quad v =< 0, dy, bdy >$$

Taking the cross product, we have

$$u \times v =< -adxdy, -bdxdy, dxdy >$$

The area is

$$\sqrt{(dxdy)^2 + (-adxdy)^2 + (-bdxdy)^2} = \sqrt{1 + a^2 + b^2}\,dxdy$$

Note dxdy is the area of the rectangle in the xy plane.

For a general surface z=f(x,y), we can approximate the area of the surface over the small rectangle in the figure above by the tangentplane through (x_0,y_0,z_0). The equation of the tangent plane is

$$z - z_0 = f_x(x_0, y_0)(x - x_0) + f_y(x_0, y_0)(y - y_0)$$

This last equation is the same as the equation for the plane with a replaced by the x derivative and b replaced by the y derivative. Hence, the area is

$$\sqrt{1} + \left[f_x(x_o, y_0)\right]^2 + \left[f_y(x_0, y_0)\right]^2 dxdy$$

In the case that the region R is not a rectangle we replace dxdy by dA, the area of a general infinitesimal region containing (x_0,y_0).

Example: To find the eigenvalues of the matrix

$$\begin{bmatrix} 2 & \frac{3}{2} \\ 2 & 0 \end{bmatrix}$$

we substitute A into the equation det(A-kI)=0 and solve for k. The matrix A-kI is given by

$$\begin{bmatrix} 2-k & \frac{3}{2} \\ 2 & 0-k \end{bmatrix}$$

which has determinant k^2-2k-3. Hence, we are looking for values k satisfying k^2-2k-3=0. So, of course, we have k=3 or k=-1. To find the eigenvectors of the eigenvalue k=3 we look for solutions v of the homogeneous system of equations (A-3I)v=0:

$$\left[\begin{array}{cc|c} -1 & \frac{3}{2} & 0 \\ 2 & -3 & 0 \end{array}\right]$$

Since the second equation is a constant multiple of the first, this system of equations reduces to the single equation -x+(3/2)y=0 or equivalently x=1.5y. This system has an infinite number of solutions. So for example, choosing y=2 yeilds the vector <3,2> which is thus an eigenvector that has eigenvalue k=3. In a general form, all eigenvectors with eigenvalue 3 have the form <2t,3t> where t is any real number. It can also be shown (by solving the system $(A + I)$ v = 0) that vectors of the form <t, – 2t> are eigenvectors with eigenvalue $k = -1$.

Example: Find the eigenvalues and corresponding eigenvalues for the matrix

$$\begin{bmatrix} 1 & -1 \\ 5 & -3 \end{bmatrix}$$

First, we must find det(A-kI):

$$|A-kI| = \begin{vmatrix} 1-k & -1 \\ 5 & -3-k \end{vmatrix} = (1-k)(-3-k)-(-1)5 = k^2+2k+2$$

This leads to the characteristic equation $k^2 + 2k + 2 = 0$ which has complex roots $k = -1+ i$ and $k = -1 - i$. To find the eigenvectors for k = $-1+ i$, we solve $(A-(-1 + i)I)$ $v = 0$ for v:

$$\left[\begin{array}{cc|c} 1-(-1+i) & -1 & 0 \\ 5 & -3-(-1+i) & 0 \end{array}\right] = \left[\begin{array}{cc|c} 2-i & -1 & 0 \\ 5 & -2-i & 0 \end{array}\right]$$

The second equation is a constant multiple of the first equation so the system reduces to the single equation $(2 - i) x - y = 0$ which implies $y = (2 - i)x$. There are once again an infinite number of eigenvectors of A of the form $< t,(2 - i)t >$ with eigenvalue $k = -1 + i$. By examining the system of equations $(A- (-1 -i)I)\ v = 0$ it can also be shown that vectors of the form $< t,(2 + i)t >$ are eigenvectors of A with eigenvalue $k = -1 -i$.

From the examples above we can infer a property of eigenvectors and eigenvalues: eigenvectors from distinct eigenvalues are linearly independent. The following examples illustrate that the situation is not so clear cut when the eigenvalues are not distinct.

Example: The matrix

$$\begin{vmatrix} 1 & 0 \\ 0 & 1 \end{vmatrix}$$

has two eigenvalues (1 and 1) but they are obviously not distinct. Since A is the identity matrix, $Av = v$ for any vector v, i.e. any vector is an eigenvector of A. We can thus find two linearly independent eigenvectors (say <-2,1> and <3,-2>) one for each eigenvalue.

Example: The matrix

$$D = \begin{vmatrix} 1 & 1 \\ 0 & 1 \end{vmatrix}$$

also has non-distinct eigenvalues of 1 and 1. All eigenvalues are solutions of (A-I)v=0 and are thus of the form <t,0>. Hence, in this case there do not exist two linearly independent eigenvectors for the two eigenvalues 1 and 1 since $<$t, 0$>$ and $<$s, 0$>$ are not linearly independent for any values of s and t.

SYMMETRIC MATRICES

There is a very important class of matrices called symmetric matrices that have quite nice properties concerning eigenvalues and eigenvectors. A symmetric matrix A is a square matrix with the property that $A_{ij}=A_{ji}$ for all i and j. The matrices

$$\begin{bmatrix} 2 & -3 \\ -3 & 5 \end{bmatrix} \text{ and } \begin{bmatrix} -1 & 3 & 0 \\ 3 & 5 & 2 \\ 0 & 2 & -4 \end{bmatrix}$$

are symmetric matrices. In symmetric matrices the upper right half and the lower left half of the matrix are mirror images of each other about the diagonal.

A nxn symmetric matrix A not only has a nice structure, but it also satisfies the following:

- A has exactly n (not necessarily distinct) eigenvalues
- There exists a set of n eigenvectors, one for each eigenvalue, that are mututally orthogonal.

Thus, the situation encountered with the matrix D in the example above cannot happen with a symmetric matrix: A symmetric matrix has n eigenvalues and there exist n linearly independent eigenvectors (because of orthogonality) *even if the eigenvalues are not distinct.*

Example: Find the eigenvalues and a set of mutually orthogonal eigenvectors of the symmetric matrix

$$\begin{bmatrix} 3 & 2 & 4 \\ 2 & 0 & 2 \\ 4 & 2 & 3 \end{bmatrix}$$

First we need det(A-kI):

$$|A-kI| = \begin{vmatrix} 3-k & 2 & 4 \\ 2 & 0-k & 2 \\ 4 & 2 & 3-k \end{vmatrix} = -k^3 + 6k^2 + 15k + 8$$

Thus, the characteristic equation is $(k-8)(k+1)^2 = 0$ which has roots $k=-1$, $k=-1$, and $k=8$. Note that we have listed $k=-1$ twice since it is a double root. We must find two eigenvectors for $k=-1$ and one for $k=8$. We now examine $(A+I)\,v=0$ to find the eigenvectors for the eigenvalue $k=-1$:

$$\left[\begin{array}{ccc|c} 4 & 2 & 4 & 0 \\ 2 & 1 & 2 & 0 \\ 4 & 2 & 4 & 0 \end{array}\right]$$

It is easily seen that this system reduces to the single equation $2x + y + 2z = 0$ since the other two equations are twice this one. There are two parameters here (x and z); thus, eigenvectors for k = –1 must have the form y = $-2x - 2z$ which corresponds to vectors of the form <s, –2s –2*t*,*t* >. We must choose values of s and t that yield two orthogonal vectors (the third comes from the eigenvalue k = 8). First, choose anything, say s =1 and t = 0: < 1, –2,0>. Now find a vector <*x*, –2*x* –2z,z> such that

$$0 =< 1,-2,0 > \cdot < x,-2x-2z,z >= x+4x+4x+0 = 5x+4z$$

An easy choice here is x = 4 and z = –5. So, we now have two orthogonal vectors <1, –2,0> and <4,2, –5> that correspond to the two instances of the eigenvalue $k = -1$. It can also be shown that the eigenvectors for $k = 8$ are of the form <2*r*,*r*,2*r*> for any value of r. It is easy to check that this vector is orthogonal to the other two we have for any choice of r. So, let's take $r = 1$. We now have the following: eigenvalues and orthogonal eigenvectors:

for $k = -1$: <1, –2,0> and <4,2, –5>

for $k = 8$: <2,1,2>

Note that since this matrix is symmetric we do indeed have 3 eigenvalues and a set of 3 orthogonal (and thus linearly independent) eigenvectors (one for each eigenvalue).

Chapter 6

Fourier Series

The Fourier series is a mathematical tool used for analyzing an arbitrary periodic function by decomposing it into a weighted sum of much simpler sinusoidal component functions sometimes referred to as normal Fourier modes, or simply modes for short. The weights, or coefficients, of the modes, are a one-to-one mapping of the original function. These modal coefficients are sometimes themselves confusingly referred to as "modes" for brevity, especially in physics literature.

Fourier series serve many useful purposes, as manipulation and conceptualization of the modal coefficients are often easier than with the original function.

Areas of application include electrical engineering, vibration analysis, acoustics, optics, signal and image processing, and data compression.

Using the tools and techniques of spectroscopy, for example, astronomers can deduce the chemical composition of a star by analyzing the frequency components, or spectrum, of the star's emitted light. Similarly, engineers can optimize the design of a telecommunications system using information about the spectral components of the data signal that the system will carry. The Fourier series is named after the French scientist and mathematician Joseph Fourier, who used them in his influential work on heat conduction.

The Fourier Series is a mathematical tool that allows the representation of any periodic signal as the sum of harmonically related sinusoids. This operation seems incredible. How, for instance, could a function that is not smooth be represented as the sum of sinusoids that are smooth? The answer lies in the number of sinusoids that are added to approximate the desired result. As more sinusoids are added, the output more closely resembles the desired output:

The first graphic shows a single wave. The y-axis represents magnitude and the x-axis represents 'n'. 'n' in discrete systems represents the sample number of the waveform. The above three graphics are actually constructed by taking 500 samples of the waveform and plotting them by a "connect the dot" method. If the sample rate is high enough, the resulting waveform will very closely resemble the actual waveform. The waveform in the middle graph is the sum

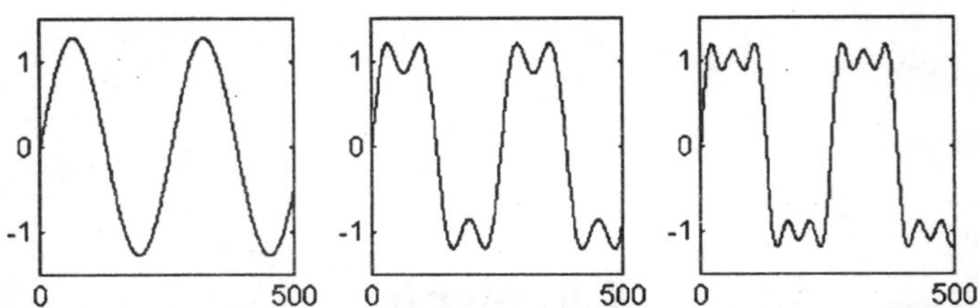

of the waveform in the first graph plus a sine wave that has a frequency that is three times that of the the original sine wave. The waveform in the last graphic contains is the sum of the middle waveform plus a sine wave that has a frequency that is six times that of the original sine wave. This waveform faintly resembles a square wave. The addition of the first ten waveforms is shown in the following animation:

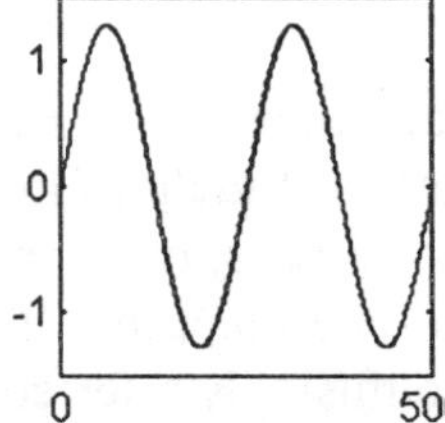

After the addition of ten and twenty sinusoids, the resultant waveform is clearly square in shape:

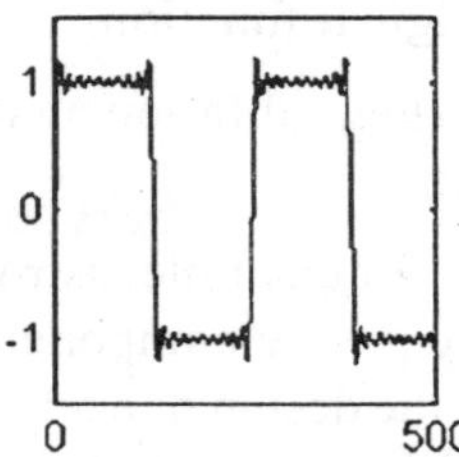

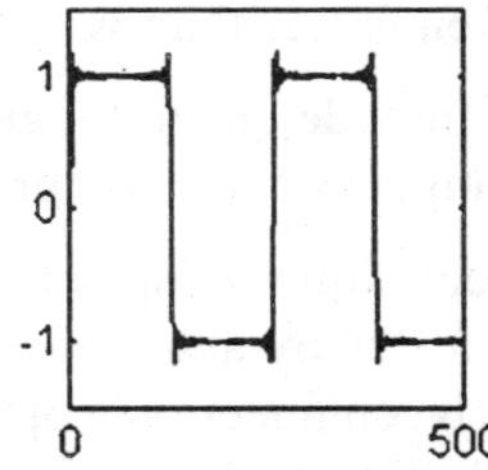

If all the intermediate graphs between the basic sine wave and the 20th iteration were placed evenly spaced on an axis, the result would be a 3-D shape:

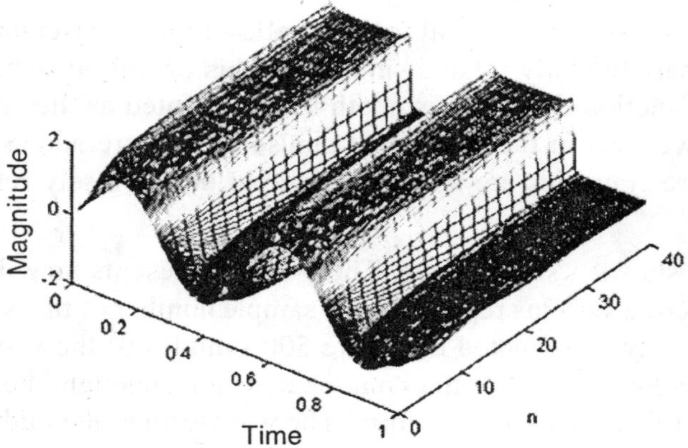

From this graph, the Fourier Series operation can be clearly seen. On the bottom left, the first waveform is the basic sine. As the graph progresses backwards and to the right into the monitor, the waveform becomes more and more square. Each line parallel to the x axis is at distinct intervals.

Definition: Given a complex-valued function f of real argument t, $f: R \rightarrow C$, where $f(t)$ is piecewise smooth and continuous, periodic with period T, and square-integrable over the interval from t_1 to t_2 of length T, that is,

$$\int_{t_1}^{t_2} |f(t)|^2 \, dt > +\infty$$

Where; $T = t_1 - t_2$, is the period, t_1 and t_2 are times to integrate between.

The Fourier series expansion of f is:

$$f(t) = \frac{a_0}{2} + \sum_{n=1}^{\infty} \left[a_n \cos(\omega_n t) + b_n \sin(\omega_n t) \right]$$

where, for any non-negative integer n:

- $\omega_n = n\dfrac{2\pi}{T}$ is the nth harmonic (in radians) of the function f,
- $a_n = \dfrac{2}{T} \int_{t_1}^{t_2} f(t)\cos(\omega_{nt})\,dt$ are the even Fourier coefficients of f, and
- $b_n = \dfrac{2}{T} \int_{t_1}^{t_2} f(t)\sin(\omega_{nt})\,dt$ are the odd Fourier coefficients of f.

Equivalently, in exponential form,

$$f(t) = \sum_{n=-\infty}^{+\infty} c_n e^{i\omega_n t}$$

$$c_n = \frac{1}{T} \int_{t_1}^{t_2} f(t) e^{-i\omega_n t} \, dt$$ i is the imaginary unit, and

where: $e^{-i\omega_n t} = \cos(\omega_n t) + i\sin(\omega_n t)$ in accordance with Euler's formula.

Canonical Form

In the special case where the period $T = 2$p, we have

$\omega_n = \text{n}$

In this case, the Fourier series expansion reduces to a particularly simple form:

$$f(t) = \frac{a_0}{2} + \sum_{n=1}^{\infty}\left[a_n \cos(nt) + b_n \sin(nt)\right]$$

where

- $a_n = \frac{1}{\pi}\int_{-\pi}^{\pi} f(t)\cos(nt)\,dt$

- $b_n = \frac{1}{\pi}\int_{-\pi}^{\pi} f(t)\sin(nt)\,dt$

for any integer n.

or, equivalently:

$$f(t) = \sum_{n=-\infty}^{+\infty} c_n e^{int}$$

where

- $c_n = \frac{1}{2\pi}\int_{-\pi}^{\pi} f(t)e^{-int}\,dt = \frac{1}{2}(a_{n-}ib_n)$

Choice of the Form

The form for period T can be easily derived from the canonical one with the change of variable defined by $x = \frac{2\pi}{T}t$. Therefore, both formulations are equivalent. However, the form for period T is used in most practical cases because it is directly applicable. For the theory, the canonical form is preferred because it is more elegant and easier to interpret mathematically, as will later be seen.

PERIODICITY OF FUNCTIONS

Let X designates the domain of a function y = $f(x)$. A function $f(x)$ is said to be *periodic* if its image values are repeated in its domain i.e. if there exists a nonzero real T such that $f(x + \text{T}) = f(x)$ for each $x \in X$.

The least positive value of those T is said to be the *period* of this function. For example for the sine function we can write sin(x + 4p) = sin x, sin(x + 24 p) = sin x, sin(x - 8p) = sin x etc., so the sine function is a periodic function.

But the least positive value for which such an equation holds is 2π and this is the period of the sine function. It is also well known that the cosine function has the period 2π and the tangent function has the period π. An example of the graph of a periodic function is illustrated in Figure

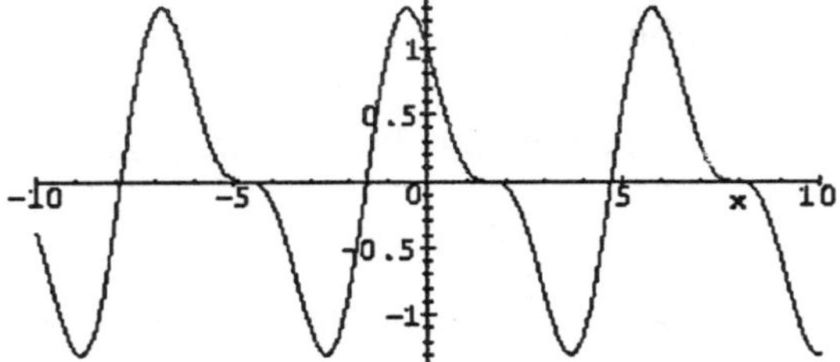

Further we need two following lemmas

Lemma: If the period of the function f(x) is T, then the period of the function f(ax) is $\frac{T}{a}$.

***Proof*:** $f\left[a\left(x+\frac{T}{a}\right)\right] = f(ax+T) = f(ax)$.

According to this lemma the period of the function sin3x is $\frac{2\pi}{3}$ and the period of the function $\cos\frac{x}{2}$ is $\frac{2\pi}{\frac{1}{2}} = 4\pi$.

Lemma: *If the* period of the function *f*(*x*) is *T*, then $\int_c^{c+T} f(x)dx = \int_0^T f(x)dx$

This result says that if we integrate the period of the function over the interval whose length equals to the period of the function then the value does not depend on the choice of the boundaries.

Proof. First we use the property of a definite integral and write $\int_c^{c+T} f(x)dx = \int_c^O f(x)dx + \int_0^T f(x)dx + \int_T^{c+T} f(x)dx$.

In the last integral we take $x = z + T$. Then $dx = dz$, $f(x) = f(z + T) = f(z)$, if $x = T$ then $z = 0$ and if $x = c + T$ then $z = c$. So we have

$$\int_c^{c+T} f(x)dx = \int_c^O f(x)dx + \int_0^T f(x)dx + \int_0^c f(z)dz$$

$$= \int_0^c f(x)dx + \int_0^T f(x)dx + \int_0^c f(z)dz$$

$$= \int_0^T f(x)dx$$

as the first and third integrals are equal.

FOURIER SERIES OF 2π-PERIODIC FUNCTIONS

In many problems in mechanics and electronics there arises a question, how to represent the behavior of a system by a combination of some simple behaviors. Mathematically – how to represent a function in the form of the functional series $f(x) = \sum_{k=1}^{\infty} c_k \varphi_k(x)$. Here $\varphi_0, \varphi_1, \varphi_2, ..., \varphi_k, ...$ form a base set of functions and $c_0, c_1, c_2, ..., ...c_k, ...,$ are said to be the *coefficients* of the expansion. The most familiar expansions are power series of the form $f(x) = \sum_{k=0}^{\infty} c_k x^k$, in which the base set comprises the power functions 1, x, x^2, ... x^k, ... Another widely used expansion is the expansion of a periodic function $f(x)$ of a period T in which the base set comprises the cosine functions, giving an expanded representation of the form

$$f(x) = A_0 + \sum_{k=1}^{\infty} A_k \cos\left(\frac{2k\pi x}{T} + \phi_k\right)$$

which is called the Fourier series of the function $f(x)$. In engineering the term *circular frequency* is used and it is defined by $\omega = \frac{2\pi}{T}$. The circular frequency is measured in radians per second. It is common to drop the term 'circular' and to refer to this simply as the frequency. Using frequency, we can rewrite in the form

$$f(x) = A_0 + \sum_{k=1}^{\infty} A_k \cos(k\omega x + \phi_k)$$

Every term $A_k \cos(k\omega x + \phi_k)$ represents a simple harmonic vibration that is called kth *harmonic*. The first harmonic $A_1 \cos(\omega x + \phi_1)$ has the same frequency ω as the parent function $f(x)$. The kth harmonic has the frequency $k\omega$, which is k times frequency of the first harmonic. A_k denotes the *amplitude* of the kth harmonic and ϕ_k is its *phase angle*. If $f(x)$ is a 2π-periodic function then its frequency $\omega = 1$, and the Fourier series has a more simple form

$$f(x) = A_0 + \sum_{k=1}^{\infty} A_k \cos(kx + \phi_k)$$

Here we shall consider the Fourier series of 2π-periodic functions. The kth harmonic can be represented in the form $A_k \cos(kx + \phi_k) = a_k \cos k_x + b_k \sin k_x$, where $a_k = A_k \cos\phi_k$ and $b_k = -A_k \sin\phi_k$. Now the expansion (ay be written as

$$f(x) = \frac{a_0}{2} + \sum_{k=1}^{\infty} a_k \cos kx + b_k \sin kx$$

Later we shall see that taking the constant term as $\frac{a_0}{2}$ enables us to make a_0 fit a general result. The expression is called *Fourier series expansion of a function f(x)* , and a_k and b_k are called the *Fourier coefficients.*

The set of functions $\{1, \cos x, \sin x, \cos 2x, ..., \cos kx, \sin kx, ...\}$ is said to be the trigonometric system of functions. It is possible to show that the functions of the trigonometric system satisfy the orthogonsality relations on an interval $[-\pi; \pi]$:

$$\int_{-\pi}^{\pi} \cos nx \cos kx dx = \begin{cases} 0 \textit{ for } n \neq k, \\ \pi \textit{ for } n = k: \end{cases}$$

$$\int_{-\pi}^{\pi} \sin nx \sin kx dx = \begin{cases} 0 \textit{ for } n \neq k, \\ \pi \textit{ for } n = k: \end{cases}$$

$$\int_{-\pi}^{\pi} \sin nx \cos kx dx = 0 \textit{ for all } \mathrm{n} \textit{ and } k;$$

$$\int_{-\pi}^{\pi} \cos kx dx = \begin{cases} 2\pi \textit{ for } k = 0, \\ 0 \textit{ for } k = 0; \end{cases}$$

$$\int_{-\pi}^{\pi} \sin kx dx = 0 \textit{ for all } k.$$

Assume that term-by-term integration of the series is permissible. Then integrating the series with respect to x over the interval $[-\pi; \pi]$ and we find that

$$\int_{-\pi}^{\pi} f(x) dx = \frac{a_0}{2} \int_{-\pi}^{\pi} dx + \sum_{k=1}^{\infty} \left(a_k \int_{-\pi}^{\pi} \cos kx dx + b_k \int_{-\pi}^{\pi} \sin kx dx \right)$$

$$= \frac{a_0}{2} .2\pi + \sum_{k=1}^{\infty} (a_k .0 + b_k .0) = \pi a_0$$

To find the coefficients an a_n (n = 1, 2, ...) we multiply both sides of equality by $\cos nx$. Integrating the result of the multiplication with respect to x over the interval $[-\pi; \pi]$ we have

$$\int_{-\pi}^{\pi} f(x) \cos nx dx = \frac{a_0}{2} \int_{-\pi}^{\pi} \cos nx dx + \sum_{k=1}^{\infty} \left(a_k \int_{-\pi}^{\pi} \cos nx \cos kx dx + b_k \int_{-\pi}^{\pi} \cos nx \sin kx dx \right)$$

The first term on the right side of the equality equals to zero. By all coefficients of b_k equal zero and we obtain the result that there is only one nonzero coefficient of a_k in the case $k = n$. We obtain

$$\int_{-\pi}^{\pi} f(x) \cos nx dx = \pi a_n$$

Analogously, multiplying both sides of the equality by $\sin nx$ integrating the result with respect to x over the interval $[-\pi;\pi]$ we find that

$$\int_{-\pi}^{\pi} f(x)\sin nx dx = \pi b_n$$

Now, dividing both sides of by , and replacing n by l we find respectively.

$$a_0 = \frac{1}{\pi}\int_{-\pi}^{\pi} \int f(x)\,dx,$$

$$a_k = \frac{1}{\pi}\int_{-\pi}^{\pi} \int f(x)\cos kx\,dx \quad for\ k=1,\ 2,\ .....$$

$$bk = \frac{1}{\pi}\int_{-\pi}^{\pi} \int f(x)\sin kx\,dx \quad for\ k=1,\ 2,\ .....$$

Evidently the formula for finding a_0 fits the formula with $k = 0$ and instead of two formulae we can write one formula.

Actually, if we compute the Fourier coefficients according to the formulae and compose the Fourier series, we don't know whether the obtained series is convergent at all, and if it is convergent then does it converge to the parent function $f(x)$ or to some different function. In this context there arises a question as to which conditions a function should satisfy so that its Fourier series expansion is convergent to the function $f(x)$. We shall consider the convergence problems in the next section . Note that the class of functions which are representable by their Fourier series expansion is quite wide and in most of the practical cases we can assume that the convergence conditions are satisfied.

Example: Find the Fourier series expansion of the function $f(x)$ with period 2π defined on the half-interval $[-\pi;\pi]$ as $f(x) = x$.

Solution: The graph of this function $f(x) = x$ for $x \in (-p, p)$ is illustrated in Figure.

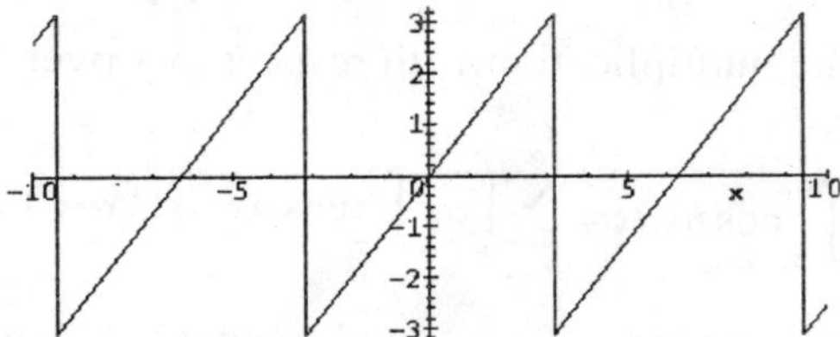

First we find a_0 using:

$$a_0 = \frac{1}{\pi}\int_{-\pi}^{\pi} x dx = \frac{1}{\pi}\frac{x^2}{2}\Bigg|_{-\pi}^{\pi} = 0$$

Using and integrating by parts we find that

$$a_k = \frac{1}{\pi}\int_{-\pi}^{\pi} x \cos kx\, dx$$

$$= \frac{1}{\pi}\left[\frac{x\sin kx}{k}\Big|_{-\pi}^{\pi} - \int_{-\pi}^{\pi} \sin kx\right] = 0$$

and integrating by parts we find that

$$b_k = \frac{1}{\pi}\int_{-\pi}^{\pi} x \sin kx\, dx$$

$$= \frac{1}{\pi}\left[-\frac{x\cos kx}{k}\Big|_{-\pi}^{\pi} + \frac{1}{k}\int_{-\pi}^{\pi} \cos kx dx\right] = (-1)^{k+1}\frac{2}{k}$$

So we have the Fourier series expansion of the function

$$f(x) \sim \left[\frac{\sin 2x}{1} - \frac{\sin 2x}{2} + \frac{\sin 3x}{3} - .... + (-1)^{k+1}\frac{\sin kx}{k} + ...\right]$$

The sign ``~" indicates that we have found the Fourier series expansion according to the formulae, but we don't know whether this expansion converges to this function, so we can't still use the sign ``=". The nth partial sum of this series expansion is

$$Sn(x) = 2\sum_{k=1}^{n(n)} (-1)^{k+1}\frac{\sin kx}{k}$$

The graph of the nth partial sum $S_n(x)$ for $n = 4, 8, 16$ is shown in Figure.

Remark: According to Lemma we can replace the interval of integration $[-\pi;\pi]$ with an arbitrary chosen interval $[c; c + 2\pi]$ i.e. we can use the formulas

$$a_0 = \frac{1}{\pi}\int_{-c}^{c+2\pi} f(x)dx,$$

$$a_k = \frac{1}{\pi}\int_{-c}^{c+2\pi} f(x)\cos kx\, dx,$$

$$b_k = \frac{1}{\pi}\int_{-c}^{c+2\pi} f(x)\sin kx\, dx,$$ where c is any real number.

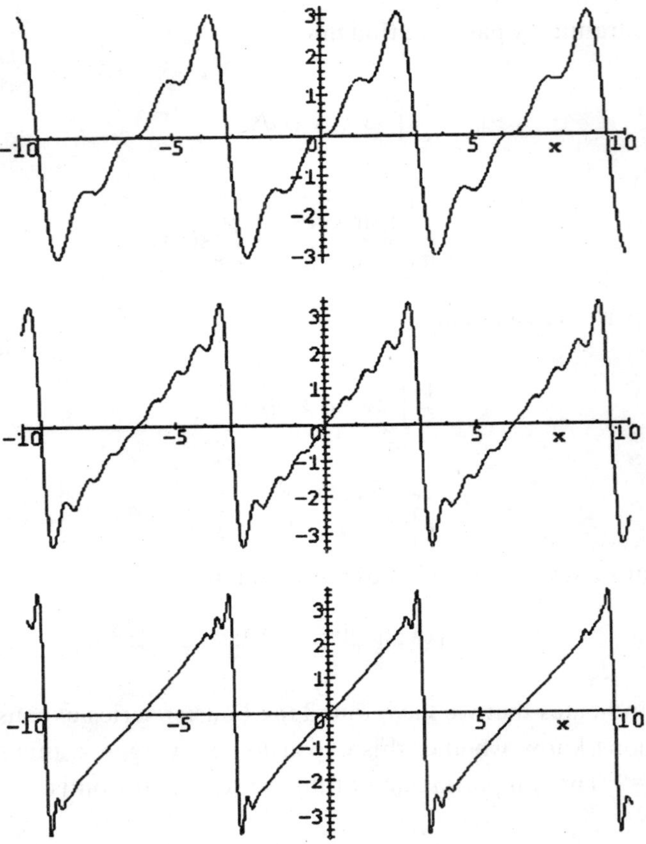

Fig. The Graph of the nth Partial sum $S_n(x)$

FOURIER SERIES OF EVEN AND ODD FUNCTIONS

If a particular function possesses certain symmetrical properties then some terms are absent from its Fourier series expansion and the expressions determining the remaining coefficients can be simplified.

A function $f(x)$ is said to be an even *function* if for each x from its domain

$$f(-x)=f(x)$$

A function $y = f(x)$ is said to be odd *function* if for each x from its domain

$$f(-x)=-f(x)$$

The graph of an even function is symmetrical about the vertical axis and the graph of an odd function is symmetrical about the origin.

The even and the odd functions satisfy the following properties:

- The product of two even functions is an even function,

- The product of two odd functions is an even function,
- The product of an even and an odd function is an odd funnction.

Lemma: *If limits of integration are symmetric with respect to zero then the integral of any odd function equals to zero, but for the integral of any even function f(x)*

$$\int_{-c}^{c} f(x)\,dx = 2\int_{0}^{c} f(x)\,dx$$

Proof: Using the additivity property of the integral we have that

$$\int_{-c}^{c} f(x)\,dx = 2\int_{0}^{c} f(x)\,dx + \int_{0}^{c} f(x)\,dx$$

Substituting in the first integral the variable $x = -y$. Then $dx = -dy$ and

$$\int_{-c}^{0} f(x)\,dx = -\int_{-c}^{0} f(-y)\,dy = -\int_{0}^{c} f(-y)\,dy$$

Denoting the variable of integration again by x , we have

$$\int_{-c}^{c} f(x)\,dy = \int_{0}^{c}\left[f(x) + f(-x)\right]dx$$

Now, if $f(x)$ is an odd function then integrand equals zero, and if $f(x)$ is an even function then integrand equals $2f(x)$ and we have proved this lemma.

Let $f(x)$ be an even function with period 2π . Then $f(x)\cos kx$ is even and $f(x)\sin kx$ is an odd function for any k =1, 2, ...Using Lemma, we obtain for the calculation of Fourier coefficients the formulas

$$a_0 = \frac{2}{\pi}\int_{0}^{\pi} f(x)\,dx,$$

$$a_k = \frac{2}{\pi}\int_{0}^{\pi} f(x)\cos kx\,dx \text{ for k} = 1, 2, \ldots.,$$

$$b_k = 0 \text{ for k} = 1, 2, \ldots.$$

Consequently the Fourier series expansion of an even periodic function with period 2π does not involve the terms with sines have the form

$$f(x) \sim \frac{a_0}{2} + \sum_{k=1}^{\infty} a_k \cos kx,$$

where the a-s are calculated using the formulae.

If $f(x)$ is an odd function then $f(x)\cos kx$ is odd and is even for any k = 1,2, ... By Lemma, we have for calculation of Fourier coefficients the formulae

$a_{k} =$ k= 0, 1, 2, ...,

$$b_{k} = \frac{2}{\pi}\int_{O}^{\pi} f(x)\sin kx\,dx \; for\; k=1,2,....$$

Thus the Fourier series expansion of the given periodic function with period consists of sine terms only and has the form

$$f(x) \sim \sum_{k=1}^{\infty} b_k \sin kx,$$

where the *b-s* are calculated.

Example: Find the Fourier series expansion of the function *f*(*x*) with period 2π, which on half-interval $[-\pi;\pi]$ is defined by *f*(*x*) = | *x* |.

Solution: A graph of this function is shown in Figure. As this is an even function we have to calculate only the coefficients a_k, c_k = 1, 2, ... First using we find

$$a_0 = \frac{2}{\pi}\int_{O}^{\pi} |x|\,dx = \frac{2}{\pi}\int_{O}^{\pi} x\,dx = \pi$$

and using and integration by parts we obtain

$$a_k = \frac{2}{\pi}\int_{O}^{\pi} |x|\cos kx\,dx = \frac{2}{\pi}\int_{O}^{\pi} x\cos kx\,dx$$

$$= \frac{2}{\pi}\left[\frac{1}{k}x\sin kx\Big|_{O}^{\pi} - \frac{1}{k}\int_{O}^{\pi}\sin kx\,dx\right]$$

$$= \frac{2}{\pi k^2 \cos kx}\Big|_{O}^{\pi} = \frac{2}{\pi k^2}\left[(-1)^k - 1\right]$$

$$= \begin{cases} 0 \frac{-4}{\pi k^2} & if\ k\ is\ odd \\ & if\ k\ is\ even. \end{cases}$$

Thus by the Fourier series expansion of this function is

$$f(x) \sim \frac{\pi}{2} - \frac{4}{\pi}\left(\cos x + \frac{1}{9}\cos 3x + \frac{1}{25}\cos 5x + ...\right)$$

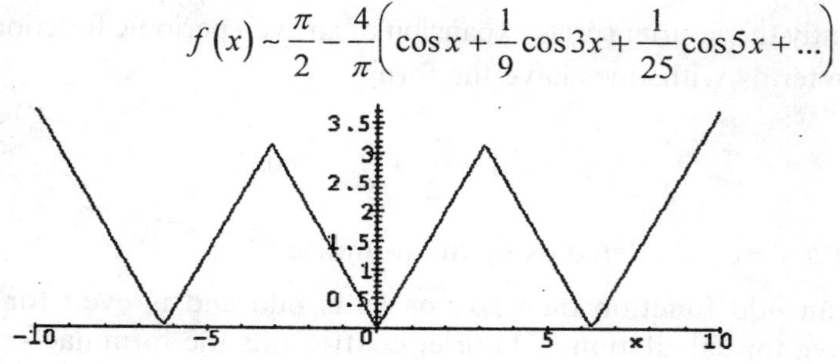

Example: Find the Fourier series expansion of the *square wave function* i.e. the periodic function $f(x)$ with period 2π defined within the period $-\pi < x < \pi$ by

$$f(x) = \begin{cases} -1 & if\ -\pi < x < 0 \\ 1 & if\ 0 < x < p \end{cases}$$

Solution: A graph of this square wave function is shown in Figure.

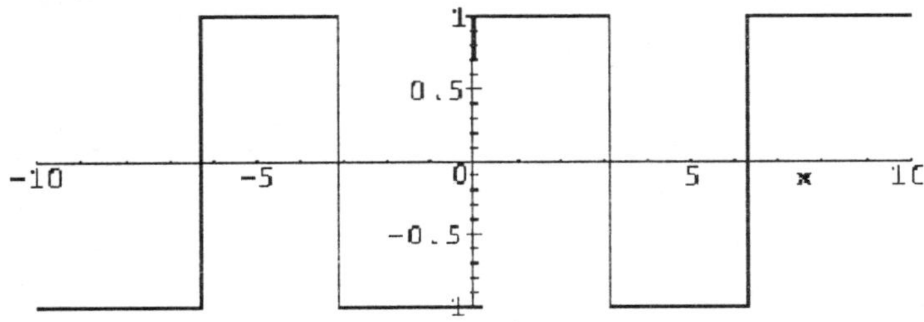

Clearly $f(x)$ is an odd function and thus its Fourier series expansion consists of sine terms only. By we calculate the coefficients

$$b_k = \frac{2}{\pi}\int_O^{\pi} \sin kx\, dx = -\frac{2}{k\pi}\cos kx\Big|_O^{\pi} =$$

$$= -\frac{2}{k\pi}\left[(-1)^k - 1\right] = \begin{cases} \dfrac{4}{k\pi} & if\ k\ is\ old \\ 0 & if\ k\ is\ even. \end{cases}$$

Thus the Fourier series expansion of $f(x)$ is

$$f(x) \sim \frac{4}{\pi}\left(\sin x + \frac{1}{3}\sin 3x + \frac{1}{5}\sin 5x + ...\right) = \frac{4}{\pi}\sum_{k=1}^{\infty}\frac{\sin(2k -}{2k -}$$

FOURIER SERIES OF FUNCTIONS WITH ARBITRARY PERIOD

Since the period 2π, i.e. the unit frequency may rarely be encountered in practice, we have to consider functions with an arbitrary period. If the period of the function $f(x)$ is T, then by Lemma the period of the function $f\left(\frac{Tx}{2\pi}\right)$ is 2π and by the Fourier series expansion of this function is

$$f\left(\frac{Tx}{2\pi}\right) \sim \frac{a_O}{2} + \sum_{k=1}^{\infty} a_k \cos kx + b_k \sin kx,$$

where

$$a_0 = \frac{1}{\pi}\int_{-\pi}^{\pi} f\left(\frac{Tx}{2\pi}\right) dx,$$

$$a_k = \frac{1}{\pi}\int_{-\pi}^{\pi} f\left(\frac{Tx}{2\pi}\right)\cos kx\,dx, k = 1, 2, \ldots$$

$$b_k = \frac{1}{\pi}\int_{-\pi}^{\pi} f\left(\frac{Tx}{2\pi}\right)\sin kx\,dx, k = 1, 2, \ldots$$

Changing the variable $t = \frac{Tx}{2\pi}$, i.e. $x = \frac{2\pi t}{T}$ and $dx = \frac{2\pi}{T}dt$, we obtain that the Fourier series expansion of the function $f(t)$ with period T is

$$f(t) \sim \frac{a_0}{2} + \sum_{k=1}^{\infty} a_k \cos\frac{2k\pi t}{T} + b_k \sin\frac{2k\pi t}{T},$$

where

$$a_0 = \frac{2}{T}\int_{-\frac{T}{2}}^{\frac{T}{2}} f(t)\,dt,$$

$$a_k = \frac{2}{T}\int_{-\frac{T}{2}}^{\frac{T}{2}} f(t)\cos\frac{2k\pi t}{T}dt \text{ for } k = 1, 2, \ldots.$$

$$b_k = \frac{2}{T}\int_{-\frac{T}{2}}^{\frac{T}{2}} f(t)\sin\frac{2k\pi t}{T}dt \text{ for } k = 1, 2, \ldots.$$

Replacing the variable t by x again, we express the Fourier series expansion in terms of

$$f(x) \sim \frac{a_0}{2} + \sum_{k=1}^{\infty} a_k \cos kwx + b_k \sin kwx,$$

where

$$a_0 = \frac{2}{T}\int_{-\frac{T}{2}}^{\frac{T}{2}} f(x)\,dx,$$

$$a_k = \frac{2}{T}\int_{-\frac{T}{2}}^{\frac{T}{2}} f(x)\cos kwx\,dx \text{ for } k = 1, 2, \ldots.$$

$$b_k = \frac{2}{T}\int_{-\frac{T}{2}}^{\frac{T}{2}} f(x)\sin kwxdx \text{ for } k = 1, 2, \ldots.$$

Remark: In formulae the interval of integration can be replaced with an arbitrary interval $c; c + T$ with length T.

Now, if the periodic function $f(x)$ with period T is even then from we obtain

$$a_0 = \frac{4}{T}\int_{0}^{\frac{T}{2}} f(x)\,dx,$$

$$a_k = \frac{4}{T}\int_0^{\frac{T}{2}} f(x)\cos kwx\,dx, \text{ for } k = 1, 2, ...,$$

$$b_k = 0 \text{ for k} = 1, 2, ...$$

and the Fourier series expansion of even function $f(x)$ is

$$f(x) \sim \frac{A_0}{2} + \sum_{K=1}^{\infty} a_k \cos kwx.$$

If $f(x)$ is a odd periodic function with period T then from we have

$$a_k = 0 \text{ for k} = 0, 1, 2, ...$$

$$b_k = \frac{4}{T}\int_0^{\frac{T}{2}} f(x)\sin kwx\,dx$$

and the Fourier series expansion of odd function $f(x)$ is

$$f(x) \sim \sum_{K=1}^{\infty} b_k \sin kwx\,dx.$$

Example: Find the Fourier series expansion of the function $f(x) = |\sin x|$.

The graph of this function is shown in Figure.

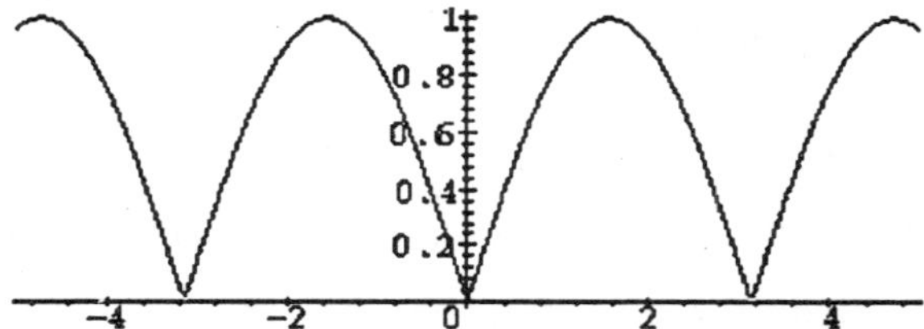

Solution: This function is clearly even and its period is π (i.e. frequency $\omega = 2$). Therefore its Fourier series expansion consists of cosine terms only and we calculate the coefficients a_k for $k = 1, 2, ...$ using the formulae. By

$$a_0 = \frac{4}{T}\int_0^{\frac{T}{2}} |\sin x|\,dx = \frac{4}{\pi}\cos x\Big|_0^{\frac{\pi}{2}} = \frac{4}{\pi}$$

$$a_k = \frac{4}{T}\int_0^{\frac{T}{2}} |\sin x|\cos 2kx dx$$

$$= \frac{4}{\pi}\frac{1}{2}\left[\int_0^{\frac{\pi}{2}} \sin(x + 2lx)\,dx + \int_0^{\frac{\pi}{2}} \sin(x - 2kx)\,dx\right]$$

$$= \frac{2}{\pi}\left[-\frac{1}{2k+1}\cos(2k+1)x\Big|_0^{\frac{\pi}{2}}\right.$$

$$+\frac{1}{2k-1}\cos(2k-1)x\Big|_0^{\frac{\pi}{2}}\Bigg]$$

$$=\frac{2}{\pi}\left[\frac{1}{2k+1}-\frac{1}{2k-1}\right]$$

$$=\frac{2}{\pi}\frac{2k-12k-1}{4k^2-1}$$

$$-\frac{4}{\pi}\frac{1}{4k^2-1}.$$

The Fourier series expansion of the given function is

$$|\sin x| \sim \frac{2}{\pi}-\frac{4}{\pi}\sum_{k=1}^{\infty}\frac{\cos 2kx}{4k^2-1}$$

FOURIER SERIES OF FUNCTIONS DEFINED OVER A FINITE INTERVAL

Suppose the given function $f(x)$ is defined only over the finite half-interval a, $a+T$ [(or over a, $a+T$)].

Then we can define the periodic extension $\phi(x)$ of $f(x)$ by

$$\phi(x) = f(x) \; if \; a \le x < a+T,$$

As the result we obtain the periodic function $\phi(x)$ which on half-interval [a, $a+T$] is identical with $f(x)$. The periodic function $\phi(x)$ has the Fourier series expansion

$$f(x) \sim \frac{a_0}{2}+\sum_{K=1}^{\infty} a_k \cos kwx + b_k \sin kwx$$

where

$$a_0 = \frac{2}{T}\int_a^{a+T} f(x)dx,$$

$$a_k = \frac{2}{T}\int_a^{a+T} f(x)\cos kwx\,dx,$$

$$b_k = \frac{2}{T}\int_a^{a+T} f(x)\sin kwx\,dx,$$

If this Fourier series expansion represents the function $\phi(x)$ then on the half-interval [a, $a+T$] it represents also the function $f(x)$.

Now we consider the case if the given function $f(x)$ is defined only over the finite interval $\left[0;\frac{T}{2}\right]$. Its even periodic extension $\phi_1(x)$ is the even periodic function with period T defined by

$$\phi_1(x) = \begin{cases} f(x) \, if \, 0 < x < \frac{T}{2}, \\ f(-x) \, if \, -\frac{T}{2} < x < 0, \end{cases}$$

$$\phi_1(x+T) = \phi_1(x)$$

The Fourier series expansion of the function $\phi_1(x)$ consists of cosine terms only and is given by

$$\phi_1(x) \sim \frac{a_0}{2} + \sum_{k=1}^{\infty} a_k \cos kwx$$

called the *cosine series expansion*, where

$$a_k = \frac{4}{T}\int_0^{\frac{T}{2}} f(x)\cos kwx dx \; for \; k = 0, 1, 2, \ldots$$

For a given function $f(x)$ defined only over the finite interval $\left[0;\frac{T}{2}\right]$, its odd periodic extension $\phi_2(x)$ is theodd periodic function with period T defined by

$$\phi_2(x) = \begin{cases} f(x) \, if \, 0 < x < \frac{T}{2}, \\ -f(-x) \, if \, -\frac{T}{2} < x < 0, \end{cases}$$

$$\phi_2(x+T) = \phi_2(x)$$

The Fourier series expansion of the function $\phi_2(x)$ consists of sine terms only and is given by

$$\phi_2(x) \sim \sum_{k=1}^{\infty} b_k \sin kwx$$

called the *sine series expansion*, where

$$b_k = \frac{4}{T}\int_0^{\frac{T}{2}} f(x)\sin kwx dx \; for \; k = 1, 2, \ldots$$

Example: The function $f(x) = x$ is defined only in the interval [0;4]. Obtain (a) the cosine series expansion and (b) sine series expansion.

Solution: (a) To obtain the cosine series expansion we define the 3 periodic extension with period.

$$\phi_1(x) = \begin{cases} f(x)\ if 0 < x < 4, \\ -x\ if\ -4 < x < 0, \end{cases}$$

$$\phi_1(x+8) = \phi_1(x)$$

Using we calculate the coefficients

$$a_0 = \frac{4}{8}\int_0^4 x\,dx = 4$$

and (as $\omega = \frac{\pi}{4}$)

$$a_k = \frac{4}{8}\int_0^4 x \,cox \frac{k\pi x}{4}dx$$

$$= \frac{1}{2}\left[\frac{4\pi}{k\pi}\sin\frac{k\pi x}{4}\bigg|_0^4 - \frac{4}{k\pi}\int_0^4 \sin\frac{k\pi x}{4}dx\right]$$

$$= \frac{1}{2}\frac{16}{k^2\pi^2}\cos\frac{k\pi x}{4}\bigg|_0^4$$

$$= \frac{8}{k^2\pi^2}\left[(-1)^k - 1\right]$$

$$= \begin{cases} -\frac{16}{k^2\pi^2}\ if\ k\ is\ odd, \\ 0\ if\ k\ is\ even \end{cases}$$

The cosine series expansion of the function $\phi_1(x)$ is

$$\phi_1(x) \sim 2 - \frac{16}{\pi^2}\sum_{k=1}^{\infty}\frac{1}{(2k-1)^2}\cos\frac{(2K-1)\pi x}{4}$$

In the interval $]0;4[$ the function $\phi_1(x)$ is identical with $f(x) = x$. Hence on the interval $]0;4[$ we have obtained also the cosine series expansion

$$\phi_1(x) \sim 2 - \frac{16}{\pi^2}\sum_{k=1}^{\infty}\frac{1}{(2k-1)^2}\cos\frac{(2K-1)\pi x}{4}.$$

(b) To obtain the sine series expansion we define the odd periodic extension $\phi_2(x)$ of $f(x)$:

$$\phi_2(x) = \text{x if} -4 < \text{x} < 4.$$

$$\phi_2(x+8) = \phi_2(x)$$

Using gives

$$b_k = \frac{4}{8}\int_0^4 x\sin\frac{k\pi x}{4}dx$$

$$= \frac{1}{2}\left[-\frac{4x}{k\pi}\cos\frac{k\pi x}{4}\Big|_0^4 + \frac{4}{k\pi}\int_0^4 \cos\frac{k\pi x}{4}dx\right]$$

$$= \frac{1}{2}\left[-\frac{16x}{k\pi}(-1) + \frac{16}{k^2\pi^2}\sin\frac{k\pi x}{4}\Big|_0^4\right]$$

$$= \frac{8}{k\pi}(-1)^{\kappa+1}$$

and the sine series expansion of the function $\phi_2(x)$ is

$$\phi_2(x) \sim \frac{8}{\pi}\sum_{k=1}^{\infty}\frac{(-1)^{k+1}}{k}\sin\frac{k\pi x}{4}.$$

On the interval [0; 4] also the sine series expansion of the given function is

$$x \sim \frac{8}{\pi}\sum_{k=1}^{\infty}\frac{(-1)^{k+1}}{k}\sin\frac{k\pi x}{4}.$$

THE COMPLEX FORM OF FOURIER SERIES

Let us assume that the periodic function $f(x)$ has the period 2π. To obtain the complex form of the Fourier series

$$f(x) \sim \frac{\alpha_0}{2} + \sum_{k=1}^{\infty} a_k \cos kx + \sin kx,$$

we use the Euler formulae

$$\cos\varphi = \frac{e^{i\varphi} + e^{-i\varphi}}{2}$$

and

$$\sin\varphi = \frac{e^{i\varphi} + e^{-i\varphi}}{2i}.$$

By these formulae we have for kth harmonic

$$a_k \cos kx + b_k \sin kx = a_k \frac{e^{ikx} + e^{-ikx}}{2} + b_k \frac{e^{ikx} - e^{-ikx}}{2i}$$

$$= a_k \frac{e^{ikx} + e^{-ikx}}{2} - ib_k \frac{e^{ikx} - e^{-ikx}}{2}$$

$$= \frac{a_k - ib_k}{2} e^{ikx} + \frac{a_k + ib_k}{2} e^{-ikx}.$$

Denoting by

$$c_O = \frac{a_O}{2},\ c_k = \frac{a_k - ib_k}{2} \text{ and } c_{-k} = \frac{a_k + ib_k}{2},$$

the Fourier series expansion takes the form

$$f(x) \sim c_O + \sum_{k=1}^{\infty} \left(c_k e^{ikx} + c_{-k} e^{-ikx}\right) = \sum_{k=-\infty}^{\infty} c_k e^{ikx}.$$

To calculate the coefficients c_k for k > 0 we use the formulae:

$$c_k = \frac{a_{k-ib_k}}{2}$$

$$= \frac{1}{2\pi} \int_{-\pi}^{\pi} f(x) \cos kx dx - \frac{i}{2\pi} \int_{-\pi}^{\pi} f(x) \sin kx dx$$

$$= \frac{1}{2\pi} \int_{-\pi}^{\pi} f(x)(\cos kx - i \sin kx)\, dx$$

$$= \frac{1}{2\pi} \int_{-\pi}^{\pi} f(x) e^{-ikx} dx.$$

An immediate extension shows that this formula is also valid for $k = 0$. If $k < 0$ then $-k > 0$ and

$$c_k = \overline{c-k} = \overline{\frac{1}{2\pi} \int_{-\pi}^{\pi} f(x) e^{ikx} dx.} = \frac{1}{2\pi} \int_{-\pi}^{\pi} f(x) \overline{e^{ikx}} dx.$$

$$= \frac{1}{2\pi} \int_{-\pi}^{\pi} f(x) e^{-ikx} dx.$$

Hence for any $k = 0, \pm 1, \pm 2, \ldots$ we have

$$c_k = \frac{1}{2\pi} \int_{-\pi}^{\pi} f(x) e^{ikx} dx.$$

Consequently, the complex form of the Fouries series of 2π-periodic function $f(x)$ is

$$f(x) \sim \sum_{k=-\infty}^{\infty} c_k e^{ikx}$$

where the coefficients c_k ($k = 0, \pm 1, \pm 2, \ldots$) are calculated

Now, if the period of the function $f(x)$ is T then by Lemma the period of the function $f\left(\frac{Tx}{2\pi}\right)$ is 2π it has the Fourier series expansion

$$f\left(\frac{Tx}{2\pi}\right) \sim \sum_{k=-\infty}^{\infty} c_k e^{ikx}$$

where the coefficients are

$$c_k = \frac{1}{2\pi}\int_{-\pi}^{\pi} f\left(\frac{Tx}{2\pi}\right) e^{-ikx} dx \text{ for } k = 0, \pm 1, \pm 2, \ldots$$

Changing in the variable $t = \frac{T_2}{2\pi}$, we have that $x = \frac{2\pi t}{T}$, $dx = \frac{2\pi}{T} dt$, and

$$f(t) \sim \sum_{k=-\infty}^{\infty} c_k e^{\frac{2k\pi it}{T}}$$

where

$$c_k = \frac{1}{T}\int_{-\frac{T}{2}}^{\frac{T}{2}} f(t) e^{\frac{-2k\pi it}{T}} dt$$

Denoting the variable by x again, we have that the complex form of the Fourier series expansion of the function $f(x)$ with period T is

$$f(x) \sim \sum_{k=-\infty}^{\infty} c_k e^{ikwx}$$

where

$$c_k = \frac{1}{T}\int_{-\frac{T}{2}}^{\frac{T}{2}} f(x) e^{-ikwx} dx \ \textit{for}\ k = 0, \pm 1, \pm 2, \ldots$$

The complex numbers c_k form an infinite sequence $(c_k)_{k=-\infty}^{\infty}$ which is called the *spectrum* of the periodic function $f(x)$. Let $|c_k|$ designates the magnitude of the complex number c_k and φk its argument, i.e. $c_k = |c_k| e^{i\varphi k}$. The sequence of real numbers $(|c_k|)_{k=-\infty}^{\infty}$ is called the *amplitude spectrum* and the sequence of real numbers $(\varphi_k)_{k=-\infty}^{\infty}$ is called the *phase spectrum* of the function $f(x)$.

Example: Find the amplitude spectrum and the phase spectrum of the 2π -periodic function $f(x)$ defined over interval $\pi; -\pi$ as $f(x) = e^x$.

Solution: First we find the spectrum, i.e. the complex Fourier coefficients, of the function $f(x)$ using

$$c_k = \frac{1}{2\pi}\int_{-\pi}^{\pi} e^x e^{-ikx}\,dx = \frac{1}{2\pi}\int_{-\pi}^{\pi} e(1-ik)^x\,dx$$

$$= \frac{1}{2\pi(1-ik)} e^{(1-ik)x}\Big|_{-\pi}^{\pi} = \frac{e^{(1-ik)\pi} - e^{(1-ik)\pi}}{2\pi(1-ik)}.$$

Since $e^{\pm ik\pi = \cos k\pi \pm i\sin k\pi = \cos k\pi = (-1)^k}$ we obtain that

$$c_k = \frac{(-1)^k\left(e^{\pi} - e^{-\pi}\right)}{2\pi(1-ik)} = \frac{(-1)^k\left(e^{\pi} - e^{-\pi}\right)}{2\pi\left(1+k^2\right)}(1+ik).$$

Since $|1+ik| = \sqrt{1+k^2}$ we find the amplitude spectrum $\left|c_k = \frac{e^{\pi} - e^{-\pi}}{2\pi\sqrt{1+k^2}}\right|$ and phase spectrum $\varphi_k = \arg c_k = \arg \tan k$.

Chapter 7

Sequences and Series of Functions

The terms of the infinite series hitherto considered have been constants, whence these series (when convergent) always represented definite numbers. But both in theory and in applications, the series of outstanding importance are those in which the terms are functions of a variable, so that the sum of the series is also a function of the variable, as in the case of Taylor series.

We shall therefore consider a series $g_1(x) + g_2(x) + g_s(x) + ...$, in which the functions $g_n(x)$ are defined in an interval $a > x > b$. We will denote the n-th partial sum $g_1(x) + g_2(x) + ... + g_n(x)$, by $f_n(x)$. Then the sum $f(x)$ of our series, where it exists, is simply the limit $\lim_{n\to\infty} f_n(x)$.

We may therefore regard the sum of an infinite series of functions as the limit of a sequence of functions $f_1(x), f_2(x), f_n(x), ...$ Conversely, for any such sequence of functions $f_1(x), f_2(x), ...$, we can form an equivalent series by setting $g_1(x) = f_1(x) ...$ and $g_n(x) = f_n(x) - f_{n-1}(x)$ for $n > 1$. Hence, when it is convenient, we can pass from the consideration of series to that of sequences and conversely.

LIMITING PROCESSES WITH FUNCTIONS AND CURVES

We shall now state exactly what we mean by saying that a function $f(x)$ is the limit of a sequence $f_1(x), f_2(x), ... , f_n(x)$, in an interval $a \geq x \geq b$. The definition is: The sequence $f_1(x)$, $f_2(x)$, ... converges in that interval to the limit function $f(x)$, if at each point x of the interval the values $f_n(x)$ converge in the usual sense to the value $f(x)$. In this case, we shall write $\lim_{n\to\infty} f_n(x) = f(x)$

According to Cauchy's test , we can express the convergence of the sequence without necessarily knowing or stating the limit function $f(x)$. In fact, our sequence of functions will converge to a limit function if, and only if, at each point x in our interval and for every positive number e the quantity $|f_n(x) - f_m(x)|$ is less than e, provided that the numbers n and m are chosen large enough, i.e., larger than a certain number $N = N(\varepsilon)$. This number $N(\varepsilon)$ usually depends on e and x and increases beyond all bounds as ε tends to zero.

We have frequently encountered cases of limits of sequences of functions. We mention only the definition of the power x^a for irrational values of α by the equation

$$x^\alpha = \lim_{n\to\infty} x^{r_n}$$

where $r_1, r_2, ..., r_n, ...$ is a sequence of rational numbers tending to α or the equation

$$e^\alpha = \lim_{n\to\infty}\left(1+\frac{x}{n}\right)^n,$$

where the functions $f_n(x)$ on the right hand side are polynomials of degree n.

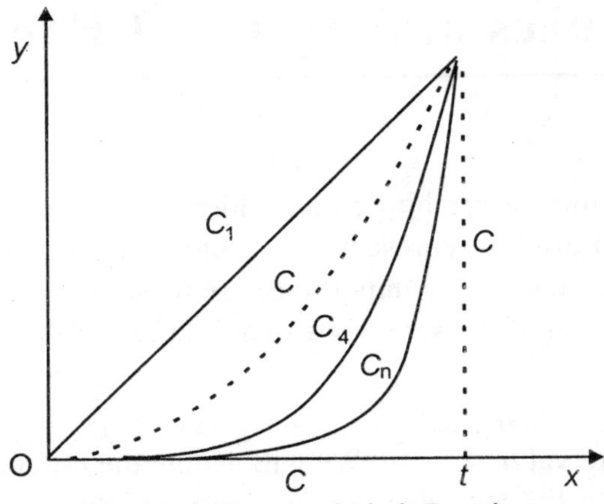

Fig. Limit Curve and Limit Function

The graphical representation of functions by means of curves suggests that we can also speak of limits of sequences of curves, saying, for example, that the graphs of the above limit functions x^a and e^x are to be regarded as the limit curves of the graphs of the functions x^{r_n} *and* $\left(1+\frac{x}{n}\right)^n$ respectively. However, there is a fine distinction between the passages to the limit with functions and with curves. Until the middle of the Nineteenth Century, this distinction was not sufficiently recognized and only by having a clear idea of it can we avoid apparent paradoxes. We shall illustrate this point by an example.

Consider the functions

$$f_n(x) = x^n \ (n = 1, 2, ...)$$

in the interval $0 \leq x \leq 1$. All these functions are continuous and the limit function

$$\lim_{n\to\infty} f_n(x) = f(x)$$

exists. However, this limit function is not continuous. On the contrary, since for all values of n the value of the function $f_n(1) = 1$, the limit $f(1) = 1$, while, on the other hand, for $0 \leq x < 1$, the limit $f(x) \lim_{n\to\infty} f_n(x) = 0$. The function $f(x)$ is therefore a discontinuous function which at $x=1$ has the value 1 and for all other values of x in the interval has the value 0.

This discontinuity becomes clear, if we consider the graphs C_n of the functions $y = f_n(x)$. These are continuous curves, all of which pass through the origin and the point $x = 1$, $y = 1$ and which draw in closer and closer to the x-axis as n increases. The curves possess a limit curve C which is not at all discontinuous, but consists of the portion of the x-axis between $x=0$ and $x=1$ and the portion of the line $x=1$ between $y=0$ and $y=1$. The curves therefore converge to a continuous limit curve with a vertical portion, while the functions converge to a discontinuous limit function. We thus recognize that this discontinuity of the limit function expresses itself by the occurrence in the limit curve of a portion perpendicular to the x-axis. Such a portion must involve a discontinuity in the limit function and, in fact, such a portion is always present when the limit function is discontinuous. This limit curve is not the graph of the limit function nor can any curve with a vertical portion be the graph of any single-valued function $y = f(x)$; in fact, corresponding to the value of x at which the vertical portion occurs, the curve gives an infinite number of values of y and the function only one. Hence the limit of the graphs of the functions $f_n(x)$ is not the same as the graph of the limit $f(x)$ of these functions.

UNIFORM AND NON-UNIFORM CONVERGENCE

The distinction between the concepts of convergence of functions and of convergence of curves introduces a phenomenon of which it is essential that the student should clearly recognize it. This is the so-called non-uniform convergence of sequences or infinite series of functions. Since it is well known that, as a rule, beginners encounter difficulties here, we should discuss the matter in some detail.

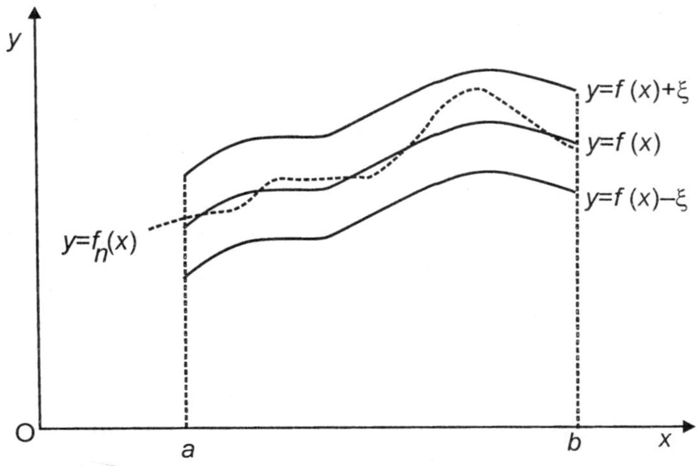

Fig. To Illustrate Uniform Convergence

By definition, a function $f(x)$ is the limit of a sequence $f_1(x), f_2(x)$, ... in an interval $a \leq x$

} *b*,only means that the usual limit relationship $f(x) = \lim_{n\to\infty} f(x)$ holds at each point x of the interval. From a naive point of view, one might expect that it would follow automatically from this concept of convergence that if we assign an arbitrary degree of accuracy, say ε=1/1000 or ε=1/100, then from a certain index N onwards all the functions $f_n(x)$ would lie between $f(x)$ + e and $f(x) - \varepsilon$ for all values of x, so that their graphs $y = f_n(x)$ will lie entirely in the strip indicated. Thus, for every positive ε , there is a corresponding number $N = N(\varepsilon)$, which, of course, will ordinarily increase beyond all bounds as $\varepsilon < 0$, such that, for $n > N$, the difference $|f(x) — f_n(x)| < \varepsilon$, no matter where x is chosen in the interval.

If the accuracy of the approximation can be made at least equal to a pre-assigned number ε everywhere in the interval at the same time, i.e., by everywhere choosing the same number $N(\varepsilon)$ independent of x, we may say that the approximation is uniform. One is at first astonished to find that the naive assumption that convergence is necessarily uniform is altogether wrong; in other words, that convergence may very well be non-uniform.

A TEST OF UNIFORM CONVERGENCE

The preceding considerations show that the uniform convergence of a sequence or series is a special property which is not possessed by all sequences and series. We shall now reformulate the concept of uniform convergence. The convergent series $g_1(x) + g_2(x) + ...$ is said to be uniformly convergent in an interval, if the sum $f(x)$ can be approximated to within ε (where ε is an arbitrarily small positive number) by taking a number of terms which is sufficiently large and which is the same throughout the interval.

First of all, we assume that the series $g_1(x) + g_2(x) + ...$ converges at every point of a certain interval to a limit function $f(x)$; we denote by $f_n(x)$ the n-th partial sum of the series $f_n(x) = g_1(x) + g_n(x)$ and by $R_n(x)$ the remainder of the series after n terms:

$$R_n(x) = f(x) - f_n(x)$$

The series $g_1(x) + g_2(x) + ...$ is said to be uniformly convergent in the interval if there corresponds to every positive number ε a number N, which only depends on e and not on x, such that for $n > N$ one has the inequality $|R(x)|=|f(x)-f_n(x)|<e$ for all values of x *in* the interval. Expressed more figuratively, the partial sum $f_n(x)$ represents the sum $f(x)$ to within an error of less than e everywhere in the interval at the same time, provided only that n is chosen large enough. By Cauchy's test, we readily see that the series converges uniformly, if, and only if the difference $|f_n(x) - f_m(x)|$ can be made less than an arbitrary quantity e everywhere in the interval by choosing n and m larger than a number N independent of x.

In fact, firstly, if the convergence is uniform, we can make both $|f_n(x) - f(x)|$ and $|f_m(x) - f(x)|$ less than $e/2$ by choosing n and m larger than a number N independent of x, whence it follows that $|f_n(x) - f_m(x)| < e$; and, secondly, if $|f_n(x) - f_m(x)| < e$, for all values of x whenever n and m are larger than N, then, on choosing any fixed value of $n > N$ and letting m increase beyond all bounds, we have

$$|f_n(x) - f(x)| = \lim_{m\to\infty} |f_n(x) - f_m(x)| \leqq \varepsilon,$$

for every value of x, whence the convergence is uniform. If we wish to speak of the uniform convergence of a sequence of functions, we have only to make minute changes in the above definition; the sequence $f_1(x), f_2(x), \ldots$ converges uniformly to $f(x)$ in an interval, if the difference $|f(x)-f_n(x)|$ can be made less than ε everywhere in the interval by choosing n larger than a number N independent of x.

As above, a necessary and sufficient condition for the uniform convergence of the sequence is that $|f_n(x) - f_m(x)| < \varepsilon$ for all values of x when both n and m are larger than a certain number N, which depends on ε but not on x.

We shall soon see that it is just this condition of uniform convergence which makes infinite series and other limiting processes with functions into convenient and useful tools of analysis. Fortunately, in the limiting processes, usually encountered in the Calculus and its applications, non-uniform convergence is a sort of exceptional phenomenon which will scarcely trouble us in our present applications of analysis.

In most cases, the uniformity of the convergence of a series is established by means of the criterion:

If the terms of the series $\sum_{\nu=1}^{\infty} g_\nu(X)$ satisfy the condition $|g_\nu(X)| \leqq a_\nu$, where the numbers a_ϖ are constants which form a convergent series $\sum_{\nu=1}^{\infty} a_\nu$ then the series $\sum_{\nu=1}^{\infty} g_\nu(X)$ converges uniformly (and, we incidentally remark, absolutely).

In fact, we have then $\left|\sum_{\nu=n}^{m} g_\nu(x)\right| \leqq \sum_{\nu=n}^{m} |g_\nu(x)| \leqq \sum_{\nu=n}^{m} a_\nu$, and since, by Cauchy's test, the sum $\sum_{\nu=n}^{m} a_\nu$, can be made arbitrarily small by choosing n and $m > n$ large enough, this expresses exactly the necessary and sufficient condition for uniform convergence. A first example is offered by the geometric series $1 + x + x + \ldots$, where x is restricted to the interval $|x| \leqq q$, a positive number less than 1. The terms of the series are then numerically less than or equal to the terms of the convergent geometric series $\sum q^\nu$

A further example is the trigonometric series $\frac{c_1 \sin(x-\delta_1)}{1^2} + \frac{c_2 \sin(x-\delta_2)}{2^2} + \frac{c_1 \sin(x-\delta_3)}{3^3}$ $+ \ldots$, provided that $|c_n| < c$, a positive constant independent of n. Because we have then

$$g_n(x) = \frac{c_n \sin(x-\delta_n)}{n^2}, \text{ so that } |g_n(x)| < \frac{c}{n^2}$$

Hence the uniform and absolute convergence of the trigonometric series follows from the convergence of the series

CONTINUITY OF THE SUM OF A UNIFORMLY CONVERGENT SERIES

As we have already hinted, the significance of the uniform convergence of an infinite series lies in the fact that a uniformly convergent series in many respects behaves exactly like the sum of a finite number of functions. For example, the sum of a finite number of continuous functions is itself continuous and, correspondingly, we have the theorem:

If a series of continuous terms converges uniformly in an interval, its sum is also a continuous function.

The proof is quite simple. We subdivide the series $f(x) = g_1(x) + g_2(x) + \ldots$ into the *m*-th partial sum $f_n(x)$ plus the remainder $R_n(x)$. As usual, $f_n(x) = g_1(x) + \ldots g_n(x)$.

If now any positive number ε is assigned, we can,by virtue of the uniform convergence, choose the number n so large that the remainder is less than $\varepsilon/4$ throughout the entire interval, whence $\left|R_n(x+h) - R_n(x)\right| < \frac{\varepsilon}{2}$ for every pair of numbers x and $x + h$ in the interval. The partial sum $f_n(x)$ consists of the sum of a finite number of continuous functions, whence it is continuous; thus, for each point x in the interval, we can choose a positive δ so small that $\left|f_n(x+h) - f_n(x)\right| < \frac{\varepsilon}{2}$, provided $|h| < \delta$ and the points x and $m + h$ lie in the interval. It then follows that

$$\begin{aligned}\left|f(x+h) - f(x)\right| &= \left|f_n(x+h) - f_n(x) + R_n(x+h) - R_n(x)\right| \\ &\leq \left|f_n(x+h) - f_n(x)\right| + \left|R_n(x+h) - R_n(x)\right| < \varepsilon,\end{aligned}$$

which expresses the continuity of our function.

The significance of this theorem becomes clear when we recall that the sums of non-uniformly convergent series of continuous functions are not necessarily continuous, as our previous examples show. We may conclude from the preceding theorem that, if the sum of a convergent series of continuous functions has a point of discontinuity, then in every neighbourhood of this point the convergence is non-uniform, whence every representation of discontinuous functions by series of continuous functions is based on the use of non-uniformly convergent limiting processes.

INTEGRATION OF UNIFORMLY CONVERGENT SERIES

A sum of a finite number of continuous functions can be integrated term by term, i.e., the integral of the sum can be found by integrating each term separately and adding the integrals. In the case of a convergent infinite series, the same procedure is possible, provided that the series converges uniformly in the interval of integration.

A series $\sum_{\nu=1}^{\infty} g_\nu(X) = f(X)$ which converges uniformly in an interval can be integrated term by term in that interval, or, more precisely, if a and x are two numbers in the interval of uniform convergence, the series $\sum_{\nu=1}^{\infty} \int_a^X g_\nu(t)\,dt$ converges and, in fact, converges uniformly with respect to x for each fixed value of a, its sum being equal to $\int_a^X f(t)\,dt$..

In order to prove this we write, as before,

$$f(x) = \sum_{\nu=1}^{\infty} g_\nu(x) = f_n(x) + R_n(x)$$

We have assumed that the separate terms of the series are continuous, the sum is also continuous and therefore integrable. Now, if ε is any positive number, we can find a number N so large that, for every $n > N$, one has $|R_n(x)| < \varepsilon$ for every value of x in the interval. By the first mean value theorem of the integral calculus, we have $\left|\int_a^x \{f(t) - f_n(t)\}\,dt\right| \leqq \epsilon\, l$,

where l is the length of the interval of integration. Since the integration of the finite sum $f_n(x)$ can be performed term by term, this yields $\left|\int_a^x f(t)\,dt - \sum_1^n \int_a^x g_\nu(t)\,dt\right| \leqq \epsilon\, l$

However, since $\varepsilon\, l$ can be made as small as we please, this yields $\sum_{\nu=1}^{\infty} \int_a^x g_\nu(t)\,dt = \lim_{n\to\infty} \sum_{\nu=1}^{n} \int_a^x g_\nu(t)\,dt = \int_a^x f(t)\,dt$, as was to be proved.

If we wish to conside instead of infinite series sequences of functions, our result can be expressed as follows:

If in an interval the sequence of functions $f_1(x)$, $f_2(x)$, ... tends uniformly to the limit function $f(x)$, then $\int_a^b f(x)\,dx = \lim_{n\to\infty} \int_a^b f_n(x)\,dx$ for every pair of numbers a and b in the interval; in other words, we can then interchange the order of the operations of integration and going to the limit.

This fact is far from being a triviality. It is true that, from a naive point of view such as prevailed in the Eighteenth Century, the interchangeabi]ity of the two processes is hardly to be doubted; however, shows that in the case of non-uniform convergence the above equation might not hold. We need only consider Example 2, in which the integral of the limit function is 0, while the integral of the function $f_n(x)$ over the interval $0 \leq x \leq 1$, i.e., the area of the triangle has the value $\int_a^b f_n(x)\,dx = n^{\alpha-2}$, and, when $\alpha \geq 2$, this does not tend to zero. We see here immediately from figure that the reason for the difference between $\int_0^1 f_n(x)\,dx = n^{\alpha-2}$, and $\lim_{n\to\infty} \int_0^1 f_n(x)\,dx$ lies in the non-uniformity of the convergence.

On the other hand, by considering values of α in the interval $1 \leq \alpha \leq 2$, we see that the equation $\lim_{n\to\infty} \int_0^1 f_n(x)\,dx = \int_0^1 f(x)\,dx$ can be valid, although the convergence is non-uniform.

As a further example, we can integrate term by term the series $\sum_0^\infty g_n(x)$, , where $g_n(x) = x^n - x^{n-1}$ for $n > 1$ and $g_0(x) = 1$, between 0 and 1, even though it does not converge uniformly. Thus, while uniformity of convergence is a sufficient condition for term-by-term integrability, it is by no means a necessary condition. Neglect of this aspect may easily lead to a misunderstanding.

DIFFERENTIATION OF INFINITE SERIES

The behaviour of uniformly converging series or sequences with respect to differentiation is quite different from that with respect to integration. For example, the sequence of functions $f_n(x) = \dfrac{\sin n^2\pi}{n}$ certainly converges uniformly to the limit function $f(x) = 0$, but the derivative $f_n'(x) = n\cos n^2x$ does not converge everywhere to the derivative of the limit function $f'(0) = 0$, as we see by considering $x = 0$. In spite of the uniformity of the convergence, we cannot therefore change the order of the process of differentiation and pass on to the limit.

Naturally, corresponding statements hold for infinite series. For example, the series $\sin x + \dfrac{\sin 2^4 x}{2^2} + \dfrac{\sin 3^4 x}{3^2} + \ldots$ is absolutely and uniformly convergent, because its terms are numerically not greater than the terms of the convergent series $\dfrac{1}{1^2} + \dfrac{1}{2^2} + \dfrac{1}{3^2} + \ldots$.

However, if we differentiate the series term by term, we obtain the series $\cos x + 2^2 \cos 2^4x + 3^2 \cos 3^4x + \ldots$, which obviously does not converge everywhere, since, for example, it diverges at $x = 0$.

The only useful criterion which assures us in special cases that term-by-term differentiation is permissible is given by the theorem:

If, on differentiating a convergent infinite series $\sum_{\nu=0}^{\infty} G_\nu(X) = F(X)$ term by term, we obtain a uniformly convergent series of continuous terms $\sum_{\nu=0}^{\infty} g_\nu(X) = f(X)$, then the sum of this last series is equal to the derivative of the sum of the first series.

Hence, this theorem expressly requires that, after differentiating the series term by term, we must still investigate whether the result of the differentiation is a uniformly convergent series or not.

The proof of the theorem is almost trivial, because, we can integrate term by term the series obtained by differentiation. Recalling that $g_n(t) = G_n'(t)$, we obtain

$$\int_a^x f(t)\,dt = \int_a^x \left(\sum_{\nu=0}^{\infty} g_\nu(t)\right) dt = \sum_{\nu=0}^{\infty} \int_a^x g_\nu(t)\,dt = \sum_{\nu=0}^{\infty} \left(G_\nu(x) - G_\nu(a)\right)$$

$$= F(x) - F(a)$$

Since this is true for every value of x in the interval of uniform convergence, it follows that $f(x) = f(a)$ as was to be proved.

POWER SERIES

Among infinite series, power series occupy the main position. We mean by a power series a series of the type

$$P(x) = c_0 + c_1x + c_2x^2 + ... = \sum_{\nu=0}^{\infty} c_\nu x^\nu$$

(a power series in x) or, more generally,

$$P(x) = c_0 + c_1x(x - x_0) + c_2(x - x_0)^2 + ... = \sum_{\nu=0}^{\infty} c_\nu (x - x_0)^\nu$$

(a power series in $(x - x_0)$), where x_0 is a fixed number. If we introduce into the last series $\xi = x - x_0$ as a new variable, it becomes a power series $\sum_{\nu=0}^{\infty} c_\nu \xi^\nu$ in the new variable ξ , whence, without any loss of generality, we can confine our attention to power series of the more special form $\sum_{\nu=0}^{\infty} c_\nu x^\nu$

We have considered the approximate representation of functions by polynomials and were thus led to the expansion of functions in Taylor series, which, in fact, are power series. Wė shall now study power series in somewhat greater detail and obtain expansions of the most important functions in series in simpler and more convenient ways than before.

CONVERGENCE PROPERTIES OF POWER SERIES

There are power series which converge for no value of x, except, of course, for $x = 0$, e.g., the series $x + 2^2x^2 + 3^3x^3 + ... + n^nx^n + ...$

In fact, if $x < 0$, we can find an integer N such that $|x| > 1/N$. Then the absolute values of all the terms n^nx^n, for which $n > N$, will be larger than 1 and, as n increases, nn^nx^n will increase beyond all bounds, so that the series fails to converge.

On the other hand, there are series which converge for every value of x; for example, the power series for the exponential function

$$e^x = 1 + x + \frac{x^2}{2!} + \frac{x^3}{3!} + ...,$$

the convergence of which for every value of x follows at once from the ratio test. The $(n + 1)$-th term, divided by the n-th term, yields x/n and, whatever number x is chosen, this ratio tends to zero as n increases. The behaviour of power series with regard to convergence is expressed in the fundamental theorem:

If a power series in x converges for a value $x = \xi$, it converges absolutely for every value x such that $|x| <|\xi|$ and the convergence is uniform in every interval $|x| \leq \eta$, where η is any positive number less than $|\xi|$. Hence h may lie as close to $|\xi|$ as we please.

The proof is simple. If the series $\sum_{\nu=0}^{\infty} c_\nu \xi^\nu$ converges, its terms tend to 0 as n increases. From this follows the weaker statement that all the terms lie below a bound M independent of n, i.e., $|c_n \xi^n| < M$. If now q is any number such that $0<q<1$ and we restrict x to the interval $|x| \leq q|\xi|$, then $|c_n x^n| < |c_n x^n| q^\nu < Mq^\nu$. Hence, in this interval, the terms of the series are smaller in absolute value than the terms of the convergent geometric series ΣMq^ν, whence follows, by the theorem, the absolute and uniform convergence of the series in the interval $-q|\xi| < x < q|x|$.

If a power series does not converge everywhere, i.e., if there is a value $x = x$ for which it diverges, it must diverge for every value of x such that $|x| > |\xi|$. In fact, if it were convergent for such a value of x, by the theorem above, it would have to converge for any numerically smaller value of ξ.

Hence, we recognize that a power series, which converges for at least one value of x other than 0 and which diverges for at least one value of x, has an interval of convergence, i.e., there exists a definite positive number r such that the series diverges for $|x| > \rho$ and converges for $|x| < \rho$. No general statement can be made for $|x| = \rho$. The limiting cases, in which the series converges only for $x=0$ and that which converges everywhere, are expressed symbolically by writing $\rho = 0$ and $\rho = \infty$, respectively.

It is possible to find this interval of convergence directly from the coefficients c_ν of the series. If the limit $\lim_{n\to\infty} \sqrt[n]{|c_n|}$ exists, then

$$\rho = \frac{1}{\lim_{n\to\infty} \sqrt[n]{|c_n|}}$$

In general, ρ is given by the formula

$$\rho = \frac{1}{\overline{\lim}_{n\to\infty} \sqrt[n]{|c_n|}}$$

where $\overline{\lim}$ is the symbol for the upper limit.

For example, $\rho = 1$ for the geometric series $1 + x + x + \ldots$; at the end-points of the interval of convergence, the series diverges. Similarly, for the series for the inverse tangent

$$\text{arc tan } x = x - x^3/3 + x^5/5 - + \ldots,$$

we have $\rho = 1$ and at both of the interval's end-points $x = 1$ the series converges, as we recognize at once from Leibnitz's test.

We deduce from the uniform convergence the important fact that within its interval of convergence (if such an interval exists) a power series represents a continuous function.

Integration and Differentiation of Power Series: On account of the uniformity of convergence, it is always permissible to integrate a power series

$$f(x) = \sum_{\nu=0}^{\infty} c_\nu x^\nu$$

term by term over any closed interval lying entirely within the interval of convergence. We thus obtain the function

$$F(x) = c + \sum_{\nu=0}^{\infty} \frac{c_\nu}{\nu+1} x^{\nu+1},$$

for which

$$F'(x) = f(x)$$

Moreover, since $|c_\nu/(n+1)| \leq |c_\nu|$ for all values of ν, the series obtained by integration converges more rapidly than the original series.

We can also differentiate a power series term by term within its interval of convergence, thus obtaining the equation

$$f'(x) = \sum_{\nu=1}^{\infty} \nu c_\nu x^{\nu-1}$$

In order to prove this statement, we need only show that the series on the right hand side converges uniformly, if x is restricted to an interval lying entirely within the interval of convergence. Now, let x be a number, lying as close to ρ as we please, for which $\sum_{\nu=1}^{\infty} c_\nu \xi^\nu$ converges; then, as we have seen before, all the numbers $|c_n \xi_n|$ lie below a bound M independently of ν, so that $\left|c_\nu \xi^{\nu-1}\right| < \frac{\text{M}}{|\xi|} = N$ Now let q be any number such that $0 < q < 1$; if we restrict x to the interval $|x| \leq q|\xi|$, the terms of the series under discussion are not larger than those of the series $\sum_{\nu=1}^{\infty} \left|\nu c_\nu q^{\nu-1} \xi^{\nu-1}\right|$, and therefore less than those of the series $\sum_{\nu=1} N\nu q^{\nu-1}$ .However, in this last series, the ratio of the $(n+1)$-th term to the n-th term is $\{(n+1)/n\}q$, which tends to q as n increases. Since $0<q<1$, it follows by Criterion III*a* that this series converges. Hence the series, obtained by differentiation, converges uniformly and by the theorem at the end of last section represents the derivative $f'_n(x)$ of the function $f(x)$, which proves our statement.

If we apply this result again to the power series

$$f'(x) = \sum_{\nu=1}^{\infty} \nu c_\nu x^{\nu-1},$$

we find on differentiating term by term that

$$f''(x) = \sum_{\nu=2}^{\infty} \nu(\nu-1)c_\nu x^{\nu-2}$$

and, by continuing the process, we arrive at the theorem:

Every function represented by a power series can be differentiated as often as we please within the interval of convergence and the differentiation can be performed term by term.

As an explicit expression for the *k*-th derivative, we obtain

$$f^{(k)}(x) = \sum_{\nu=k}^{\infty} \nu(\nu-1)...(\nu-k+1)c_\nu x^{\nu-k},$$

or in a slightly different form

$$\frac{f^{(k)}(x)}{k!} = \sum_{\nu=k}^{\infty} \binom{\nu}{k} c_\nu x^{\nu-k} = \sum_{\nu=0}^{\infty} \binom{k+\nu}{k} c_{k+\nu} x^{\nu}$$

These two formulae are frequently useful.

OPERATIONS WITH POWER SERIES

The preceding theorems on the behaviour of power series are our justification for operating in the same way with power series as with polynomials. Obviously, two power series can be added or subtracted by adding or subtracting the corresponding coefficients. It is also clear that a power series, like any other convergent series, can be multiplied by a constant factor by multiplying each term by that factor. We merely mention here without a proof that two power series $f(x) = \sum_{\nu=0}^{\infty} a_\nu x^\nu$ and $g(x) = \sum_{\nu=0}^{\infty} b_\nu x^\nu$ can be multiplied by each other like polynomials. In order to be specific, we have the theorem:

Throughout the common part of the interval of convergence of two power series, their product is given by the convergent power series $\sum_{\nu=0}^{\infty} c_\nu x^\nu$ with the coefficients

$$c_0 = a_0 b_0,$$
$$c_1 = a_0 b_1 + a_1 b_0,$$
$$c_2 = a_0 b_2 + a_1 b_1 + a_2 b_0,$$

....

$$c_n = a_0 b_n + a_1 b_{n-1} + ... + a_n b_0,$$

....

Theorem of Uniqueness of Power Series: In the theory of power series, the following fact is important:

If both power series $\sum_{\nu=0}^{\infty} a_\nu x^\nu$ and $\sum_{\nu=0}^{\infty} b_\nu x^\nu$ converge in an interval which contains the point

$x = 0$ inside and if in that interval the two series represent the same function $f(x)$, then they are identical, that is, the equation $a_n = b_n$ is true for every value of n.

In other words:

If at all, a function $f(x)$ can be represented by a power series in x in only one way,

Briefly speaking, the representation of a function by a power series is unique.

In order to prove this, we need only note that the difference of the two power series, i.e., the power series $\phi(x) = \sum_{\nu=0}^{\infty} c_\nu x^\nu$ with the coefficients $c_\nu = a_\nu - b_\nu$ represents the function $\phi(x) = f(x) - f(x) = 0$ in the interval, i.e., this last power series converges to the limit 0 everywhere in the interval. In particular, for $x = 0$, the sum of the series must be 0, i.e., $c_0 = 0$, so that $a_0 = b_0$. We now differentiate the series in the interior of the interval and obtain $\phi'(x) = \sum_{\nu=1}^{\infty} \nu c_\nu x^{\nu-1}$ However, $\phi'(x)$ is also 0 throughout the interval, whence, in particular, we have for $x = 0$: $c_1 = 0$ or $a_1 = b_1$. Continuing this process of differentiation and then setting $x = 0$, we find successively that all the coefficients c_n are equal to zero, which proves the theorem.

Moreover, we see that we can draw from the above discussion the conclusion:

If we take the ν-th derivative of a series $f(x) = \sum a_\nu x^\nu$ and then set $x = 0$, we at once obtain

$$a_\nu = \frac{1}{\nu!} f^{(\nu)}(0)$$

i.e., Every power series which converges for points other than $x = 0$ is the Taylor series of the function which it represents. The uniqueness of the expansion is here expressed by the fact that the coefficients are uniquely determined by the function itself.

EXPANSION OF GIVEN FUNCTIONS IN POWER SERIES

Within its interval of convergence, every power series represents a continuous function with continuous derivatives of all orders. We shall now discuss the converse problem of the expansion of a given function in a power series. In theory, we can always do this by means of Taylor's theorem, however, in practice, we often encounter difficulties with the actual calculation of the n-th derivative and with the estimation of the remainder. But we can often reach our goal more simply by means of the device:

We first write down the statement $f(x) = \sum_{\nu=0}^{\infty} c_\nu x^\nu$, where, to begin with, the coefficients c_ν are unknown. Then we determine by some known property of the function $f(x)$ the coefficients and prove subsequently the convergence of the series. The series represents a function and there remains only to prove that this function is identical to $f(x)$. Due to the uniqueness of the expansion in power series, we know that no other series than the one just found can be the required expansion.

We shall now consider some examples of the use of this method. In fact, we have already obtained the series for artan x and log $(1+x)$ by a method, which forms part of the range of ideas. We simply integrated term by term the series for the derivatives of these functions, which we knew to be geometric series.

The Exponential Function: Our problem is to find a function for which $f'(x) = f(x)$ and $f(0) = 1$. If we write down the series with undetermined coefficients $f(x) = c_0 + c_1x + c_2x^2 + ...$, and differentiate it, we obtain $f'(x) = c_1 + 2c_2x + 3c_3x^2 + ...$

Since, by assumption, these two power series must be identical, we have the equation $nc_n = c_{n-1}$, which is true for all values of $n \geq 1$. If we observe that, due to the relation $f(0)=1$, the coefficient c_0 must have the value 1, we can compute all the coefficients successively and obtain the power series $f(x) = 1 + \frac{x}{1!} + \frac{x^2}{2!} + \frac{x^3}{3!} + ...$

As is easily shown by the ratio test, this series converges for all values of x and therefore represents a function for which the relations $f'(x) = f(x)$, $f(0) = 1$ are actually fulfilled.

Now, the function e^x certainly possesses these properties; we readily deduce that the function $\phi(x)$ is identical to e^x. In fact, if we form the quotient $f(x)=f(x)/e^x$ and differentiate it, we find

$$\varphi(\xi) = \frac{e^{\alpha} f'(x) - e^{\alpha} f'(x)}{e^{2x}} = 0$$

Hence, the function $\phi(x)$ is a constant and since it has the value 1 for $x = 0$, it must be identically equal to 1, thus proving that our power series and the exponential function are identical.

THE BINOMIAL SERIES

We can now return to the binomial series, this time using the method of undetermined coefficients. We wish to expand the function $f(x) = (1+x)^\alpha$ in a power series, whence we write $f(x) = (1 + x)^\alpha = c_0 + c_1x + c_2x^2 + ...$,

where the coefficients c_n are undetermined. We now note that obviously our function satisfies the relation $(1 + x) = f'(x) = \alpha f(x) = \sum_{\nu=0}^{\infty} \alpha c_\nu x^\nu$

On the other hand, if we differentiate the series for $f(x)$ term by term and multiply by $(1+x)$, we obtain $(1 + x)\, f'(x) = c_1 + (2c_2 + c_1)\, x + (3c_3 + 2c_2)\, x^2 + ...$; since these two power series for $f'(x)$ must be identical, we have $\alpha c_0 = c_1$, $\alpha c_1 = 2c_2 + c_1$, $\alpha c_2 = 3c_3 + 2c_2$, ...

It is now certain that $c_0 = 1$, since our series must have the value 1 for $x = 0$; we thus find in succession the expressions

$$c_1 = \alpha,\ c_1 = \frac{(\alpha-1)\alpha}{2},\ c_1 = \frac{(\alpha-2)(\alpha-1)\alpha}{3\cdot 2},...,$$

for the coefficients, and, in general, as is readily established,

$$c_\nu = \frac{(\alpha-\nu-1)(\alpha-\nu+2)\ldots(\alpha-1)\alpha}{\nu(\nu-1)\ldots2.1} = \binom{\alpha}{\nu}$$

Substituting these values for the coefficients, we have the series $\sum_{\nu=0}^{\infty}\binom{\alpha}{\nu}x^\nu$;

we must yet investigate the convergence of this series and show that it actually represents $(1 + x)^n$.

By the ratio test, we find that, when α is not a positive integer, the series converges if $|x| < 1$ and diverges if $|x| > 1$; in fact, then the ratio of the $(n + 1)$-th term to the n-th term is $\{(\alpha-n+1)/n\}x$ and the absolute value of this expression tends to $|x|$ as n increases beyond all bounds.

We state here without proof the exact conditions under which this series converges. If the index a is an integer, the series terminates and is therefore valid for all values of x (as it becomes the ordinary binomial theorem). For all other values of α, the series is absolutely convergent for $|x|<1$ and diverges for $|x|>1$. For $x = +1$, the series converges absolutely if $\alpha > 0$, conditionally if $-1 < \alpha < 0$, and it diverges if $\alpha = -1$. Finally, at $x = -1$, the series is absolutely convergent if $a > 0$, divergent if $a < 0$.

Hence, if $|x| < 1$, our series represents a function $f(x)$ which satisfies the condition $(1 + x)f'(x) = af(x)$, as follows from the method of forming the coefficients. Moreover, $f(0) = 1$. But these two conditions ensure that the function $f(x)$ is identical to $(1 + x)^\alpha$. In fact, setting

$$\phi(x) = f(x)/(1 + x)^\alpha,$$

we find that

$$\phi(x) = \frac{(1+x)^\alpha f'(x)-\alpha(1+x)^{\alpha-1}f(x)}{(1+x)^{2\alpha}}$$

hence $f(x)$ is a constant and is always equal to 1, since $\phi(0) = 1$. We have thus proved that for $|x| < 1$

$$(1 + x)^\alpha = \sum_{\nu=0}^{\infty}\binom{\alpha}{\nu}x^\nu$$

which is the binomial series.

We present here the following special cases of the binomial series:

The geometric series: $\frac{1}{1+x} = (1+x)^{-1} = 1-x+x^2-x^3+x^4-+\ldots = \sum_{\nu=0}^{\infty}(-1)^\nu x^\nu$; the series $\frac{1}{(1+x)^2} = (1 + x)^{-2} = 1 - 2x + 3x^2 - 4x^3 + - \ldots = \sum_{\nu=0}^{\infty}(-1)^\nu(\nu+1)x^\nu$, which may otherwise be obtained from the geometric series by differentiation and the

$$\sqrt{(1+x)} = (1+x)^{\frac{1}{2}} = 1+\frac{1}{2}x-\frac{1}{2.4}x^2+\frac{1.3}{2.4.6}x^3-\frac{1.3.5}{2.4.6.8}x^4+-...,$$

$$\frac{1}{\sqrt{(1+x)}} = (1+x)^{-\frac{1}{2}} = 1+\frac{1}{2}x+\frac{1.3}{2.4}x^2-\frac{1.3.5}{2.4.6}x^3\frac{1.3.5.7}{2.4.6.8}x^4-+...,$$

series the first two or three terms of which are useful expressions.

The Series for arsin x: This series can be established readily by expansion of the expression $1/\sqrt{(1-t^2)}$ as a binomial series

$$\left(1-t^2\right)^{-\frac{1}{2}} = 1+\frac{1}{2}t^2+\frac{1.3}{2.4}t^4+...$$

This series converges if $|t| \rangle 1$, whence it converges uniformly for $|t| < q < 1$. On integrating term by term between 0 and x, we obtain

$$\text{arc sin}x = x+\frac{1}{2}\frac{x^3}{3}+\frac{1.3}{2.4}\frac{x^5}{5}+...;$$

by the ratio test, we find that this series converges if $|x| <1$ and diverges if $|x|>1$.

The derivation of this series by Taylor's theorem would be decidedly less convenient owing to the difficulty of estimating the remainder.

The Series for arsinh $x = \log \{x + (1 + x)\}$: We obtain this expansion by a similar method. By the binomial theorem, we write down the series for the derivative of arsinh x:

$$\frac{1}{\sqrt{1+x^2}} = 1-\frac{1}{2}x^2+\frac{1.3}{2.4}x^4-\frac{1.3.5}{2.4.6}x^6+-...,$$

and then integrate term by term to obtain the expansion

$$\text{arc sinh}x = x-\frac{1}{2}\frac{x^3}{3}+\frac{1.3}{2.4}\frac{x^5}{5}-+...;$$

with the interval of convergence $-1 < x < 1$.

MULTIPLICATION OF SERIES

The expansion of the function $\frac{\log(1+x)}{1+x}$ is a simple example of the application of the rule for the multiplication of power series. We have only to multiply the logarithmic series $\log(1 + x) = x - \frac{x^2}{2} + \frac{x^3}{3} + \frac{x^4}{4} + - ...$ by the geometric series $\frac{1}{1+x} = 1-x+x^2-x^3+x^4-+...$; the reader should verify that this yields the remarkable expansion

$$\frac{\log(1+x)}{1+x} = x-\left(1+\frac{1}{2}\right)x^2+\left(1+\frac{1}{2}+\frac{1}{3}\right)x^3-\left(1+\frac{1}{2}+\frac{1}{3}+\frac{1}{4}\right)x^4+-... \text{ for } |x| < 1.$$

Example of Term-by-Term Integration (Elliptic Integral):. In previous applications, we have encountered the elliptic integral

$$K = \int_0^{\pi/2} \frac{d\varphi}{\sqrt{\left(1-k^2\sin^2\varphi\right)}} (k^2 < 1)$$

(the period of oscillation of a pendulum . In order to evaluate the integral, we can first expand the integrand by the binomial theorem:

$$\frac{1}{\sqrt{\left(1-k^2\sin^2\varphi\right)}} = 1+\frac{1}{2}k^2\sin^2\varphi+\frac{1.3}{2.4}k^4\sin^4\varphi+\frac{1.3.5}{2.4.6}k^6\sin^6\varphi+...,$$

Since k } sin } φ is never larger than k} , this series converges uniformly for all values of φ and we may integrate term by term:

$$K = \int_0^{\pi/2} \frac{d\varphi}{\sqrt{1-k^2\sin^2\varphi}} = \int_0^{\pi/2} d\varphi + \frac{1}{2}k^2 \int_0^{\pi/2} \sin^2\varphi \, d\varphi + \frac{1.3}{2.4}k^4 \int_0^{\pi/2} \sin^4\varphi \, d\varphi + ...$$

Substitution of their values yields

$$K = \int_0^{\pi/2} \frac{d\varphi}{\sqrt{1-k^2\sin^2\varphi}} = \frac{\pi}{2}\left\{1+\left(\frac{1}{2}\right)^2 k^2 + \left(\frac{1.3}{2.4}\right)^2 k^4 + \left(\frac{1.3.5}{2.4.6}\right)k^6 + ...\right\}$$

POWER SERIES WITH COMPLEX TERMS

COMPLEX TERMS INTO POWER SERIES

The similarity between certain power series representing functions which are apparently unrelated led Euler to set up a purely formal link between them, found by giving complex values, in particular, pure imaginary values to the variable x. We shall first do this formally, unhindered by questions of rigour and investigate the results of the process.

The first striking relation of this sort is obtained, if we replace the quantity x in the series for e^x by a pure imaginary $i\phi$, where ϕ is a real number. If we recall the fundamental equation for the imaginary unit i, that is, $i = -1$, from which it follows that $i^3 = -i$, $i^4 = 1$, $i^5 = i$, ..., then, on separating the real and imaginary terms of the series, we obtain

$$e^{1\phi} = \left(1-\frac{\phi^2}{2!}+\frac{\phi^4}{4!}+\frac{\phi^6}{6!}+-...\right)+i\left(\phi-\frac{\phi^3}{3!}+\frac{\phi^5}{5!}-\frac{\phi^7}{7!}+-...\right),$$

or, in another form, $e^{1\phi} = \cos\phi + i\sin\phi$

This is the well-known and important Euler formula. which is as yet purely formal. It is consistent with De Moivre's theorem which is expressed by the equation $(\cos\phi + i\sin\phi)(\cos\psi + i\sin\psi) = (\cos\phi+\psi) + i\sin(\phi+\psi)$.

By virtue of Euler's formula, this equation merely states that the relation $e^x . e^v = e^{x+v}$ continues to hold for pure imaginary values $x = i\phi$, $y = i\psi$.

If we replace the variable x in the power series for $\cos x$ by the pure imaginary ix, we obtain at once the series for $\cosh x$; this relation can be expressed by the equation

$$\cosh x = \cos ix.$$

In the same way, we obtain

$$\sinh x = \frac{1}{i}\sinh ix.$$

Since Euler's formula also yields

$$e^{-if} = \cos\phi - i\sin\phi,$$

we arrive at the exponential expressions for the trigonometric functions

$$\sin x = \frac{e^{ix} - e^{-ix}}{2i}, \qquad \cos x = \frac{e^{ix} + e^{-ix}}{2}$$

These are exactly analogous to the exponential expressions for the hyperbolic functions and are, in fact, transformed into them by the relations

$$\cosh x = \cos ix, \qquad \sinh x = \frac{1}{i}\sin ix$$

Naturally, corresponding formal relations can also be obtained for the functions $\tan x$, $\tanh x$, $\cot x$ and $\coth x$, which are interrelated by the equations

$$\tanh x = \frac{1}{i}\tan ix \qquad \coth x = i\cot ix$$

Finally, similar relations can also be found for the inverse trigonometric and hyperbolic functions. For example, we immediately find from

$$y = \tan x = \frac{e^{ix} - e^{-ix}}{i\left(e^{ix} + e^{-ix}\right)} = \frac{e^{2ix} - 1}{i\left(e^{2ix} + 1\right)}$$

that

$$e^{2ix} = \frac{1+iy}{1-iy}$$

It we take the logarithms of both sides of this equation and then write x instead of y and artan x instead of x, we obtain the equation

$$\text{artan}\,x = \frac{1}{2i}\log\frac{1+ix}{1-ix},$$

which expresses a remarkable connection between the inverse tangent and the logarithm.

If we replace in the known power series for $\log(1+x)/(1-x)$ x by ix, we actually obtain the power series for artan x:

$$\text{artan} x = \frac{1}{i}\left(ix + \frac{(ix)^3}{3} + \frac{(ix)^5}{5} + \ldots\right) = x - \frac{x^3}{3} + \frac{x^5}{5} - + \ldots$$

The above relations have as yet a purely formal character and naturally demand a more exact statement as to the meaning which they are intended to convey. In the next sub-section, we shall indicate how this can be given with the help of function theory.

However, we shall later on only require Euler's formula $e^{i\phi} = \cos\phi + i\sin\phi$ This being so, we can avoid here a thorough analysis. We need only regard the symbol $e^{i\phi}$ as a formal abbreviation $\cos\phi + i\sin\phi$, for in which case De Moivre's formula $e^{i\phi} \cdot e^{i\psi} = e^{i(\phi+\psi)}$ appears merely as a consequence of the elementary addition theorem of trigonometry. From this formal point of view, in order to make the relation $e^x \cdot e^y = e^{x+y}$ e*. e* remain valid for any complex arguments, we set up the further definition

$$e^x = e^{\xi}(\cos\eta + i \text{ xin}\eta)$$

where $x = \xi + i\eta$ (ξ, η real).

GENERAL THEORY OF FUNCTIONS OF A COMPLEX VARIABLE

Although the purely formal point of view, indicated above, is by itself free from objection, it is still desirable to recognize in the above formula something more than a mere formal connection. In order to pursue this aim, we are led to the general theory of functions, as (for the sake of brevity) we call the general theory of the so-called analytic functions of a complex variable. In this, we may use as our starting point a general discussion of the theory of power series with complex variables and complex coefficients. The construction of such a theory of power series presents no difficulties once we have defined the concept of limit in the domain of complex numbers; in fact, it follows almost exactly the theory of real power series . But as we shall not make any use of these matters in what follows, we shall content ourselves here by stating certain facts and omit the proofs.

If a power series converges for any complex value $x = \xi$ whatever, then it converges absolutely for every value x for which $|x| < |\xi|$; if it diverges for a value $x = x$, then it diverges for every value x for which $|x| > |\xi|$. A power series, which does not converge everywhere, but does converge for some other point in addition to $x = 0$, has a circle of convergence, that is, there exists a number $\rho > 0$ such that the series converges absolutely for $|x| < \rho$and diverges for $|x| > \rho$.

Having established the concept of functions of a complex variable, represented by power series, and having developed the rules for operating with such functions, we can think of the functions e^x, sin x, cos x, artan x, etc., of the complex variable x as simply defined by the power series which represent them for real values of x. Then, all the above formal relationships reduce to trivialities. We shall merely indicate by two examples, how this introduction of complex variables helps us to understand the elementary functions.

The geometric series for l/(l + xýÿ) ceases to converge when x leaves the interval -1 } x } l and so does the series for artan x, although there are no peculiarities in the behaviour of these functions at the ends of the interval of convergence; in fact, they and all their derivatives are continuous for all real values of x.

On the other hand, we can readily understand that the series for l/(l – x) and log (1 – x) cease to converge as x passes through the value 1, since they become infinite there. But this divergence of the series for the inverse tangent and the series $\sum_{\nu=0}^{\infty}(-1)^{\nu} x^{2\nu}$ *for* $|x|>1$ becomes clear immediately, if we also consider complex values of x. In fact, we find that when $x = i$, the sum functions become infinite and so cannot be represented by a convergent series.

Hence, by our theorem about the circle of convergence, the series must diverge for all values of x such that $|x| > |i|$; in particular, for real values of x, the series diverge outside the interval -1 } x } 1. Another example is given by the function $f(x) = e^{-1/x2}$ for x ψ0, which, in spite of its apparently regular behaviour, cannot be expanded in a Taylor series.

As a matter of fact, this function ceases to be continuous, if we take pure imaginary values of $x = i\xi$ into account. The function then takes the form e^{1/ξ^2} and increases beyond all bounds as ξ } 0. It is therefore clear that no power series in x can represent this function for all complex values of x in a neighbourhood of the origin, no matter how small a neighbourhood we choose. These remarks on the theory of functions and power series of a complex variable must be sufficient for us here.

Chapter 8

Integration of Elementary Functions

In this chapter we have developed several methods for evaluating indefinite integrals, such as the Sum and Constant Rules, change of variables, integration by parts, and partial fractions. In the integration problems up to this point, the method to be used was usually given. But in a real life integration problem, one will have to decide which method to use on his own.

This section has two purposes. First, to review all the methods of integration. Second, to explain how one might decide which method to use for a given problem. Almost all the examples and problems in this book involve what are called elementary functions. A real function f(x) is called an elementary function if f(x) is given by a term T(x) which is built up from constants, sums, differences, products, quotients, powers, roots, exponential functions, logarithmic functions, and trigonometric functions and their inverses. These are the functions for which we have introduced names. Given an elementary function f(x), an indefinite integral $\int f(x)$ dx may or may not be an elementary function. For example, it turns out that the integrals

$$\int e^{-x^2} dx, \int \sqrt{1-x^4}\, dx$$

are not elementary functions. What is meant by the problem "evaluate the indefinite integral $\int f(x)dx$? The problem is really the following. Given an elementary function f(x), find another elementary function F(x) (if there is one) such that

$$\int f(x)dx = F(x) + C.$$

This is a hard problem. Sometimes the integral is not an elementary function at all. Sometimes the integral is an elementary function but it can be found only by guesswork. There is no routine way to evaluate an indefinite integral. However, one can often find clues which will cut down on the guesswork. We shall point out some of these clues here. The corresponding problem for differentiation is much easier. Given an elementary function f(x), the derivative f'(x) is always another elementary function. It can be found in a routine way using the rules for differentiation and the Chain Rule.

RULES FOR INTEGRATION

The fundamental theorem of and the definition of the indefinite integral reveal the possibility of writing down an integration formula corresponding to each differentiation formula. The following rules of integration are completely equivalent to the first three rules of differentiation.

Multiplication by a Constant

If c is a constant, then

$$\int cf(x)dx = c\int f(x)dx$$

Integration of a Sum

It is always true that

$$\int\{f(x)+g(x)\}dx = \int f(x)dx + \int g(x)dx$$

To the third rule of differentiation corresponds the rule for the integration of a product or, as it is usually called, the rule for integration by parts. On integration, the product rule yields

$$\int\{f(x)g(x)\}'dx = \int f(x)g'(x)dx + \int g(x)f'(x)dx$$

The indefinite integral on the left hand side is obviously $f'(x)g(x)$ (except possibly for an additive constant), whence we can write the rule for integration by parts in the form

$$\int f(x)g'(x)dx = f(x)g(x) - \int g(x)f'(x)dx$$

This last integration formula, the counterpart of the rule for the differentiation of a product, has been given here only for the sake of completeness.

INTEGRATION OF THE SIMPLEST FUNCTIONS

Corresponding to the differentiation formulae for special functions which we have found above, we now set down the equivalent integration formulae. The formula

$$\frac{d}{dx}x^n = nx^{n-1}$$

expressed as an integration formula, becomes

$$\int x^{n-1}dx = \frac{x^n}{n}, n \neq 0$$

In fact, this formula merely means that the derivative of the right hand side is equal to the integrand on the left hand side. If we replace n by $n+1$, we obtain the formula

$$\int x^{n-1}dx = \frac{x^n}{n}, n \neq 0$$

This formula holds for every integral index n (when $n < 0$, it holds of course only if $x \succ 0$) except when $n = -1$, when the denominator $n + 1$ would vanish. This exceptional case will be studied in detail.

The fundamental theorem of the integral calculus allows at once to use our integral formulae for the determination of areas, that is, the values of definite integrals. We immediately obtain

$$\int_a^b x^n dx = \frac{1}{n+1}\left(b^{n+1} - a^{n+1}\right), n \neq -1$$

where, if n is negative, we assume that a and b have the same sign, since otherwise the integrand would be discontinuous in the interval of integration.

There correspond to the differentiation formulae for $\sin x$, $\cos x$ the integration formulae:

$$\int \cos x dx = \sin x, \int \sin x dx = -\cos x,$$

$$\int \frac{1}{\cos^2 x} dx = \tan x, \int \frac{1}{\sin^2 x} dx = -\cot x$$

They yield by way of the fundamental rule of the value of the definite integral between any limits, an only restriction being that when the last two formulae are used, the interval of integration must not contain any point of discontinuity of the integrand. For example,

$$\int_a^b \cos x dx = \sin x\Big|_a^b = \sin b - \sin a$$

It hardly need be emphasized that, with the help of the first two rules of integration, we are now in a position to integrate any polynomial in x and, in fact, any linear combination with arbitrary constant coefficients of the functions integrated here.

However, note that, according to the fundamental theorem, rules of integration and differentiation must be equivalent to each other; it is therefore possible to prove the general integration rules of this section and then to read off the differentiation rules of the preceding section. The reader is well advised to carry out this suggestion.

THE CORRESPONDING INTEGRAL FORMULAE

Expressed in the language of the indefinite integral, the formulas which we have just derived become

$$\int \frac{1}{\sqrt{\left(1 - x^2\right)}} dx = ar\sin x, \int \frac{1}{\sqrt{\left(1 - x^2\right)}} dx = -ar\cos x$$

$$\int \frac{1}{1+x^2} dx = ar \tan x, \quad \int \frac{1}{1+x^2} dx = -ar \cot x$$

There arises no contradiction between the pair of formulae on the left and on the right hand sides, which express each indefinite integral in the form of two functions, which appear to be entirely different. We must remember that, in the case of the indefinite integral, an arbitrary additive constant remains at at our disposal. If we choose these constants so that they differ by $\pi/2$ and recall that $\pi/2$ – arcos x = arsin$x$ and likewise $\pi/2$ – arcot x = artan x, this formal disagreement is immediately removed. The indefiniteness simply depends on the fact that the indefinite integral is not a single definite function, but an entire family of functions which differ from each other by arbitrary additive constants. The equation for an indefinite integral does not specify the value, but only one value of it. As we have already remarked, it would be more correct to express this fact by always including the undetermined constant, thus not writing

$$\int f(x) dx = F(x)$$

but

$$\int f(x) dx = F(x) + c$$

However, for the sake of convenience, this more detailed form is usually avoided; the reader should therefore be all the more careful to bear in mind the indefiniteness which is always associated with the shorter form.

There follow immediately from the formulae for indefinite integration formulae for definite integration. In particular,

$$\int_a^b \frac{dx}{1+x^2} = ar \tan x\Big|_a^b = ar \tan b - ar \tan a$$

If we set $a = 0$, $b = 1$ and recall that tan 0 = 0 and tan $\pi/4 = 1$, we obtain the remarkable formula

$$\frac{\pi}{4} = \int_0^1 \frac{1}{1+x^2} dx$$

The number π, which originally arose from the study of the circle, is by this formula brought into a very simple relationship to the rational function $1/(1+x)$ and is expressed and defined by the area.

As a very simple example, we consider the function $y = x^\alpha$, where we $\alpha = p/q$, q being a positive integer and p a positive or negative integer, so that α is an arbitrary, positive or negative rational number. Let x be positive. By the chain rule with

$$y = \varphi^p, \varphi = x^{1/q}$$

we have the formula

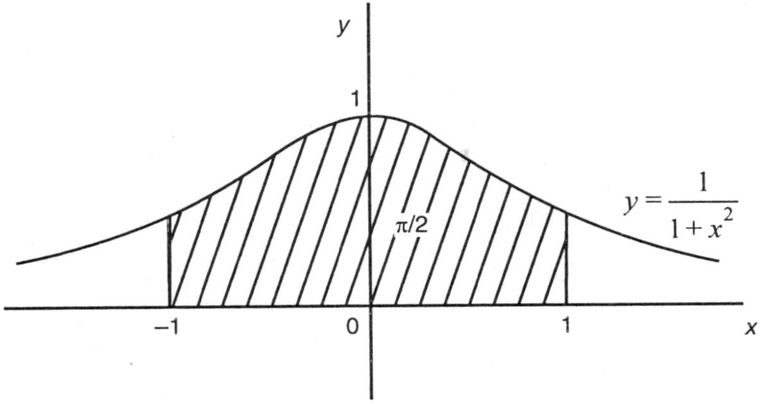

Fig. π/2 Illustrated by an Area

$$y' = p\varphi^{p-q} \cdot \frac{1}{q} x^{(1-q)/q} = \frac{p}{q} x^{p/q-1}$$

so that for arbitrary, rational values of α we obtain the differentiation formula

$$\frac{d}{dx} x^{\alpha} = \alpha x^{\alpha-1}$$

in agreement with the result a!ready found in another way. *As* a second example, we consider

$$y = \sqrt{\left(1-x^2\right)} \text{ or } y = \sqrt{\varphi}$$

where $\varphi = I - x$ and $-1 < x < 1$. The chain rule yields

$$y' = \frac{1}{2\sqrt{\varphi}} \cdot (-2x) = -\frac{x}{\sqrt{\left(1-x^2\right)}}$$

Further examples are given in brief calculations:

1. $y = \arcsin\sqrt{\left(1-x^2\right)}$

$$\frac{dy}{dx} = \frac{1}{\sqrt{\left\{1-\left(1-x^2\right)\right\}}} \cdot \frac{d\sqrt{\left(1-x^2\right)}}{dx}$$

$$= \frac{1}{|x|} \cdot \frac{-x}{\sqrt{\left(1-x^2\right)}} = \mp \frac{1}{\sqrt{\left(1-x^2\right)}}$$

2. $y=\sqrt{\left(\frac{1+x}{1-x}\right)}$

$$\frac{dy}{dx} = \frac{1}{\sqrt[2]{\left(\frac{1+x}{1-x}\right)}} \cdot \frac{a\left(\frac{1+x}{1-x}\right)}{dx}$$

$$= \frac{\sqrt{(1-x)}}{\sqrt[2]{(1+x)}} \cdot \frac{2}{(1-x)^2} = \frac{1}{(1+x)^{1/2}(1-x)^{3/2}}$$

The chain rule for differentiation can also be expressed in the form of an integration formula in agreement with the fact that there corresponds to each differentiation formula a completely equivalent integration formula.

INTEGRATION AND DIFFERENTIATION OF x^a, WHEN A IS IRRATIONAL

In view of the elementary definition of the power x^α by the equation

$$x^\alpha = \lim x^{r_n}$$

where the numbers r_n form a sequence of rational numbers with the limit α, we might be tempted to effect the differentiation of x^α by direct passage to the limit in the differentiation formula

$$\frac{d}{dx} x^r n = r_n x^{r_n - 1}$$

We are not entitled to do so unless we have the might conclude that there follows from the relation $x^{r_n} \to x^\alpha$ the relation $\frac{d}{dx}x^{r_n} \to \frac{d}{dx}x^\alpha$. However, there is a very serious objection to such a passage to the limit. In fact, in any arbitrarily small neighbourhood of a given curve one may draw other curves the direction of which at arbitrarily selected points differs from the direction of the original curve by any desired amount; for example, we may approximate to a straight line by a wave lying arbitrarily near to it, the angle between the wave and the line reaching a value as large as 45. In other words, the above example shows that from the fact that two functions differ only very little from each other, we cannot immediately conclude that their derivatives also are everywhere nearly equal to one another. This objection forbids us to perform the apparently obvious passage to the limit in the absence of further justification.

However, in this respect, the integral behaves quite differently from the derivative. We have already observed that, if two functions differ by less than ε throughout the interval from a to b, their integrals must differ by less than $\varepsilon(b\text{-}a)$. We used there this result to establish the validity of the differentiation formula

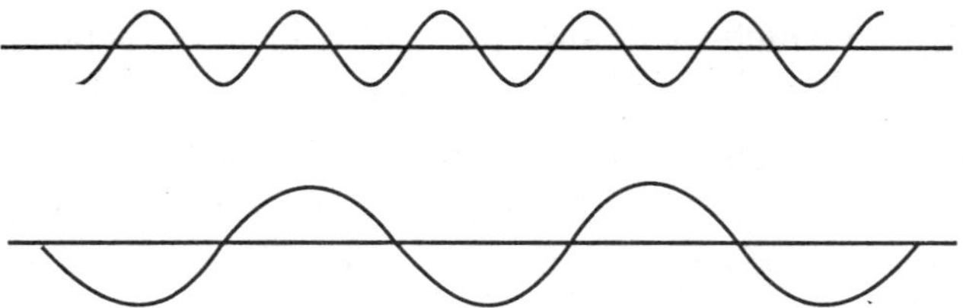

Fig. Approximation to a Straight line by Wavy Curves

$$\frac{1}{\alpha+1}\frac{d}{dx}x^{\alpha+1} = x^{\alpha}$$

or, replacing $\alpha+1$ by α,

$$\frac{d}{dx}x^{\alpha} = ax^{\alpha-1}.$$

Hence, in this indirect way, there is verified the relation $\frac{d}{dx}x^{r_n} \to \frac{d}{dx}x^{a}$ given above. This discussion is a characteristic example of the interrelation of the differential and integral calculus. Yet, in principle, it is preferable to replace the elementary definition of x^{α} by another, essentially simpler definition which will lead us once more to the same result and this time directly.

THE LOGARITHM AND THE EXPONENTIAL FUNCTION

The systematic relations between the differential and the integral calculus lead naturally to a convenient method of approach to the exponential function and the logarithm. Although we have already investigated these functions, we now define them afresh and redevelop their theory without making any use of our previous definition and the results based on it. We shall begin with the logarithm and then obtain the exponential function as its inverse.

Definition of the Logarithm

We have seen that, in general, indefinite integration of the power x^n for integral indices n leads to a power of x. The only exception is the function $1/x$, which does not appear as the derivative of any of the functions which we have dealt with so far. It is natural to assume that the indefinite integral of the function $1/x$ represents a new sort of function; thus, pursuing this idea, we will investigate the function

$$y = \int_1^x \frac{d\xi}{\xi} = f(x)$$

for $x > 0$. We call it the logarithm of x or, more accurately, the natural logarithm of x, and write it $y = \log_e x$ or $y = \ln x$. We have denoted the variable of integration by ξ, in order to avoid confusion with the upper limit x.

The choice of the number 1 as lower limit is arbitrary; however, it will soon prove its convenience. During the following argument, it will appear that the logarithm, thus defined, is the same as the logarithm which we defined previously in an elementary way. But, as we once more emphasize, the results of the following study are independent of those obtained earlier.

Geometrically, our logarithmic function means the area shown shaded Fig.; it is bounded above by the rectangular hyperbola $y = 1/\xi$, below *by* the ξ-axis, and on the sides by the lines $\xi = 1$ and $\xi = x$. This area is to be reckoned positive, if $x > 1$, negative if $x < 1$. For $x = 1$, the area vanishes, whence we have $\log 1 = 0$.

According to the above definition, the derivative of the logarithm is given by the formula

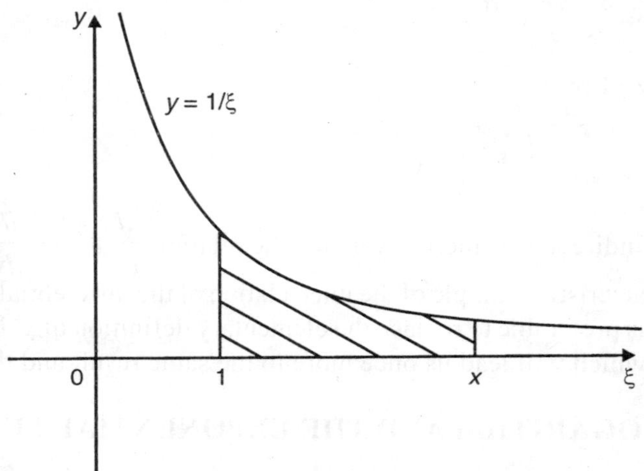

Fig. Log x Illustrated by an Area

$$\frac{d(\log x)}{dx} = \frac{1}{x}$$

We emphasize specially that we assume throughout that the argument x is positive; the logarithm of 0 or of any negative value cannot be formed in accordance with the formula above, because the integrand $1/\xi$ becomes infinite when $\xi = 0$. On the other hand, if we choose some negative number, say -1, as the lower limit, we can form the integral with a negative upper limit x, i.e. we can consider the expression

$$\int_{-1}^{x} \frac{d\xi}{\xi} \quad (x < 0)$$

Owing to the significance of the integral as the limit of a sum or as an area, we see that for $x < 0$

$$\int_{-1}^{x} \frac{d\xi}{\xi} = \int_{1}^{-x} \frac{d\xi}{\xi} = \int_{1}^{|x|} \frac{d\xi}{\xi} = \log|x|$$

Hence we can, in general, write the formula for indefinite integration as

$$\int \frac{dx}{x} = \log|x|$$

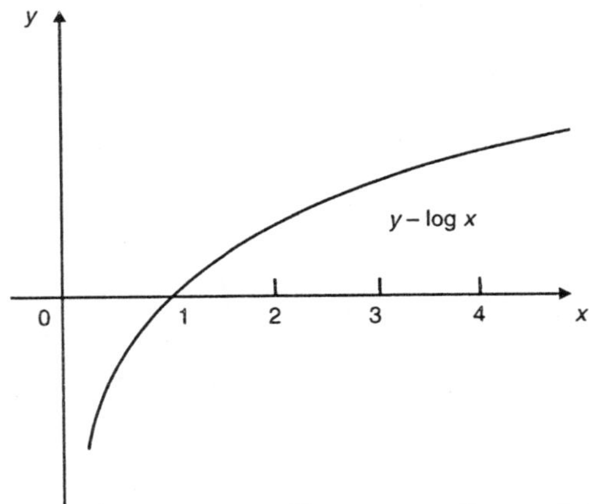

Naturally, the logarithm can be represented by means of a graph. This graph, the logarithmic curve.

The Addition Theorem

The logarithm defined as above obeys the fundamental law:

$$\log(ab) = \log a + \log b$$

The proof of this addition theorem follows directly from the differentiation formula. In fact, writing $z = \log(ax)$ and applying the chain rule, we obtain

$$\frac{dz}{dx} = \frac{1}{ax}.a = \frac{1}{x}$$

However,

$$\frac{d}{dx}\log x = \frac{1}{x}$$

and, since the functions z and $\log x$ have the same derivative, they differ only by a constant, so that $z = \log x + c$ or

$$\log ax = \log x + c$$

This being true for all positive values of x, we first put $x = 1$ to find c; since $\log 1 = 0$, this yields

$$\log a = c$$

Substituting this value for c, we have

$$\log ax = \log x + \log a$$

whence, for $x = b$,

$$\log ab = \log a + \log b$$

which was to be proved.

For arbitrary positive numbers $a_1, a_2, \ldots, a_n$, the equation

$$\log(a_1 a_2 \ldots a_n) = \log a_1 + \log a_2 + \ldots + \log a_n$$

follows from the addition theorem for the logarithm.

In particular, if all the numbers $a_1, a_2, \ldots, a_n$ are equal to one and the same number a, we have

$$\log a^n = n \log a$$

Similarly, it follows that

$$\log a + \log \frac{1}{a} = \log 1 = 0$$

so that

$$\log a = -\log \frac{1}{a}$$

Moreover, if we set $\sqrt[n]{a} = a$, it follows that $\log a = n \log \alpha$, or

$$\log \sqrt[n]{a} = \log a^{1/n} = \frac{1}{n} \log a$$

Hence, by repeated use of the addition theorem, we find that, when m is a positive integer,

$$\frac{m}{n} \log a = \log \sqrt[n]{a^m} = \log a^{m/n}$$

The equation

$$\log a^r = r \log a$$

is thus proved for all positive rational values of r, and it is obviously correct for $r = 0$. It is also valid for negative rational values of r, because then

$$\log a^r = \log \frac{1}{a^{-r}} = -\log a^{-r} = r \log a$$

Monotonic Character and Values of the Logarithm

Obviously, the value of the logarithm increases and decreases with x, whence the logarithm is a monotonic function. Since the derivative $1/x$ becomes smaller and smaller as x increases, the function increases more and more slowly as x increases.

Nevertheless, as x increases beyond all bounds, the function $\log x$ does not tend to a positive limit, but becomes infinite, that is, for every positive number A, no matter how large, there are values of x for which $\log x > A$.

This fact follows very readily from the addition theorem. In fact, $\log 2^n = n \log 2$, and since $\log 2$ is a positive number, we can, by taking $x = 2^n$ with sufficiently large values of n. make log

x as large as we please. Since log $(1/2^n)$ = – n log 2, we see that, as x tends to zero through positive values, log x is negative and increases numerically beyond all bounds.

In summary, we have: The function log x is a monotonic function which assumes all values between – and + as the independent variable x ranges over the continuum of positive numbers.

The Inverse Function of the Logarithm

Since the function $y = \log x$ $(x > 0)$ is a monotonic function of x which assumes all real values, its inverse function, which we shall at first denote by $x = E(y)$, must be a single-valued monotonic function defined for every real value of y; it is differentiable, since log x itself is differentiable. We interchange the notation for the dependent and independent variables and study the function $E(x)$ in detail. In the first place, obviously, it must be positive for every value of x. Moreover, we must have

$$E(0) = 1$$

In fact, this equation is equivalent to the statement that log 1 = 0.

From the addition theorem for the logarithm follows immediately the multiplication theorem

$$E(\alpha)E(\beta) = E(\alpha+\beta)$$

In order to prove this, we merely need note that the equations

$$E(\alpha)=a, E(\beta)=b, E(\alpha+\beta)=c$$

are equivalent to

$$E(\alpha)=a, E(\beta)=b, E(\alpha+\beta)=c$$

Since, by the addition theorem for the logarithm, $\alpha+\beta = \log ab$, it must be true that $c = ab$, which proves the multiplication theorem. We derive from this theorem a fundamental property of the function $y = E(x)$, which gives us the right to call our function the exponential function and to write it symbolically in the form

$$y = e^x$$

In order to obtain this property, we observe that there must be a number - which we shall call e - for which

$$\log e = 1$$

This is equivalent to the definition

$$E(1) = e$$

Using the multiplication theorem for the function $E(x)$, we have

$$E(n) = e^n$$

and, in the same way, for positive integers m and n,

$$E\left(\frac{m}{n}\right) = e^{m/n}$$

which we could also have found directly from the addition theorem for the logarithm.

The equation $E(r) = e^r$ thus proved for positive rational numbers r holds also for negative rational numbers by virtue of the equation

$$E(r)E(-r) = E(0) = 1$$

The function $E(x)$ is therefore a function which is continuous for all values of x and for rational values of x coincides with e^x. These facts give us the right to call our function e^x also for arbitrary irrational values of x. (It should be noted that here the continuity of e^x is an immediate consequence of its definition as the inverse function of a continuous, monotonic function, while, if the elementary definition is adopted, the continuity must be proved.)

The exponential function is differentiated according to the formula

$$\frac{d}{dx}e^x = e \text{ or } y' = y$$

This formula expresses the important fact that the derivative of the exponential function is the function itself. The proof is very simple. In fact, we have $x = \log y$, whence, by the formula for the differentiation of the logarithm, we have $dx/dy = 1/y$, and then, by the rule for inverse functions,

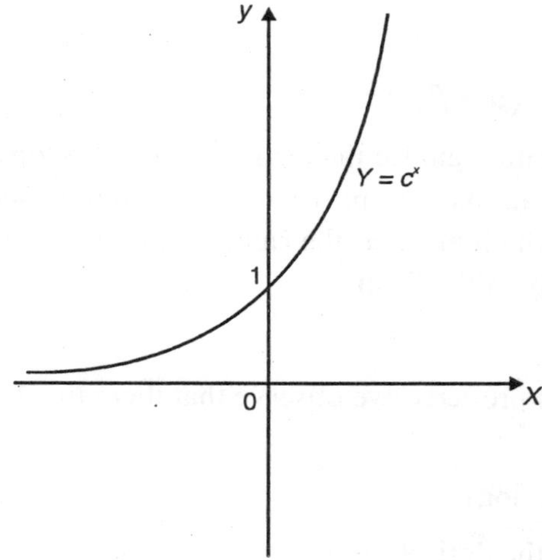

Fig. The Exponential Function

$$\frac{dy}{dx} = y = e^x$$

as stated.

The graph of the exponential function e^x - the so-called exponential curve - is obtained by reflection of the logarithmic curve in the line which bisects the first quadrant. The General Exponential Function a^x and the General Power x^α:

The exponential Function a^x for an arbitrary positive base a *is* now simply defined by the equation

$$y = a^x = e^{x\log a}$$

which agrees with the earlier definition by virtue of the relation

$$e^{\log a} = \text{a}$$

By the chain rule, we immediately obtain

$$\frac{d}{dx}a^x = \frac{d}{dx}e^{x\log} = e^{x\log a}\cdot\log a$$

$$= a^x \log \alpha$$

The inverse function of the exponential function y = a^x is called the logarithm to the base a and is written

$$x = \log_\alpha y$$

while the logarithmic function previously introduced, when it is required to distinguish it, is spoken of as the natural logarithm, or the logarithm to the base *e.*

It follows immediately from the definition that

$$\log y = x \log a = \log_\alpha yK.\log\alpha$$

which shows that the logarithm of y for an arbitrary positive base $a < 1$ is obtained by multiplying the natural logarithm of y by the reciprocal of the natural logarithm of a, the modulus of the system of logarithms to the base a.

If we take $a = 10$, we obtain the ordinary Briggian logarithms, which have already been encountered in elementary mathematics and which are advantageous for use in numerical computations. Instead of our previous definition of the general power $x^{\alpha}{}'$ $(x > 0)$, we shall now define this power by the equation

$$x^\alpha = e^{\alpha\log x}$$

The rule for differentiating the power x^α follows immediately from the definition, using the chain rule; for

$$\frac{d}{dx}x^\alpha = e^{\alpha\log x}, \frac{\alpha}{x} = ax^{\alpha-1}$$

in agreement with our previous result.

The Exponential Function and the Logarithm Represented as Limits

We are now in a position to state important limiting relations referring to the quantities introduced above. We begin with the formula for differentiating the function $f(x) = \log x$:

$$\frac{1}{x} = f'(x) = \lim_{h\to 0}\frac{f(x+h)-f(x)}{h} = \lim_{h\to 0}\frac{\log(x+h)-\log x}{h}$$

$$= \lim_{h\to 0}\frac{1}{h}\log\left(1+\frac{h}{x}\right)$$

If we set $1/x = z$, this becomes

$$\lim_{h\to 0}\frac{1}{h}\log(1+zh) = x$$

Since the function e^x is continuous for all values of x, this implies that

$$e^x = \lim_{h\to 0}{}^{e\{\log(1+zh)/h\}} = \lim_{h\to 0}{}^{(1+zh)^{1/h}}\dots$$

In particular, if we give h the sequence of values 1, 1/2, 1/3,..., $1/n$,..., we have

$$\lim_{h\to\infty}\left(1+\frac{z}{n}\right)^n = e^x\dots.$$

If we assign to z the value 1, Equation states the important fact:

As h tends to zero, the expression $(1 + h)^{1/h}$ tends to the number e:

$$\lim_{h\to 0}(1+h)^{1/h} = e$$

Equation yields

$$\lim_{h\to\infty}\left(1+\frac{1}{n}\right)^n = e$$

which proves that the number e is the same as the number denoted earlier by the symbol e.

It follows from the differentiation formula for a^x

$$a^a \log\alpha = \lim_{h\to 0}\frac{\alpha^{x+h}-\alpha^n}{h}$$

that for $x = 0$

$$\log\alpha = \lim_{h\to 0}\frac{\alpha^n - 1}{h}$$

which expresses the logarithm of a directly as a limit.

We append to this equation the remark that we can complete by it the earlier obtained relation

$$\int_a^b x^4 dx = \frac{1}{\alpha+1}\left(b^{\alpha+1}-\alpha^{\alpha+1}\right)$$

We have always been obliged to exclude the case $a = -1$. However, we can now discover what happens when the number a tends to the limit -1. If we put a=1, the left-hand side will, by our definition of the logarithm, have the limit

$$\int_1^b \frac{dx}{x} = \log b$$

whence the right-hand side has the same limit when $a < -1$. We have here carried out the passage to the limit -1 under the integral sign without further investigation; cf. the earlier discussion.

Moreover, this fact agrees with the formula

$$\log b = \lim_{h \to 0} \frac{b^n - 1}{h}$$

we only need write $a + 1 = h$.

We have thus cleared up the exceptional case $a = -1$ in the integration formula which we have so often used. The formula above is still meaninglesS when $a = -1$, but it retains as a limit formula its significance as $a < -1$.

We review here briefly the train of thought followed in this section, We have first defined the natural logarithm $y = \log x$ as for $x > 0$ by means of an integral, whence we immediately deduced the differentiation formula, the addition theorem, and the existence of an inverse. We then investigated the inverse function $y = e^x$, where the number e was seen to be the number the logarithm of which is 1, and derived its differentiation formula as well as limit expressions for it and for the logarithm. The introduction of the functions $y = x^\alpha = e^{\alpha \log x}$ and $y = a^x = e^{x \log a}$ followed naturally. In the discussion given here, in contrast to the elementary treatment, the question of continuity causes no difficulty, since the logarithm is defined as an integral and therefore as a continuous and differentiable function, the inverse function of which is also continuous.

APPLICATIONS OF THE EXPONENTIAL FUNCTION

We shall consider now some different problems involving the exponential function and thus gain insight into the fundamental importance of this function for all kinds of applications.

Definition of the Exponential Function by Means of a Differential Equation

We can define the exponential function by a simple theorem the use of which will save us many detailed investigations of particular cases:

If a function $y = f(x)$ satisfies an equation of the form

$$y' = \alpha y$$

where a is a constant other than zero, then y has the form

$$y = f(x) = ce^{ax}$$

where c if also a constant; conversely, every function of the form ce^{ax} satisfies the equation $y' = ay$.

The last equation is usually briefly referred to as a differential equation, since it expresses a relation between the function and its derivative. In order to make the theorem clear, we note first of all that in the simplest case $a = 1$ the above equation becomes $y' = y$. We know that $y = e^x$ satisfies this equation and it is clear that the same is also true of $y = ce^x$, where c is an arbitrary constant. Conversely, we can easily see that no other function satisfies the differential equation. For if y is such a function, we consider the function $u = ye^{-x}$. We must then have

$$u' = y'e^{-x} - ye^{-x} = e^{-x}\left(y' - y\right)$$

But the right-hand side vanishes, since we have assumed that $y' = y$, whence $u' = 0$, u is a constant c and $y = ce^x$, as we wished to prove. The case of any non-zero value of α can be treated in exactly the same manner as the special case $\alpha = 1$. If we introduce the function $u = ye^{-ax}$, we obtain the equation $u' = y'e^{-ax} — ae^{-ax}$. Hence we find from the assumed differential equation that $u' = 0$, so that $u = c$ and $y = ce^{ax}$. The converse is clear.

We will now apply this theorem to a number of examples and thereby make it more intelligible. Interest, Compounded Continuously. Radio-active Disintegration: A capital sum, or principal, which has its interest added to it at regular periods of time, increases by jumps at these interest periods in the following manner. If 100α is the rate of interest per cent and moreover the interest accrued is added to the principal at the end of each year, then after x years the accumulated amount of an original principal of 1 will be

$(1 + \alpha)^x$

However, if the principal had the interest added to it not at the end of each year, but at the end of each n-th part of a year, then after x years the principal would amount to

$$\left(1+\frac{\alpha}{n}\right)^{nx}$$

Taking $x = 1$, for the sake of simplicity, i.e, reckoning the interest at 100α per cent for one year, we find that, if the interest is computed in this latter way, the principal 1 amounts after one year to

$$\left(1+\frac{\alpha}{n}\right)^{n}$$

If we now let n increase beyond all bounds, i.e., if we let the interest be calculated at shorter and shorter intervals, the limiting case will signify in a sense that the interest is compounded continuously, at each instant; and we see that the total amount after one year will be e^{α} times the original principal. Similarly, if the interest is calculated in this manner, an original principal of 1 will have grown after x years to an amount $e^{\alpha x}$, where x may be any number, integral or otherwise.

We consider a quantity y, which increases (or decreases) with time. Let the rate at which this quantity changes be proportional to the total quantity. Then, if we take time as the independent variable x, we obtain a law of the form $y' = \alpha y$ for the rate of increase, where α, the factor of proportionality, is positive or negative, according to as the quantity is increasing or decreasing, The quantity y itself will be given by a formula

$$y = ce^{qx}$$

where the meaning of the constant c is immediately obvious, if we consider the instant $x =$ 0. At that instant, $e^{\alpha x} = 1$ and we find that $c = y_0$ is the. quantity at the beginning of the time under consideration, so that we may write

$$y = y_0^{ex}$$

A characteristic example of the use of these ideas is the case of radioactive disintegration. The rate at which the total quantity y of a radioactive substance is diminishing at any instant is proportional to the total quantity present at that instant; this is *a priori* plausible, as each portion of the substance decreases as rapidly as every other portion. Hence the quantity y of the substance, expressed as a function of the time, satisfies a relation of the form $y' = -ky$, where k is to be taken as positive, since we are dealing with a diminishing quantity. The quantity of substance is thus expressed as a function of the time by $y = 3.8y_0e^{-kx}$, where y_0 is the amount of the substance at the beginning of the time (time $x = 0$). After a certain time τ, the radioactive substance will be diminished to half its original quantity. This so-called half-period is given by the equation

$$\frac{1}{2}y_0 = y_0e^{-kr}$$

whence we immediately obtain $\tau = \dfrac{\log 2}{k}$

Making and Breaking of an Electric Circuit

As a final example, consider the growth of a (direct) electric current when a circuit is completed (or its decay, when it is broken). If R is the resistance of the circuit and I the impressed electro-motive force (Voltage), the current I will gradually increase from its original value 0 to the steady final value E/R. Hence we must consider I as a function of the time. The growth of the current depends on the self-induction of the circuit; the circuit has a characteristic constant L - the coefficient of self-induction - of such a nature that as the current increases, an electro-motive force of magnitude LdI/dx, opposed to the external electromotive force E, is developed. We obtain from Ohm's law, according to which the product of the resistance and the current is at each instant equal to the actual effective voltage, the relation

$$IR = E - L\frac{dI}{dx}$$

Hence, we put

$$f(x) = I(x) - \frac{E}{R}$$

we find immediately that $f'(x) = -R/Lf(x)$, so that, $f(x)=f(0)e^{-Rx/L}$. Recalling that $I(0) = 0$, we see that $f(0) = -E/R$ and thus obtain the expression for the current as a function of the time

$$I = f(x) + \frac{E}{R} = \frac{E}{R}\left(1 - e^{-Rx/L}\right)$$

We see from this expression that as the circuit is closed the current tends asymptotically to its steady value E/R.

Cooling or Heating of a Body by a Surrounding Medium

Another typical example of the occurrence of the exponential function is offered by the cooling of a body, e.g., a metal plate, which is immersed in a very large bath of a given temperature. In considering this cooling, we assume that the surrounding bath is so large that its temperature is unaffected by the cooling process.

Moreover, we assume that at each instant all parts of the immersed body are at the same temperature and that the rate at which the temperature changes is proportional to the difference between the temperature of the body and that of the surrounding medium (Newton's law of cooling).

If we denote the time by x and the temperature difference by $y = y(x)$, this law of cooling is expressed by the equation

$$y' = -ky$$

where k is a positive constant the value of which depends on the body itself. From this instantaneous relationship, which expresses the effect of the cooling process at a given instant, we now wish to derive an integral law which will allow us to find the temperature at an arbitrary time x from the temperature at an initial time $x = 0$. The theorem immediately gives us this integral law in the form

$$y = ce^{-kx}$$

where k is the above-mentioned constant depending on the body. This shows that the temperature decreases exponentially and tends to become equal to the external temperature. The rapidity with which this happens is expressed by the number k. As before, we find the meaning of the constant c by considering the instant $x = 0$; this yields $y_0 = c$, so that our law of cooling can finally be written is the form

$$y = y_0 e^{-kx}$$

Obviously, the same discussion will also apply to the heating of a body. The only difference is that the initial difference of the temperature y_0 is in this case negative instead of positive.

Variation of the Atmospheric Pressure with the Height above the Surface of the Earth

As another example of the occurrence of the exponential formula we shall derive the law according to which the atmospheric pressure varies with the height. We use here:

1 The physical fact that the atmospheric pressure is equal to the weight of the column of air vertically above a surface of area 1.

2. Boyle's law, according to which the pressure of the air (p) at a given constant temperature is proportional to the density of the air (σ). Boyle's law, expressed in symbols, is $p = \alpha\sigma$, where α is a constant which depends on a specific physical property of the air and, in addition, it is proportional to the absolute temperature—here, we are not concerned with this, as we shall assume that the temperature is constant. Our problem is to determine $p = f(h)$ as a function of the height (h) above the surface of the Earth.

If we denote by p_0 the atmospheric pressure at the surface of the Earth, i.e., the total weight of the air column supported by unit area, and by $\sigma(\lambda)$ the density of the air at the height λ above the Earth, the weight of the column up to the height h will he given by the integral $\int_0^h \sigma(\lambda)d\lambda$. Hence the pressure at height h will be

$$p = f(h) = p_0 - \int_0^h \sigma(\lambda)d\lambda$$

By differentiation, this yields the relationship between the pressure $p = f(h)$ and the density $\sigma(h)$:

$$\sigma(h) = -f'(h) = -p'$$

We now use Boyle's law to eliminate the quantity σ from this equation and obtain an equation

$$p' = -\frac{1}{a}p$$

which only involves the unknown pressure function.

$$p = f(h) = p_0 e^{-h/a}$$

If we denote, as above, the pressure at the Earth's surface, i.e., $f(0)$, by p_0, it follows immediately that $c = p_0$, and consequently

$$h = a\log\frac{p_0}{p}$$

These two formulae find frequent application. For example, if the constant a is known, they enable us to find the height of a place from the barometric pressure or to find the difference of the heights of two locations by measuring the local atmospheric pressures. Again, if the atmospheric pressure and the height h are known, we can determine the constant a, which is of great importance in gas theory.

Progress of a Chemical Solution

We will now consider an example from chemistry, namely the so-called molecular reaction. Let a substance be dissolved in a relatively large amount of solvent, say a quantity of cane sugar in water. If a chemical reaction takes place, the chemical law of mass action in this simple case states that the rate of reaction is proportional to the quantity of reacting substance present. If we assume that the cane-sugar is being transformed by catalytic action into invert sugar and if we denote by $u(x)$ the quantity of cane sugar which at time x is still unchanged, the velocity of reaction will be $-du/dx$, and, in accordance with the law of mass action, there holds an equation of the form

$$\frac{du}{dx} = -ku$$

where k is a constant which depends on the reacting substance. From this instantaneous law, we immediately obtain, an integral law, which gives us the amount of cane sugar as a function of the time

$$u(x) = ae^{-kx}$$

This formula show us clearly how the chemical reaction tends asymptotically to its final state $u = 0$, that is, the complete transformation of the reacting substance. The constant a is obviously the quantity present at $x = 0$.

THE HYPERBOLIC FUNCTIONS

ANALYTICAL DEFINITION

$$y = ce^{qx}$$

In many applications, the exponential function does not enter by itself, but in combinations of the form

$$\frac{1}{2}\left(e^{x} + e^{-x}\right) \text{ or } \frac{1}{2}\left(e^{x} - e^{-x}\right)$$

It is convenient to introduce these and similar combinations as special functions with the notation:

$$\sin x = \frac{e^{x} - e^{-x}}{2}, \cos x = \frac{e^{x} + e^{-x}}{2}$$

$$\tanh x = \frac{e^{x} - e^{-x}}{e^{x} + e^{-x}}, \coth x = \frac{e^{x} + e^{-x}}{e^{x} - e^{-x}}$$

and call them the hyperbolic sine, hyperbolic cosine, hyperbolic tangent, and hyperbolic cotangent, respectively. The functions sinh x, cosh x and tanh x are defined for all values of x, while in the case of coth x the point $x = 0$ must be excluded. This notation is designed to express a certain analogy with the trigonometric functions; it is this analogy, which we are about to study in detail. as it justifies special consideration of these new functions.

The graphs of the hyperbolic functions; the dotted lines in figure are the graphs of $y = e^x/2$ and $y = e^{-x}/2$, from which the graphs of sinh x and cosh x may readily be constructed. We see that cosh x is an even function, i.e., a function which remains unchanged when x is replaced by $-x$, while sinh x is an odd function, i.e., a function which changes sign when x is replaced.

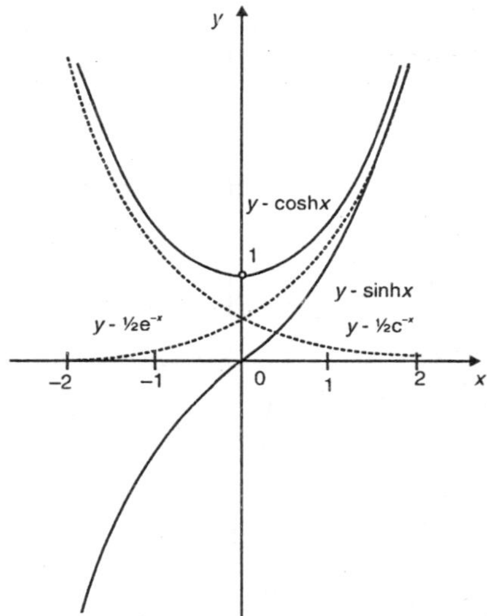

The function

$$\cosh x = \frac{e^x + e^{-x}}{2}$$

is (by its definition) positive for all values of x. It has its minimum when $x = 0$, while cosh $0 = 1$.

There exists between cosh x and sinh x the fundamental relation

$$\cosh^2 x - \sinh^2 x = 1$$

which follows immediately from the definitions of these functions. If we now denote the independent variable by t instead of x and write

$$x = \cosh t, \qquad y = \sinh t$$

we have

$$x^2 - y^2 = 1$$

thus, the point with the coordinate $x = \cosh t$, $y = \sinh t$ moves along the rectangular hyperbola $x - y = 1$ as t runs through the whole scale of values from –} to +}. According to the defining equation, x } 1 and we may easily convince ourselves that y•runs through the whole scale of values –} to +} with t; in fact, if t tends to infinity so does e^t, while e^{-r} tends to zero.

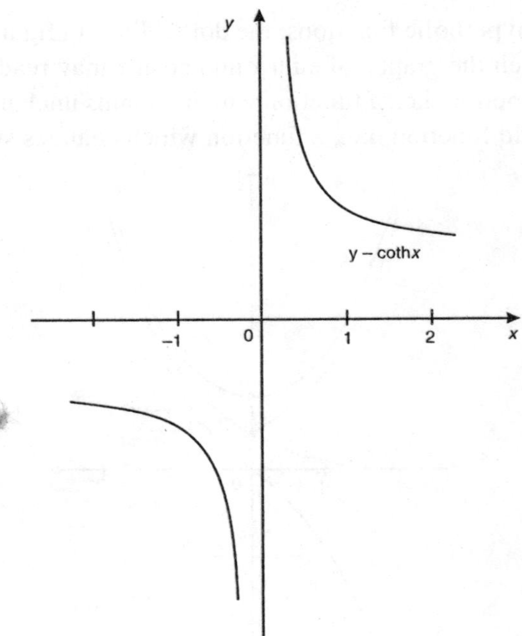

We may therefore state more exactly that as t runs from –} to +}, the equations $x = \cosh t$, $y = \sinh t$ give us one branch, namely, the right-hand one, of the rectangular hyperbola.

Addition Theorems and Formulae for Differentiation

From the definitions of our functions follow the formulae, known as addition theorems:

$$\cosh(a+b) = \cosh a \cosh b + \sinh a \sinh b,$$

$$\sinh(a+b) = \sinh a \cosh b + \cosh a \sinh b$$

The proofs are obtained at once by writing

$$\cosh(a+b) = \frac{e^{a}e^{b} + e^{-a}e^{-b}}{2}, \sinh(a+b) = \frac{e^{a}e^{b} - e^{-a}e^{-b}}{2}$$

and setting in these equations

$$e^{a} = \cosh a + \sinh a, \qquad e^{-a} = \cosh a - \sinh a,$$

$$e^{b} = \cosh b + \sinh b, \qquad e^{-b} = \cosh b - \sinh b$$

The analogy between these formulae and the corresponding trigonometrical formulae is clear. The only difference in the addition theorems is one sign in the first formula.

A corresponding analogy holds for the differentiation formulae. Remembering that $d(e^x)/dx = e^x$, we readily find

$$\frac{d}{dx}\cosh x = \sinh x, \frac{d}{dx}\sinh x = \cosh x$$

$$\frac{d}{dx}\tanh x = \frac{1}{\cosh^2 x}, \frac{d}{dx}\coth x = -\frac{1}{\sinh^2 x}$$

At times, it is convenient to introduce the functions sech $x = 1/\cosh x$, cosech $x = 1/\sinh x$.

The Inverse Hyperbolic Functions

To the hyperbolic functions $x = \cosh t$, $y = \sinh t$ correspond the inverse functions

$t = ar\cosh x, \qquad t = ar\sinh y$

The notation $\cosh^{-1}x$, etc. is also used.

Since the function sinh t is monotonic increasing throughout the interval $-1 < t < +1$, its inverse function is uniquely determined for all values of y; on the other hand, we learn from a glance at the graph that $t =$ ar cosh x is not uniquely determined, but has an ambiguity of sign, because there corresponds to a given value of x not only the number t, but also the number $-t$. Since cosh $t<1$ for all values of t, its inverse arcosh x is defined only for $x<1$.

We can express these inverse functions very easily in terms of the logarithm by regarding the quantity $e^t = u$ in the definitions

$$x = \frac{e^t + e^{-t}}{2}, y = \frac{e^t - e^{-t}}{2}$$

as unknowns and solving these (quadratic) equations for u. Then

$$u = x \pm \sqrt{(x^2 - 1)}, u = y + \sqrt{(y^2 + 1)};$$

since $u = e^t$ can have only positive values, the square root in the second equation must be taken with the positive sign, while in the first equation either sign is possible. In the logarithmic form

$$t = \log\left(x \pm \sqrt{(x^2 - 1)}\right) = ar\cosh x$$

$$t = \log\left(y + \sqrt{(y^2 + 1)}\right) = ar\sinh y$$

In the case of arcosh x, the variable x is restricted to the interval $x \geq 1$, while arsinh y is defined for all values of y.

The formula gives us two values, $\log\left\{x + \sqrt{(x^2 - 1)}\right\}$ and $\log\left\{x - \sqrt{(x^2 - 1)}\right\}$ for arcosh x, corresponding to the two branches of arcosh x. Since

$$\left\{x+\sqrt{\left(x^{2}-1\right)}\right\}\left\{x-\sqrt{\left(x^{2}-1\right)}\right\}=1$$

the sum of these two values of arcosh x is zero, which agrees with an earlier remark.

The inverses of the hyperbolic tangent and hyperbolic cotangent can be defined analogously and can also be expressed in terms of logarithms. We denote these functions by artanh x and arcotanh x; expressing the independent variable everywhere by x, we readily obtain

$$ar\tanh x = \frac{1}{2}\log\frac{1+x}{1-x} \text{ in the interval } -1 < x < 1$$

$$\text{arcoth} x = \frac{1}{2}\log\frac{x+1}{x-1} \text{ in the intervals } x < -1,\ x > 1$$

The differentiation of these inverse functions should be carried out by the reader himself; he may use either the rule for differentiating an inverse function or the chain rule in conjunction with the above expressions for the inverse functions in terms of logarithms. If x is the independent variable, the results are

$$\frac{d}{dx}ar\cosh x = \pm\frac{1}{\sqrt{\left(x^{2}-1\right)}},\ \frac{d}{dx}ar\sinh x = \frac{1}{\sqrt{\left(x^{2}+1\right)}}$$

$$\frac{d}{dx}ar\tanh x = \frac{1}{1-x^{2}},\ \frac{d}{dx}ar\coth x = \frac{1}{1-x^{2}}$$

The last two formulae do not contradict each other, since the first holds only for $-1 < x < 1$ and the second only for $x < -1$ and $1 < x$. The two values of

$d ar \cosh x/dx$, expressed by the sign in the first formula, correspond to the two different branches of the curve

$$y = ar\cosh x = \log\left\{x \pm \sqrt{\left(x^{2}-1\right)}\right\}$$

Analogies

In the above representation of the rectangular hyperbola by the quantity t, we did not attempt to bring out any geometrical meaning of the parameter t itself. We shall now return to this matter and thus gain still more insight into the analogy between the trigonometric and hyperbolic functions.

If we represent the circle by the equation $x + y = 1$ by means of a parameter t in the form $x = \cos t$, $y = \sin t$, we can interpret the quantity t as an angle or a length of arc measured along the circumference; however, we may also regard f as twice the area of the circular sector, corresponding to that angle, the area being reckoned positive or negative a cording *to as* the angle is positive or negative.

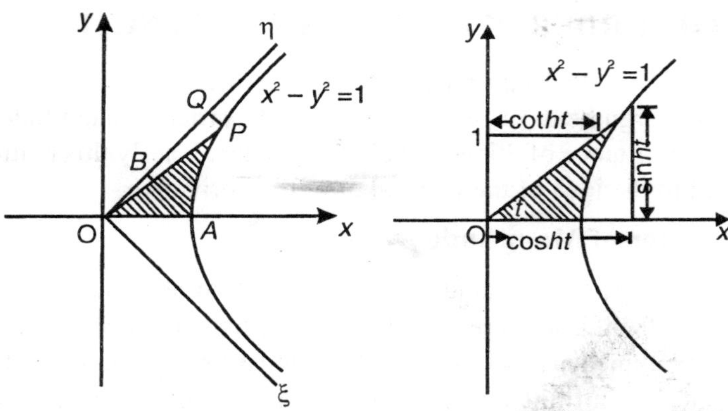

Fig. Parametric Representation of the Hyperbola and Hyperbolic Functions

We now make the analogous statement that for the hyperbolic functions the quantity t is twice the area of the hyperbolic sector.

Just as the notation $t = \text{arcos}\, x$ recalls that t is an arc of a circle of reference, so $t = \text{ar cosh}\, x$ recalls that t is a certain area connected with a rectangular hyperbola. The proof is obtained without difficulty, if we refer the hyperbola to its asymptotes as axes by means of the coordinate transformation

$$x - y = \sqrt{2}\xi, x + y = \sqrt{2}\eta$$

or

$$x = \frac{1}{\sqrt{2}}(\xi + \eta), y = \frac{1}{\sqrt{2}}(r_1 - \xi)$$

with these new coordinates, the equation of the hyperbola becomes $\xi\eta = 0$. We thus see immediately that the area in question is equal to the area of the figure *ABQP;* in fact, the two right-angled triangles *OPQ* and *OAB* have the same area, according to the equation of the hyperbola. ' The two points *A* and *P* obviously have the coordinates $\xi = \frac{1}{\sqrt{2}}, \eta = \frac{1}{\sqrt{2}}$

and $\xi = \frac{x-y}{\sqrt{2}}, \eta = \frac{x+y}{\sqrt{2}}$

respectively, and we thus obtain for double the area of our figure

$$2\int_{1/\sqrt{2}}^{(x+y)/\sqrt{2}} (1/2\eta)d\eta = \log(x+y) = \log\left\{x \pm \sqrt{(x^2-1)}\right\}$$

A comparison of this expression with the formula for the inverse function $t = \text{ar cosh}\, x$ shows us that our statement about the quantity t is true. In conclusion, it may be pointed out that above, the hyperbolic functions can be diagrammatically represented on the circle.

THE ORDER OF MAGNITUDE OF FUNCTIONS

The various functions which exhibit very important differences as regards their behaviour for large values of their arguments or, as we also say, of the order of magnitude of their increase. In view of the great importance of this, we shall discuss here briefly this matter, even though it is not directly linked to the idea of the integral or of the derivative.

The Concept of Order of Magnitude

The simplest Cases: If the variable x increases beyond all bounds, then, for $a > 0$, the functions a, $\log x$, e^x, e^{ax} will also increase beyond all bounds. However, as regards the manner of this increase, we can immediately point out an essential difference between them. For example, the function x will become infinite to a higher order than x; this means that, as x increases, the quotient x/x itself increases beyond all bounds. Similarly, we shall say that the function x^α becomes infinite to a higher order than x^β, if $\alpha > \beta > 0$, etc.

Quite generally, we shall say of two functions $f(x)$ and $g(x)$, the absolute values of which increase with x beyond all bounds, that $f(x)$ becomes infinite of a higher order than $g(x)$, if, as x increases, the quotient $|f(x)/g(x)|$ increases beyond all bounds; we shall say that $f(x)$ becomes infinite of a lower order than $g(x)$, if the quotient $|f(x)/g(x)|$ tends to zero as x increases; and we shall say that the two functions become infinite of the same order of magnitude if as x increases the quotient $|f(x)/g(x)|$ has a limit other than 0 or at least remains between two fixed positive bounds. For example, the function $ax + bx + c=f(x)$, where $a > 0$, will be of the same order of magnitude as the function $x=g(x)$; in fact, the quotient $\left|\frac{f(x)}{g(x)}\right| = \left|\frac{ax^3 + bx^2 + c}{x^3}\right|$ has the limit $|a|$.

On the other hand, the function $x + x + 1$ becomes infinite of a higher order of magnitude than the function $x + x + 1$. A sum of two functions $f(x)$ and $\phi(x)$, where $f(x)$ is of higher order of magnitude than $\phi(x)$, has the same order of magnitude as $f(x)$. In fact,

$$\left|\frac{f(x)+\phi(x)}{f(x)}\right| = \left|1+\frac{\phi(x)}{f(x)}\right|$$

and, by hypothesis, this expression tends to 1as x increases.

We might be tempted to measure the order of magnitude of functions by a scale, assigning to the quantity x the order of magnitude 1 and to the power x^α $(\alpha > 0)$ the order of magnitude α. Obviously, then a polynomial of the n-th degree has the order of magnitude n; a rational function, the degree of the numerator of which is greater than that of the denominator by h, has then the order of magnitude h.

The Order of Magnitude of the Exponential Function and the Logarithm

However, it turns out that any attempt to fix the order of magnitude of an arbitrary function by the above scale must end in failure. In fact, there are functions which become infinite of higher order than the power x^α of x, no matter however large is the chosen α; again, there are functions which become infinite of lower order than the power x^α, no matter how small the positive number α is chosen. Hence, these functions will not fit in anywhere into our scale. Without entering into a detailed theory of order of magnitude we shall prove the theorem: If a is an arbitrary number greater than 1, then the quotient a^x tends to infinity as x increases. In order to prove this statement, we construct the function

$$\phi(x) = \log\frac{a^x}{x} = x\log a - \log x$$

obviously, it is sufficient to show that this increases beyond all bounds if x tends to $+1$. For this purpose, we consider the derivative

$$\phi'(x) = \log a - \frac{1}{x}$$

and note that for $x \nsucc 0$ this is not less than the positive number $1/2 \log a$, whence, for $x< c$

$$\phi(x)-\phi(c) = \int_c^x \phi'(t)\,dt \geqq \int_c^x \frac{1}{2}\log a\,dt \geqq \frac{1}{2}(x-c)\log a$$

$$\phi(x) \geqq \phi(c) + \frac{1}{2}(x-c)\log a$$

and the right-hand side becomes infinite as x increases.

We shall give a second proof of this important theorem. If we write $\sqrt{a} = b = 1+h$, we have $b > 1$ and $h > 0$. Let n be an integer such that $n< \xi < n+1$; we may take $x > 1$, so that $n <$ 1. Applying the Lemma, we have

$$\sqrt{\left(\frac{a^x}{x}\right)} = \frac{b_x}{\sqrt{x}} = \frac{(1+h)^x}{\sqrt{x}} > \frac{(1+h)^n}{\sqrt{(n+1)}} > \frac{1+nh}{\sqrt{(n+1)}} > \frac{nh}{\sqrt{2n}} = \frac{h}{\sqrt{2}}\sqrt{n}$$

so that

$$\frac{a^x}{x} > \frac{h^2}{2}\cdot n$$

whence it tends to infinity with x.

There follow many results from the fact just proved. For example, for every positive index α and every number $a > 1$, the quotient a^x/x^α tends to infinity as x increases, that is: The exponential function becomes infinite of a higher order of magnitude than any power of x.

In order to see this, we need only show that the α-th root of the expression, that is,

$$\frac{a^{x/\alpha}}{x} = \frac{1}{\alpha}\cdot\frac{a^{x/a}}{x/a} = \frac{1}{a}\cdot\frac{a^x}{y}\left(y = \frac{x}{a}\right)$$

tends to infinity. However, this follows immediately from the preceding theorem, when x is replaced by $y = x/\alpha$.

We can prove in a similar fashion the theorem: For every positive value of α, the quotient $(\log x)/x^{\alpha}$ tends to zero when x tends to infinity, that is, the logarithm becomes infinite at a lower order of magnitude than any arbitrarily small positive power of x. The proof follows immediately, if we set $\log x = y$, so that our quotient is transformed into $y/e^{\alpha y}$. We then set $e^{\alpha} = a$; then a is a number > 1 and our quotient y/a^{y} approaches 0 as y increases. Since y approaches infinity with x, our theorem is proved.

Another very simple proof maybe suggested: For $x > 1$ and $\varepsilon > 0$,

$$\log x = \int_1^x \frac{d\xi}{\xi} < \int_1^x \xi^2 - 1 d\xi = \frac{1}{\in}\left(x^2 - 1\right)$$

if we choose ε smaller than α and divide both members of the inequality by x^{α}, then as x it follow that $(\log x)/x^{\alpha} < 0$.

Based on these results, we can construct functions of an order of magnitude by far higher than that of the exponential function and other functions e^{ex} of an order of magnitude far lower than that of the logarithm. For example, the function is of higher order than the exponential function, and the function $\log \log x$ of lower order than the logarithm; obviously, we can repeat these iteration processes as often as we like, piling up the symbols e or log to any extent we please.

These considerations show that it is not possible to assign by systematic reasoning to all functions definite numbers as orders of magnitude in such a way that, when two functions are compared, the function of the higher order of magnitude has the higher number. For example, if the function x is of the order of magnitude 1 and the function $x^{1+\varepsilon}$ of the order of magnitude $1 + \varepsilon$, then the function $x \log x$ must be of an order of magnitude which is greater than I and less than $1 + \varepsilon$ no matter how small ε is chosen. But there exists no such number. Moreover, it is readily seen that functions need not possess a clearly defined order of magnitude.

For example, the function $\dfrac{x^2(\sin x)^2 + x + 1}{x^2(\cos x)^2 + x}$

approaches no definite limits as x increase; on the contrary, for $x = n\pi$ (where n is an integer) its value is $1/n\pi$, while, for $x = (n + 1/2)\pi$, it is $(n + 1/2)\pi + 1 + 1/(n +)\pi$. Although both the numerator and denominator become infinite, the quotient neither remains between positive bounds, nor tends to zero, nor tends to infinity. Hence, the numerator is neither of the same order as the denominator, nor of lower order, nor of higher order. This apparently startling situation merely means that our definitions are not designed in such a way that we can compare any pair of functions. This is not a defect; we have no desire to compare the orders of such functions like the numerator and denominator above, since knowledge of the value of one of them gives us no useful information about the other one.

The Order of Magnitude of a Function Tending to Zero

Just as we seek to describe the approach of a function to infinity more definitely by means of the concept of order of magnitude, we may also specify the way in which a function approaches zero.

$$\lim_{x \to 0} \left(x^{-a} \cdot \log x \right) = 0$$

applies for every positive α.

In the same way, we say that, for $x = \xi$, the quantity $x - \xi$ vanishes to the first order, the quantity $|x-\xi|^{\alpha}$ to the order α. With the above results, it is easy to prove the relations

$$\lim_{x \to 0} \left(|x|^{\alpha} \cdot \log |x| \right) = 0$$

$$\lim_{x \to 0} \left(|x|^{-\alpha} \cdot e^{-1/|x|} \right) = 0$$

which are usually expressed as follows:

The function $1/\log |x|$ vanishes as $x < 0$ to a lower order than any power of x; the exponential function $e^{-1/|x|}$ vanishes to a higher order than any power x.

The Order of Magnitude of a Function in the Neighbourhood of an Arbitrary Point

Just as we can investigate the behaviour of a function when x increases without limit, we may also ask ourselves whether and how functions which become infinite at the point $x = \xi$ may be distinguished with regard to their behaviour at that point. Moreover, we state that the function $f(x) = 1/|x - \xi|$ becomes infinite at the first order at the point $x = \xi$, and correspondingly that the function $1/|x - \xi|^{\alpha}$ becomes infinite at the order α, provided that a is positive. We then recognize that the function $e^{1/|x - \xi|}$ becomes infinite at higher order, and the function $\log |x - \xi|$ infinite at lower order than all these powers, i.e., that there hold the limiting relations

$$\lim_{x \to \xi} \left(|x - \xi|^{\alpha} \cdot e^{1/|x-\xi|} \right) = \infty$$

and

$$\lim_{x \to \xi} \left(|x - \xi|^{\alpha} \cdot \log |x - \xi| \right) = 0$$

In order to see this, we merely set $1/|x - \xi| = y$; our statements then reduce to the known theorem, since

$$|x - \xi|^{a} \cdot e^{1/|x-\xi|} = e^{v} / y^{\alpha} \text{ and}$$

$$|x - \xi|^{\alpha} \cdot \log |x - \xi| = -(\log y) / y^{\alpha}$$

and y increases beyond all bounds as x tends to ξ The method of reduction of the behaviour at a finite point to the behaviour at infinity by the substitution $1/|x - \xi| = y$ frequently proves useful.

Chapter 9

Integrating Differential Equations Numerically

POINT-SLOPE METHODS - EULER'S METHOD

If you don't like Laplace transforms you might prefer using a numerical method of integrating the differential equations that you might find in pharmacokinetic studies. Actually a computer program such as Boomer will do this for you but doing this once or twice using a calculator might be informative. There are a number of different types of numerical integration methods but the easiest to understand is probably the point-slope methods. Euler's method is the simplest of these point-slope type of numerical integration. If we have an initial amount or concentration (the point) and a differential equation (the slope) Euler's method could be used. At time zero we have a Dose ($X_1(0)$) and a slope expressed as the differential equation for a one compartment linear model:

$$\frac{dX_1}{dt} = -\text{kel} \bullet X_1$$

This first step of Euler's method can be illustrated graphically

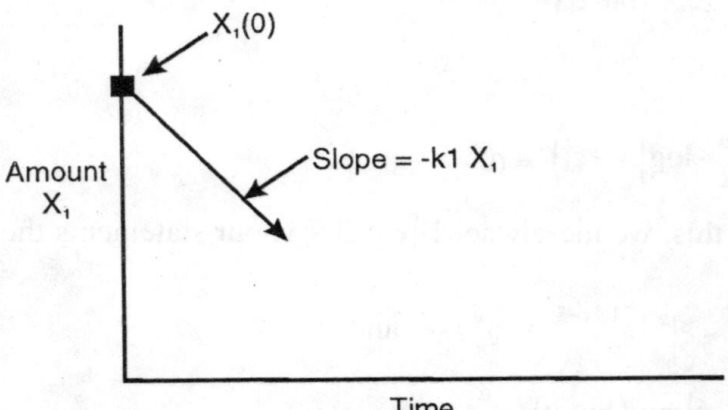

Fig. A first Step Using Euler's Method

In practice the method involves using this initial slope for a finite (small) stepsize (in the time direction). The method is then repeated for another stepsize using the result from the previous step as the new starting point. If we start with a dose of 100 mg, k1 of 0.25 hr^{-1} and a step size of 0.1 hour the amount of drug remaining is 97.5 mg.

$$
\begin{aligned}
X_1(new) &= X_1(\text{old}) + \text{slope} \bullet ss \\
&= X_1(\text{old}) + \left(\text{k1} \bullet X_1(\text{old})\right) \bullet ss \\
&= 100 + (-0.25' \ 100)' \ 0.1 \\
&= 100 - 2.5 \\
&= 97.5
\end{aligned}
$$

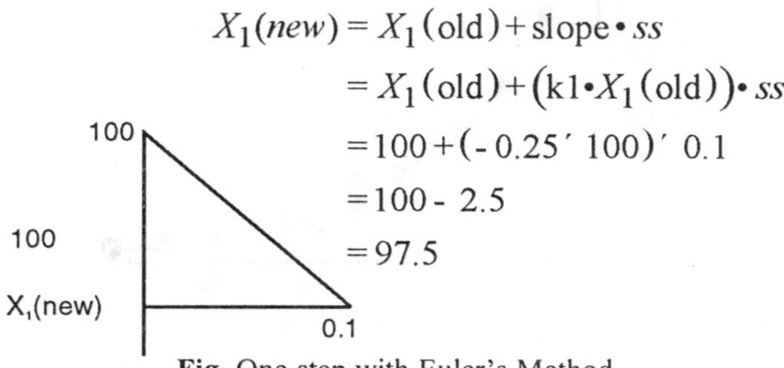

Fig. One step with Euler's Method

If this process is continued we can estimate the amount remaining at the time or times of interest.

AREA UNDER THE PLASMA CONCENTRATION TIME CURVE (AUC)

The area under the plasma (serum, or blood) concentration versus time curve (AUC) has an number of important uses in toxicology, biopharmaceutics and pharmacokinetics. ToxicologyAUC can be used as a measure of drug exposure.

It is derived from drug concentration and time so it gives a measure how much - how long a drug stays in a body.

A long, low concentration exposure may be as important as shorter but higher concentration. Some drugs are dosed using AUC to quantitate the maximum tolerated exposure (AUC Dosing).

BiopharmaceuticsThe AUC measured after administration of a drug product is an important parameter in the comparison of drug products. Studies can be performed whereby different drug products may be given to a panel of subject on separate equations. These bioequivalency or bioavailability studies can be analysed by comparing AUC values.

Pharmacokinetics Drug AUC values can be used to determine other pharmacokinetic parameters, such as clearance or bioavailability, F. Similar techniques can be used to calculate area under the first moment curve (AUMC) and thus mean resident times (MRT).

CALCULATION OF AUC USING THE TRAPEZOIDAL RULE

The area under the plasma concentration time curve (AUC) is very useful for calculating the relative efficiency of different drug products. It can used to calculate the total body clearance (CL) and the apparent volume of distribution.

If we have a smooth line for concentration versus time or an equation for Cp versus time from a pharmacokinetic model we could slice the area into vertical segments. Each segment would be very thin, Δt and in extreme dt, in width . The total AUC is calculated by adding these segments together. In calculus this would be the integral. Each very narrow segment has an area = Cp*dt. Thus the total area (AUC) is given by Equation:

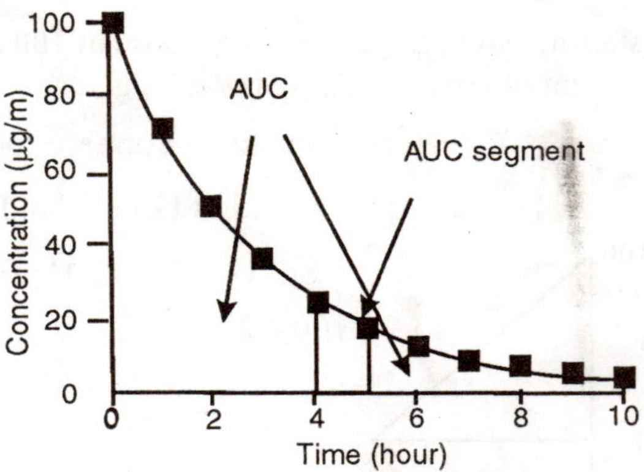

Fig. Linear Plot of Cp Versus Time Showing AUC and AUC Segment

$$AUC = \int_{t=0}^{t=\infty} Cp_t \bullet dt$$

Equation: Total AUC calculated from Very Narrow Segment

Moving ahead a little to Chapter 5 the integrated equation for plasma concentration as a function of time is:

$$Cp_t = Cp_0 \bullet e^{-kel \bullet t}$$

$$= \frac{Dose}{V} \bullet e^{-kel \bullet t}$$

Equation: Cp versus Time after an IV Bolus dose

This is essentially the same as the integrated equation we derived for amount of drug remaining using Laplace transforms.

The difference here is the use of V, apparent volume of distribution, to convert amount into concentration. We can substitute this equation for Cp_t into Equation to derive an analytical equation for AUC:

$$AUC = \int_{t=0}^{t=\infty} \frac{Dose}{V} \bullet e^{-\text{kel} \bullet t} \bullet dt$$

$$= \int_{t=0}^{t=\infty} e^{-\text{kel} \bullet t} \bullet dt$$

Equation: AUC calculated as the integral of Cp versus time

From Math Table we would find that:

$$\int_{t=t_1}^{t=t_2} e^{-a \bullet t} \bullet dt = \left[\frac{e^{-a \bullet t}}{-a} \right]_{t=t_1}^{t=t_2} = \frac{e^{-a \bullet t_2}}{-a} - \frac{e^{-a \bullet t_1}}{-a}$$

With a = kel, $t_1 = 0$, and $t_2 = \infty$,

At t = 0, $e^{-kel*t} = 1$ and at t = ", $e^{-kel*t} = 0$

Therefore:

$$AUC_{0-\infty} = \frac{Dose}{V} \bullet \left[\frac{e^{-\text{kel} \bullet \infty}}{-\text{kel}} - \frac{e^{-\text{kel} \bullet 0}}{-\text{kel}} \right] = \frac{Dose}{V} \bullet \left[0 - \frac{1}{-kel} \right]$$

or

$$AUC_{0-\infty} = \frac{Dose}{V \bullet \text{kel}} = \frac{Cp_0}{\text{kel}}$$

Equation. AUC Calculated from Concentration and kel

This is analytical integration (exact solution, given exact values for V and kel)

Note: For t = 0 to ∞, AUC = Cp_0/kel. With t = t to ∞, AUC is calculated as Cp_t/kel. We will use this result further down on this page

That is:

$$AUC_{0-\infty} = \frac{Dose}{V} \bullet \left[\frac{e^{-\text{kel} \bullet \infty}}{-\text{kel}} - \frac{e^{-\text{kel} \bullet 0}}{-\text{kel}} \right] = \frac{Dose}{V} \bullet \left[0 - \frac{e^{-kel \bullet t}}{-kel} \right]$$

Rearranging Equation to solve for V gives:

V = Dose/(AUC * kel)
Equation: Volume of Distribution Calculated from Dose, AUC and kel

We could use Equation to calculate the AUC value if we knew DOSE, kel, and V but usually we don't do this.

We can calculate AUC directly from the Cp versus time data. We need to use a different approach. The simplest, most common approach is a numerical approximation method called the trapezoidal rule.

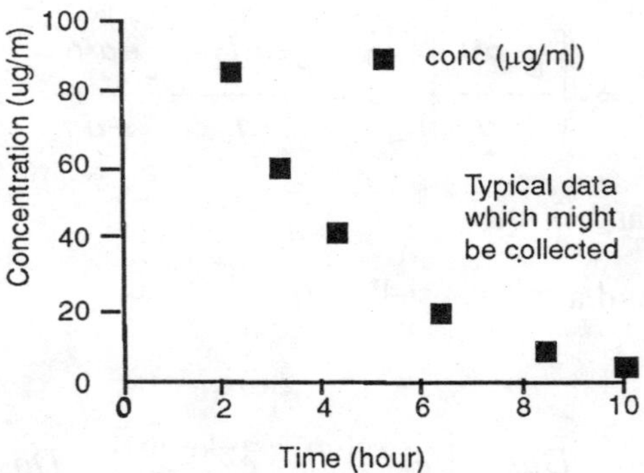

Fig. Linear Plot of Cp versus Time showing Typical Data Points

We can calculate the AUC of each segment if we consider the segments to be trapezoids. The area of each segment can be calculated by multiplying the average concentration by the segment width. For the segment from Cp_2 to Cp_3:

For example: $AUC_{2-3} = \dfrac{Cp_2 + Cp_3}{2} \bullet (t_3 - t_2)$

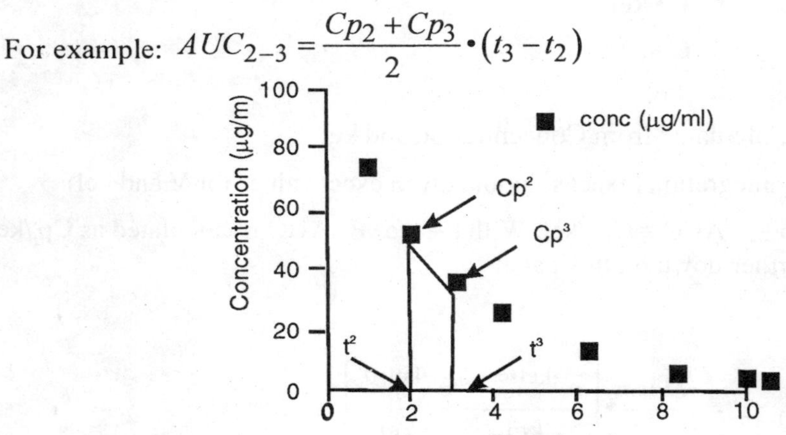

Fig. Linear Plot of Cp versus Time showing One Trapezoid

The area from the first to last data point can then be calculated by adding the areas together.

$$AUC_{1-n} = \sum\left\{\frac{Cp_1 + Cp_2}{2} \bullet (t_2 - t_1)\right\} + \left\{\frac{Cp_2 + Cp_3}{2} \bullet (t_3 - t_2)\right\} + \ldots$$

This gives:

To finish this calculation we have two more areas to consider. The first and the last segments.

After a rapid IV bolus, the first segment can be calculated after determining the zero plasma concentration Cp^0 by extrapolation.

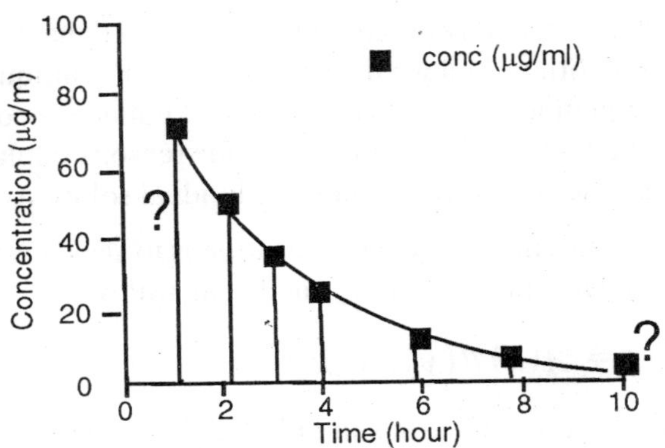

Fig. Linear Plot of Cp Versus Time Showing Areas from Data 1 to Data n

Thus

$$AUC_{0-1} = \left\{\frac{Cp_0 + Cp_1}{2} \bullet (t_1)\right\}$$

If we assume that the last data points follow a single exponential decline (a straight line on semi-log graph paper) the final segment can be calculated from the equation above from t_{last} to infinity:

$$AUC_{t_{last}-\infty} = \int_{t=\infty}^{t=\infty} Cp \bullet dt = \frac{Cp_{last}}{kel}$$

Thus the total AUC can be calculated as:

$$AUC_{0-\infty} = AUC_{0-1} + AUC_{1-last} + AUC_{last-\infty}$$

$$= \frac{Cp_0 + Cp_1}{2} \bullet t_1 + \frac{Cp_1 + Cp_2}{2} \bullet (t_2 - t_1)$$

$$+ \frac{Cp_2 + Cp_3}{2} \bullet (t_3 - t_2) + \ldots + \frac{Cp_{last}}{\text{kel}}$$

THE LAPLACE EQUATION AND HARMONIC FUNCTIONS

Given a scalar field φ, the Laplace equation in Cartesian coordinates is

$$\frac{\partial^2 \varphi}{\partial x^2} + \frac{\partial^2 \varphi}{\partial y^2} + \frac{\partial^2 \varphi}{\partial z^2} = 0$$

This equation arises in many important physical applications, such as potential fields in gravitation and electro-statics, velocity potential fields in fluid dynamics, etc. Typically we are given a set of boundary conditions and we need to solve for the (unique) scalar field φ that is a solution of the Laplace equation and that satisfies those boundary conditions. A solution of Laplace's equation is called a "harmonic function" (for reasons explained below). Since the Laplace equation is linear, the sum of two or more individual solutions is also a solution.

The usual approach to solving the Laplace equation is to seek a "separable" solution given by the product of independent function of *x*, *y*, and *z*, as follows

$$\varphi(x,y,z) = A(x)B(y)C(z)$$

Every solution can be expressed as a sum of (possibly infinitely many) solutions of this form. Inserting this into the Laplace equation and evaluating the derivatives gives

$$\frac{d^2A(x)}{dx^2}B(y)C(z)+A(x)\frac{d^2B(x)}{dy^2}C(z)+A(x)B(y)\frac{d^2C(z)}{dz^2}=0$$

Dividing through by the product $A(x)B(y)C(z)$, this can be written in the form

$$\frac{1}{A(x)}\frac{d^2A(x)}{dx^2}+\frac{1}{B(y)}\frac{d^2B(y)}{dy^2}+\frac{1}{C(x)}\frac{d^2C(z)}{dz^2}=0$$

Since *x*, *y*, and *z* can be varied independently, this equation can be identically satisfied only if each of the three terms is a constant, and these three constants sum to zero. It will be convenient to denote these constants by squared quantities, α^2, β^2, γ^2 respectively. Thus we have three separate ordinary differential equations

$$\frac{1}{A(x)}\frac{d^2A(x)}{dx^2} = \alpha^2$$

$$\frac{1}{B(y)}\frac{d^2B(y)}{dy^2} = \beta^2$$

$$\frac{1}{C(x)}\frac{d^2C(z)}{dz^2} = \gamma^2$$

with the condition $\alpha^2+\beta^2+\gamma^2=0$. Taking just the first of these, we can re-write it in the form

$$\frac{d^2A}{dx^2}-\alpha^2A(x) = 0$$

which the familiar equation of harmonic motion. This is the reason solutions of Laplace's equation are called "harmonic functions". The characteristic polynomial is

$$s^2 - \alpha^2 = 0$$

whose roots are $\pm\alpha$, so the general solution of the differential equation is

$$A(x) = a_1 e^{\alpha x} + a_2 e^{-\alpha x}$$

for constants a_1 and a_2. We arrive at similar solutions for $B(y)$ and $C(z)$, so the general solution of the Laplace equation of the form $A(x)B(y)C(z)$ is

$$\varphi(x,y,z) = \left(a_1 e^{\alpha x} + a_2 e^{-\alpha x}\right)\left(b_1 e^{\beta y} + b_2 e^{-\beta y}\right)\left(c_1 e^{\gamma z} + c_2 e^{-\gamma z}\right)$$

where the a_i and b_i are arbitrary constants, and α, β, γ are any complex constants such that $\alpha^2 + \beta^2 + \gamma^2 = 0$. By summing terms of this general form using the techniques of Fourier analysis we can determine the values of these constants necessary to match any given boundary conditions.

The above analysis is useful when the boundary conditions are easily expressible in terms of Cartesian coordinates, but we often encounter problems that are easier to express in terms of other, possibly curvilinear, coordinate systems. The general form of the Laplacian operator for an arbitrary (differentiable) coordinate system x^k, $k = 1,2,3$, with the metric tensor g_{ij} is

$$\nabla^2\varphi = \frac{1}{\sqrt{g}} \sum_{m=1}^{3} \sum_{m=1}^{3} \frac{\partial}{\partial x^m}\left(\sqrt{g}\, g^{mn} \frac{\partial\varphi}{\partial x^n}\right)$$

where g is the determinant of the covariant metric tensor. If the basis vectors of the coordinate system are mutually orthogonal at each point, the fundamental line element has the diagonal form

$$(ds)^2 = \left(h_1 dx^1\right)^2 + \left(h_2 dx^2\right)^2 + \left(h_3 dx^3\right)^2$$

where $h_i = \sqrt{g_{ii}}$. In this case we have $\sqrt{g} = h_1 h_2 h_3$ and $g_{ij} = 0$ for all $i \neq j$. Also, the non-zero components of the contravariant metric are $g^{ii} = 1/g_{ii}$, so the Laplacian expressed in terms of orthogonal coordinates has the form

$$\nabla^2\varphi = \frac{1}{h_1 h_2 h_3} \frac{\partial}{\partial x^1}\left[\left(\frac{h_2 h_3}{h_1}\frac{\partial\varphi}{\partial x^1}\right) + \frac{\partial}{\partial x^2}\left(\frac{h_1 h_3}{h_2}\frac{\partial\varphi}{\partial x^2}\right) + \frac{\partial}{\partial x^3}\left(\frac{h_1 h_3}{h_3}\frac{\partial\varphi}{\partial x^3}\right)\right]$$

For example, ordinary spherical coordinates r, θ, ϕ where

$$x = r\sin(\theta)\cos(\phi)$$

$$y = r\sin(\theta)\cos(\phi)$$

$$z = r\cos(\theta)$$

are orthogonal, i.e., the basis vectors in the r, θ and ϕ directions are mutually orthogonal at each point, so we can express the Laplacian in terms of spherical coordinates by means of the preceding expression. The line element for spherical coordinates is found by simply evaluating the derivatives ds/dr, etc., along the coordinate axes. This gives

$$(ds)^2 = (dr)^2 + r^2(d\theta)^2 + r^2\sin(\theta)^2(d\phi)^2$$

and we have $h_r = 1$, $h_\theta = r$, $h_\phi = r\sin(\theta)$. Inserting these into the general expression for the Laplacian in orthogonal coordinates gives

$$\nabla^2\varphi = \frac{1}{r^2}\frac{\partial}{\partial r}\left(r^2\frac{\partial\varphi}{\partial r}\right) + \frac{1}{r^2\sin(\theta)}\frac{\partial}{\partial\theta}\left(\sin(\theta)\frac{\partial\varphi}{\partial\theta}\right) + \frac{1}{r^2\sin(\theta)^2}\left(\frac{\partial^2\varphi}{\partial\phi}\right)$$

We are often interested in situations with axial symmetry, such as when determining the potential flow around a sphere. In such cases we can take the axis of symmetry over the angle ϕ, so we have $\partial\varphi/\partial\phi = 0$, and Laplace's equation reduces to

$$\frac{\partial}{\partial r}\left(r^2\frac{\partial\varphi}{\partial r}\right) + \frac{1}{\sin(\theta)}\frac{\partial}{\partial\theta}\left(\sin(\theta)\frac{\partial\varphi}{\partial\theta}\right) = 0$$

As in the case of Cartesian rectangular coordinates, we can focus on separable solutions, i.e., solutions of the form $\varphi(r,\theta) = A(r)B(\theta)$. Evaluating the Laplacian of this, equating this to zero, and dividing through by $A(r)B(\theta)$ gives

$$r^2\frac{\ddot{A}(r)}{A(r)} + 2r\frac{\dot{A}(r)}{A(r)} + \frac{\ddot{B}(\theta)}{B(\theta)} + \frac{\cos(\theta)}{\sin(\theta)}\frac{\dot{B}(\theta)}{B(\theta)} = 0$$

where dots signify derivatives of the function with respect to the argument. Since r and θ are independently variable, this equation can be identically satisfied only if the sum of the first two terms is constant, and the sum of the last two terms is constant. Denoting these constant by μ,ν, we have the two separate ordinary linear differential equations

$$r^2A''(r) + 2rA'(r) - \mu A(r) = 0$$

$$\sin(\theta)B''(\theta) + \cos(\theta)B'(\theta) - \nu\sin(\theta)B(\theta) = 0$$

with the condition $\mu + \nu = 0$. We could solve these explicitly, but notice that the first is satisfied by setting $A(r) = r^n$, provided we set $\mu = n(n+1)$. Also, if we set $B(\theta) = \cos(\theta)$ the second equation is satisfied with $\nu = 0$. Hence the function $\varphi(r,\theta) = r^n \cos(\theta)$ is a solution of Laplace's equation provided $n(n+1) = 0$, which implies either n = 0 or n = -1. More generally, we can consider solutions that are linear combinations of functions of the form

$$\varphi(r,\theta) = r^n p_n(\cos(\theta))$$

where $P_n(x)$ is a polynomial. Inserting this into the axially symmetric form of Laplace's equation gives

$$n(n+1)r^n P_n(\cos(\theta)) + r^n\left(\sin(\theta)^2 \ddot{P}_n(\cos(\theta))\right) = 0$$

where dotted functions signify derivatives of $P_n(x)$ with respect to x. Dividing through by r^n, replacing $\cos(\theta)$ with "x", and noting that $\sin(\theta)^2 = 1-\cos(\theta)^2$, this becomes

$$\left(1-x^2\right)\ddot{P}_n(x) - 2x\dot{P}_n(x) + n(n+1)P_n(x) = 0$$

There is a unique (up to a constant factor) polynomial P_n for each value of $n(n+1) = k$, which implies that each such polynomial is compatible with two distinct values of n, i.e., the two roots n and n' of $n^2 + n = k$. Of course, n need not be an integer, but if it is, we can easily determine the polynomial $P_n(x)$ that satisfies the above equation, simply by inserting a polynomial with undetermined coefficients into the equation and solving for the coefficients. The results for the first few values of n are as shown below:

n	n'	$p(x)$
0	–1	1
1	–2	x
2	–3	$3x^2–1$
3	–4	$5x^3–3$
4	–5	$35x^4–30x^2+3$

Up to a scale factor, these are the Legendre polynomials. We can express an axially symmetrical solution of the Laplace equation as a linear combination of these individual solutions as follows

$$\varphi(r,\theta) = \left(a_0 + b_0\frac{1}{r}\right) + \left(a_1 + b_1\frac{1}{r^2}\right)\cos(\theta)$$

$$\left(a_2 r^2 + b_2\frac{1}{r^3}\right)\left(3\cos(\theta)^2 - 1\right) + \ldots$$

where the a_i and b_i are arbitrary constants. The factors of the form $P_n(\cos(\theta))$ in this expansion are called zonal harmonics. As an example, the velocity potential of the flow of an ideal, irrotational, frictionless fluid past a stationary sphere of radius R is given by the second zonal harmonic with $a_1 = U$ and $b_1 = UR^3/2$, where U is the free-stream velocity of the fluid moving in the positive x direction (assuming the x axis is our axis of symmetry). Thus we have

$$\varphi(r,\theta) = Ur\cos(\theta) + \frac{UR^3}{2r^2}\cos(\theta)$$

The first term is obviously just Ux in Cartesian coordinates, which represents a uniform flow field of constant velocity U in the positive x direction. The second term is a dipole potential, proportional to x/r^3. This is the form of the "far field" potential associated with two nearby oppositely charged particles in electrostatics (aligned along the axis of symmetry). It's interesting that the superposition of these two simple fields gives the potential flow field around a sphere, as shown by the fact that the radial velocity vanishes at $r = R$.

One of the interesting properties of harmonic functions, i.e., continuous functions $\varphi(x,y,z)$ that satisfy Laplace's equation, is that the mean value of φ on *any* sphere centered on the point p is equal to the value of φ at the point p. This is called the *mean value theorem*, and a formal proof of it is presented in Differential Operators and the Divergence Theorem. This theorem is often invoked to prove that a continuous differentiable harmonic function cannot contain a local maximum or minimum, because by definition such points have values of φ properly greater then (resp. less than) the values of φ for all the surrounding points, which would make it impossible for the average on the surface of a sphere surrounding the point to equal the value of φ at that point.

In fact, by the same reasoning, if the value of a harmonic function at any given point is increasing in one direction then it must be decreasing in some other direction. (Of course, we don't actually need the mean value theorem to prove these facts; we can see at a glance that the Laplace equation automatically precludes local maxima and minima, because at any such point all three of the second partial derivatives with respect to x, y, and z would have the same sign, making it impossible for their sum to be zero.)

From this it follows that if the value of a harmonic function φ has the constant value c over an entire surface that completely encloses a region of space, then the value of φ must be c throughout that region. This is because if the value was anything other than c at any point in the interior of the region, then the region must contain a properly highest (or lowest) value, so there must be a point from which the function is decreasing (or increasing) in some directions but not increasing (or decreasing) in any directions, which is impossible.

Having established this, we can also see that the solution of Laplace's equation satisfying a complete set of boundary conditions is unique. To prove this, suppose two distinct harmonic functions φ_1 and φ_2 have the same values on a closed surface, but have different values in the interior of the enclosed region. We know that harmonic functions are additive, so $\varphi_1 - \varphi_2$ is also a harmonic function, and it's value is zero over the closed surface. But this implies the value of this difference is zero throughout the interior region enclosed by the surface (by the preceding proposition), so φ_1 and φ_2 are identical throughout the region.

It's also interesting to consider the mean value theorem in the context of the general power series of a harmonic function. If we expand $\varphi(x,y,z)$ into a power series about the point p (after shifting the coordinates to place p at the origin), we get an expression of the form

$$\varphi(x,y,z) = c_{000} + c_{100}x + x_{010}y + c_{001}z + c_{110}xy$$

$$+x_{101}xz + c_{011}yz + c_{200}x^2 + c_{020}y^2 + c_{002}z^2 + ..$$

$$= \sum c_{\alpha\beta\gamma}x^{\alpha}y^{B}z^{\gamma}$$

Recall that if θ denotes the "latitude" (zero at the equator) and ϕ the "longitude", then the mean value of a function f(θ,ϕ) over the surface of a sphere of radius R is given by the double integral

$$f_{mean} = \frac{1}{4\pi}\int_{-\pi}^{\pi}\int_{-\pi/2}^{\pi/2} f(\theta,\varphi)\cos(\theta)\,d\theta d\varphi$$

Making the substitutions $x = R\cos(\phi)\cos(\theta)$, $y = R\sin(\phi)\cos(\theta)$, $z = R\sin(\theta)$ into the expression for φ and evaluating the mean value over this sphere gives

$$\varphi_{mean} = c_{000} + \frac{1}{3}(c_{200} + c_{020} + c_{002})R^2$$

$$+\frac{1}{5}\left(c_{400} + c_{040} + c_{004} + \frac{1}{3}(c_{220} + c_{202} + 022)\right)R^4$$

$$+\frac{1}{7}\left(c_{600} + c_{060} + c_{006} + \frac{1}{5}\left((c_{420} + c_{402} + c_{042} + c_{240} + c_{024} + c_{204}) + \frac{1}{3}(c_{222})\right)\right)$$

$$R^6 + ...$$

In general we find that only even indices appear, and the mean value can be written as

$$\varphi_{mean} = \sum_{\alpha,\beta,\gamma\geq 0} \frac{(\alpha+\beta+\gamma)!}{\alpha!\beta!\gamma!}\frac{(2\alpha)!+(2\beta)!+(2\gamma)!}{((\alpha+\beta+\gamma)+1)!}c_{2\alpha,2\beta,2\gamma}R^{2(\alpha+\beta+\gamma)}$$

The constant coefficient in this expression is c_{000}, which equals the value of φ at the central point p, corresponding to $R = 0$. For all other values of R, the mean value theorem asserts that the mean value remains constant, so the coefficients of all non-zero powers of R must vanish. To show that they do, we can substitute the original power series for φ into the Laplace equation and collect terms by powers of x,y,z. We find the coefficient of $x^q y^r z^s$ is

$$C_{qrs} = (q+2)(q+1)c_{q+2,r,s} + (r+2)(r+1)c_{q,r+2,s} + (s+2)(s+1)c_{q,r,s+2}$$

Each of these coefficients must vanish. With $q = r = s = 0$ this gives

$$2(c_{200} + c_{020} + c_{002}) = 0$$

which proves that the coefficient of R^2 in the series for φ_{mean} is zero. Likewise for the (q,r,s) combinations (2,0,0), (0,2,0), and (0,0,2) we have the three conditions

$$12c_{400} + 2c_{220} + 2c_{202} = 0$$
$$12c_{040} + 2c_{220} + 2c_{022} = 0$$
$$12c_{004} + 2c_{022} + 2c_{202} = 0$$

Summing these three conditions and dividing by 12 gives

$$c_{400} + c_{040} + c_{004} + \frac{1}{3}(c_{220} + c_{202} + c_{022}) = 0$$

which proves that the coefficient of R^4 in the double integral for φ_{mean} is zero. Proceeding on to the next term, with the (q,r,s) combinations (4,0,0), (0,4,0), (0,0,4), (2,2,0), (2,0,2), and (0,2,2) we get the six conditions

$$30c_{600} + 2c_{420} + 2c_{402} = 0$$
$$30c_{060} + 2c_{240} + 2c_{042} = 0$$
$$30c_{006} + 2c_{024} + 2c_{204} = 0$$
$$12c_{420} + 12c_{240} + 2c_{222} = 0$$
$$12c_{402} + 12c_{204} + 2c_{222} = 0$$
$$12c_{042} + 12c_{024} + 2c_{222} = 0$$

Multiplying the first three by 3 and then adding all these expressions together and dividing by 90 gives

$$\left(c_{600} + c_{060} + c_{006} + \frac{1}{5}\left((c_{420} + c_{402} + c_{042} + c_{240} + c_{024} + c_{204}) + \frac{1}{3}(c_{222})\right)\right) = 0$$

which proves that the coefficient of R^6 in the double integral for φ_{mean} is zero. This process can be continued indefinitely, corresponding to the fact that the coefficient of each positive power of R is zero, and hence $\varphi_{mean} = c_{000}$ for all R.

FRACTIONAL CALCULUS

Differentiation and integration are usually regarded as discrete operations, in the sense that we differentiate or integrate a function once, twice, or any whole number of times. However, in some circumstances it's useful to evaluate a *fractional derivative*. In a letter to L'Hospital in 1695, Leibniz raised the possibility of generalizing the operation of differentiation to non-integer orders, and L'Hospital asked what would be the result of half-differentiating x. Leibniz replied "It leads to a paradox, from which one day useful consequences will be drawn". The paradoxical aspects are due to the fact that there are several different ways of generalizing the differentiation operator to non-integer powers, leading to inequivalent results.

In some ways the most natural and appealing generalization is based on the exponential function $f(x) = e^{ax}$, whose nth derivative is simply $a^n e^{ax}$. This immediately suggests defining the derivative of order ν (not necessarily an integer) as

$$\frac{d^{\nu}}{dx^{\nu}} e^{ax} = a^{\nu} e^{ax}$$

Negative values of ν represent integrations, and we can even extend this to allow complex values of ν. Any function expressible as a sum of exponential functions can then be differentiated in the same way. For example, the generalized derivative of the cosine function according to this approach is given by

$$\frac{d^{\nu}}{dx^{\nu}} \cos(x) = \frac{d^{\nu}}{dx^{\nu}} \left[\frac{e^{ix} + e^{-ix}}{2} \right] = \frac{(i)^{\nu} e^{ix} + (-i)^{\nu} e^{-ix}}{2}$$

Since $(\pm i)^{\nu} = e^{\pm \nu\pi/2}$, we have the nice result

$$\frac{d^{\nu}}{dx^{\nu}} \cos(x) = \cos\left(x + \nu \frac{\pi}{2} \right)$$

Thus the generalized differential operator simply shifts the phase of the cosine function (and likewise the sine function) in proportion to the order of the differentiation. Needless to say, this approach can be applied to the exponential Fourier representation of an arbitrary function to define the generalized derivative and integral. Recall that the exponential Fourier representation of a function $f(x)$ is defined as

$$f(x) = \frac{1}{\sqrt{2\pi}} \int_{-\infty}^{\infty} g(\alpha) e^{-i\alpha x} d\alpha$$

where

$$g(\alpha) = \frac{1}{\sqrt{2\pi}} \int_{-\infty}^{\infty} f(x) e^{-i\alpha x} dx$$

The functions f and g are Fourier transforms of each other. Given the representation of $f(x)$ (meaning that we are given the "coefficient function" $g(\alpha)$), the generalized derivatives (and integrals) are simply

$$\frac{d^{\nu}}{dx^{\nu}} f(x) = \frac{1}{\sqrt{2\pi}} \int_{-\infty}^{\infty} g(\alpha)(-i\alpha)^{\nu} e^{-i\alpha x} d\alpha$$

Hence if $g(\alpha)$ is the Fourier transform of $f(x)$, then the Fourier transform of the generalized νth derivative of $f(x)$ is $(-i\alpha)^{\nu} g(\alpha)$.

The exponential approach seems to give a very satisfactory way of defining fractional derivatives... but we have yet to answer L'Hospital's question, which was to determine the half-derivative of $f(x) = x$. There is no Fourier representation of this open-ended function, so it has no well-defined spectral decomposition. Of course, we can find the Fourier representation of x over some finite interval, but what interval should we choose? This ambiguity gives a hint of why Leibniz considered the subject to be paradoxical. Leibniz was well aware that the result of integrating a function is neither unique nor local, because it depends on how the function behaves over the range for which the integration is performed, not just at a single point. But he was used to thinking of differentiation as both unique and local, because whole derivatives happen to possess both of those attributes.

The apparent paradoxes of fractional derivatives arise from the fact that, in general, differentiation is non-unique and non-local, just as is integration. This shouldn't be surprising, since the generalization essentially unifies integrals and derivatives into a single operator. If anything, we ought to be surprised at how this operator takes on uniqueness and locality for positive integer arguments.

To get a clearer idea of the ambiguity in the concept of a generalized derivative, it's useful to examine a few other approaches, and compare them with the exponential approach described above. The most fundamental approach may be to begin with the basic definition of the whole derivative of a function $f(x)$

$$\frac{d}{dx} f(x) = \lim_{\varepsilon \to 0} \frac{f(x) = f(x-\varepsilon)}{\varepsilon}$$

Repeated composition of this operation leads to

$$\frac{d^n}{dx^n} f(x) = \lim_{\varepsilon \to 0} \frac{1}{\varepsilon^n} \sum_{j=0}^{n} (-1)^j \binom{n}{j} f(x - j\varepsilon)$$

for any positive integer n. To illustrate, this formula gives the second derivative of the function $f(x) = x^4$ as

$$\frac{d^2}{dx^2}\left(x^4\right) = \lim_{\varepsilon \to 0} \frac{1}{\varepsilon^2} \sum_{j=0}^{2} (-1)^j \binom{2}{j} (x - j\varepsilon)^4$$

$$= \lim_{\varepsilon \to 0} \frac{1}{\varepsilon^2}\left[x^4 - 2(x-\varepsilon)^4 + (x-2\varepsilon)^4\right]$$

$$= \lim_{\varepsilon \to 0} \left(12x^2 - 24x\varepsilon + 14\varepsilon^2\right) = 12x^2$$

We can generalize equation for non-integer orders, but to do this we must not only generalize the binomial coefficients, we also need to determine the appropriate generalization of the upper summation limit, which we wrote as n in equation. To clarify the situation, let us go back and derive "from stratch" the operations of differentiation and integration in a unified context. Consider an arbitrary smooth function $f(x)$.

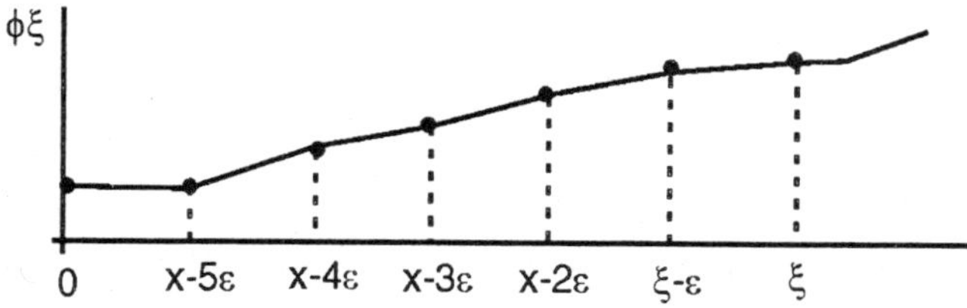

In addition to the point at x, we've also marked six other equally-spaced values on the interval from 0 to x, each a distance ε from its neighbours. The number k of these points is related to the values of x and ε by $x = k\varepsilon$. For convenience, we define a shift operator σ_ε such that $\sigma_\varepsilon f(x) = f(x-\varepsilon)$. Now we consider a general operation D defined by

$$D^n[f(x)] = \lim_{\varepsilon \to 0}\left(\frac{1-\sigma_\varepsilon}{\varepsilon}\right) f(x)$$

Recalling the geometric series expansion $1/(1-z) = 1 + z + z^2 + z^3 + \ldots$, the effects of this operator with $n = +1$ or -1 are

$$D^1 f(x) = \left(\frac{1-\sigma_\varepsilon}{\varepsilon}\right)^1 f(x)$$

$$= \frac{f(x)-f(x-\varepsilon)}{\varepsilon}$$

$$D^1 f(x) = \left(\frac{1-\sigma_\varepsilon}{\varepsilon}\right)^{-1} f(x)$$

$$= \varepsilon\left[f(x)+f(x-\varepsilon)+f(x-2\varepsilon)+\ldots+f(0)\right]$$

in the limit as ε goes to zero. Thus we see that D is simply the differentiation operator, and its inverse, D^{-1}, is the integration operator. This reproduces the ordinary whole derivatives. For example, the second derivative of $f(x)$ is

$$D^2 f(x) = \left(\frac{1-\sigma_\varepsilon}{\varepsilon}\right)^{2} f(x)$$

$$= \frac{f(x)-2f(x-\varepsilon)+f(x-2\varepsilon)}{\varepsilon^2}$$

in the limit as ε goes to zero, which illustrated how we recover the binomial equation for any whole number of differentiations. However, strictly speaking, this context makes it clear that we should actually write the second derivative as

$$\left(\frac{1-\sigma_\varepsilon}{\varepsilon}\right)^{2} f(x) = \frac{1f(x)-2f(x-\varepsilon)+1f(x-2\varepsilon)+0f(x-3\varepsilon)+\ldots+0f(0)}{\varepsilon^2}$$

It just so happens that, if n is a positive integer, all the binomial coefficients after the first $n+1$ are identically zero (i.e., we have $C(n;j) = 0$ for all j greater than n), so we can truncate the series, but for any negative or fractional positive values of n, the binomial coefficients are non-terminating, so we must include the entire summation over the specified range. Consequently, the upper summation limit should actually be $(x-x_0)/\varepsilon$, where x_0 is the lower bound on the range of evaluation. We often choose $x_0 = 0$ by convention, but it is actually arbitrary, and we will see below some circumstances in which the lower bound is not zero. In any case, we can re-write equation in the more correct form that does not rely on n being a positive integer

$$\frac{d^n}{dx^n} f(x) = \lim_{\varepsilon\to 0} \frac{1}{\varepsilon^n} \sum_{j=0}^{\left[\frac{x-x_0}{\varepsilon}\right]} (-1)^j \binom{n}{j} f(x-j\varepsilon)$$

To define the binomial coefficient for non-integer values of n, recall that for integer arguments these coefficients are defined as

$$\binom{n}{j} = \frac{n!}{j!(j-j)!}$$

so we need a way of evaluating the factorial function for non-integer arguments. Notice that for any positive integer n we have the definite integral

$$\int_{-1}^{+1} \left(1-x^2\right)^n dx = \frac{2^{2n+1}(n!)^2}{(2n1)!}$$

Solving for $n!$ gives

$$n! = \sqrt{\frac{(2n+1)!}{2^2 n+1} \int_{-1}^{+1} \left(1-x^2\right)^n dx}$$

The argument of the factorial on the right side is $2n+1$, so the right hand expression is well-defined for half-integer value of n such that $2n+1$ is non-negative. Hence this is a well-defined expression for the factorial of any such half-integer argument. For example, setting n = –1/2, we get

$$(-1/2)! = \sqrt{\pi}$$

Furthermore, now that the factorial of all (positive) half-integers is defined, the above formula allows us to compute the factorial of any quarter-integer, and then every sixteenth, and so on. Hence, using the binary representation of real numbers, and using the identity $(x+1)! = (x+1)x!$, we now have a well-defined factorial function for any real number. This is traditionally called the gamma function, with the argument offset by 1 relative to the factorial notation, so we have

$$\Gamma(n) = (n-1)!$$

for any positive integer n. The fundamental recurrence formula for the gamma function is therefore $\Gamma(x+1) = x\,\Gamma(x)$. Thus we have the following values for positive half-integer arguments

$$\Gamma\left(-\frac{3}{2}\right) = \frac{4}{3}\sqrt{\pi} \quad \Gamma\left(-\frac{1}{2}\right) = 2\sqrt{\pi} \quad \Gamma\left(\frac{1}{2}\right) = \sqrt{\pi} \quad \Gamma\left(\frac{3}{2}\right) = \frac{1}{2}\sqrt{\pi} \quad \Gamma\left(\frac{5}{2}\right) = \frac{3}{4}\sqrt{\pi}$$

Note the reflection relation $\Gamma(x)\,\Gamma(1-x) = \pi / \sin(\pi x)$. Now that we have a general way of expressing "factorials" for non-integers, we can re-write equation in generalized form, replacing each appearance of the integer n with the real number ν. This gives

$$\frac{d^{\nu}}{dx^{\nu}} f(x) = \lim_{\varepsilon\to 0} \frac{1}{\varepsilon^{\nu}} \sum_{j=0}^{\left[\frac{x-x_0}{\varepsilon}\right]} (-1)^j \frac{\Gamma(\nu+1)}{j!\Gamma(\nu+1-j)} f(x-j\varepsilon)$$

If ν is an integer *n*, the vanishing of the binomial coefficients for all j greater than n implies that we don't really need to carry the summation beyond j = n, and in the limit as ε goes to zero the n values of $f(x-k\varepsilon)$ with non-zero coefficients all converge on *x*, so the derivative is local. However, in general, the binomial expansion has infinitely many non-zero coefficients, so the result depends on the values of *x* all the way down to x_0. We typically choose $x_0 = 0$, so we are effectively evaluating the "derivative" (which is not the same as the "slope") for the interval from 0 to *x*.

Thus, as mentioned previously, the generalized derivative is a non-local operation, just as is integration. The general derivative depends on the value of the function f over the whole range from x_0 to *x*. This can be seen from the factor $f(x-j\varepsilon)$ in the summation in equation, showing that as j ranges from zero to $(x-x_0)/\varepsilon$ the argument of f ranges from *x* down to zero. It just so happens that this non-locality disappears for positive whole derivatives.

Choosing $x_0 = 0$ as the low end of our differentiation interval, the formula for the general derivative becomes

$$\frac{d^{\nu}}{dx^{\nu}} f(x) = \lim_{\varepsilon\to 0} \frac{1}{\varepsilon^{\nu}} \sum_{j=0}^{[x/\varepsilon]} (-1)^j \frac{\Gamma(\nu+1)}{j!\Gamma(\nu+1-j)} f(x-j\varepsilon)$$

With this, we are finally equipped to attempt to answer L'Hospital's question. Taking the function $f(x) = x$ with $\nu = 1/2$, signifying the half-derivative, this formula gives

$$\frac{d^{1/2}}{dx^{1/2}}(x) = \lim_{\varepsilon\to 0} \frac{1}{\sqrt{\varepsilon}}\left[1(x) - \frac{1}{2}(x-\varepsilon) - \frac{1}{8}(x-2\varepsilon) - \frac{1}{16}(x-3\varepsilon)\right.$$

$$\left. - \frac{5}{128}(x-4\varepsilon) - \ldots - \binom{1/2}{[x/\varepsilon]-2}(2\varepsilon) - \binom{1/2}{[x/\varepsilon]-1}(\varepsilon)\right]$$

In this equation the binomial coefficients symbol is understood to denote the generalized function, with the factorials expressed in terms of the gamma function. As explained previously, the coefficients in the above expression are just the coefficients in the binomial expansion of $(1-\sigma_{\varepsilon})^{1/2}$. Evaluating this expression in the limit as ε goes to zero, we find that the half-derivative of *x* is

$$\frac{d^{1/2}}{dx^{1/2}}(x) = 2\sqrt{\frac{x}{\pi}}$$

This is exactly what we would expect based on a straightforward interpolation of the derivatives of a power of *x*. Recall that the first few (whole) derivatives of x^m are

$$\frac{d\left(x^{m}\right)}{dx} = mx^{m-1}$$

$$\frac{d^{2}\left(x^{m}\right)}{dx^{2}} = m(m-1)x^{m-2}$$

$$\frac{d^{3}\left(x^{m}\right)}{dx^{3}} = m(m-1)(m-2)x^{m-3}$$

Thus we expect to find that the general form of the nth derivative of x^m is

$$\frac{d^{n}\left(x^{m}\right)}{dx^{n}} = \frac{m!}{(m-n)!}x^{m-n}$$

Replacing the integer n with the general value ν, and using the gamma function to express the factorial, this suggests that the a fractional derivative of x^n is simply

$$\frac{d^{\nu}\left(x^{m}\right)}{dx^{\nu}} = \frac{m!}{\Gamma(m+1-\nu)}x^{m-\nu}$$

which is exactly the same answer to L'Hospital's question as we got previously, i.e., the half-derivative of x is given. Now, since analytic functions can be expanded into power series, we can use equation to determine the fractional derivatives of all such functions.

Furthermore, applying this formula with negative values of ν gives a plausible expression for the integral of a power of x. For example, to find the whole integral of x^3 we set m = 3 and ν = -1, and then compute

$$\frac{d^{-1}\left(x^{3}\right)}{dx^{-1}} = \frac{3!}{\Gamma(3+1-(-1))}x^{3}-(-1) = \frac{3!}{\Gamma(5)}x^{4} = \frac{6}{24}x^{4} = \frac{1}{4}x^{4}$$

So, in a sense, equation is an algebraic expression of the fundamental theorem of calculus, i.e., the inverse relationship between the operations of differentiation and integration, since the nth derivative of the *–nth* derivative is the identity. The unification of these two operations makes it even less surprising that generalized differentiation is non-local, just as is integration.

Incidentally, the generalized derivative as developed so far gives some slightly surprising results. For example, the half-derivative of any constant function cx^0 is

$$\frac{d^{k}(c)}{dx^{k}} = c\frac{0!}{\Gamma(1/2)}x^{-1/2} = \frac{c}{\sqrt{\pi x}}$$

Thus, not only is the half-derivative of a constant with (respect to x) non-zero, it is infinite at $x = 0$. Nevertheless, equations are agreeably consistent with each other, giving some confidence in the significance of this generalization of the derivative. Given this equivalence, one might wonder about the value of the elaborate derivation of equation when it seems to be so much easier and more direct to arrive at equation. In answer to this there are two points to consider. First, equation applies to fairly arbitrary functions, whereas equation applies only to functions expressible as power series. Still, a very large class of functions can be expressed as power series, so this in itself is not an overriding factor.

More important is the fact that equation gives no hint of the non-locality of the generalized derivative, i.e., the dependence on the function over a finite range rather than just at a single point, and the need to specify (implicitly or explicitly) the chosen range. The importance of this can be seen in several different ways. Perhaps the most significant reason for taking care of the derivative interval is brought to light when we try to apply equation to the simple exponential function. We previously proposed that the general nth derivative of e^{ax} is simply $(a)^n\, e^{ax}$, and yet if we expand the exponential function e^x into a power series

$$e^{ax} = 1+\frac{a}{1!}x+\frac{a^2}{2!}x^2+\frac{a^3}{3!}x^3+..$$

and apply equation to determine the half-derivative, term by term, we get

$$\frac{d^{1/2}}{dx^{1/2}}\left(e^x\right) = \frac{1}{\sqrt{\pi x}}\left(1+2x+\frac{4}{3}x^2+\frac{8}{15}x^3+\frac{16}{105}x^4....\right)$$

A plot of this function, along with ex, is shown in the figure below

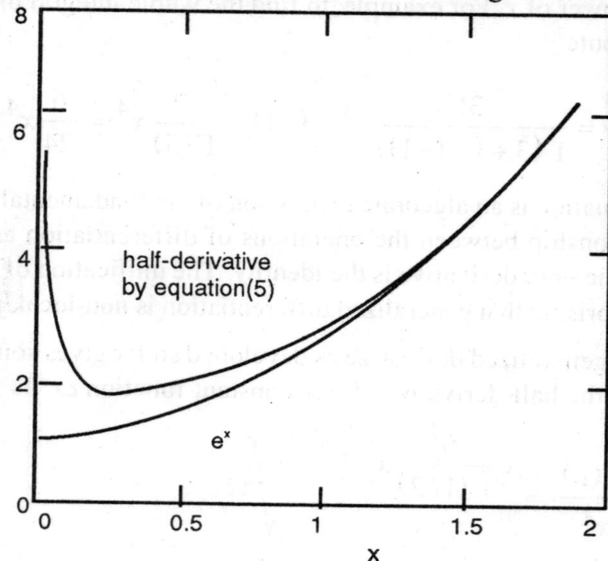

Here we see one of the paradoxes that might have intrigued Leibniz. Accordi reasonable general definition we expect any derivative (including fractional derivativ exponential function to equal itself, and the exponential goes to 1 as x goes to zero, and y carefully-derived formulas for the half-derivative of the exponential function goes to infinity a = 0. Clearly something is wrong. Must we abandon the elegant exponential approach, along with it's beautiful explanation of the trigonometric derivatives as simple phase shifts, etc? No, we can reconcile our results, provided we recognize that the derivative is non-local, and therefore depends on the chosen range of differentiation. Consider the two anti-differentiations shown below

$$\int_{x-0}^{x} u^3 du = \frac{x^4}{4} - \frac{x_0^4}{4}$$

$$\int_{x-0}^{x} e^u du = e^x - e^{x_0}$$

The left-hand integral shows that when we say x^3 is the derivative of $x^4/4$ we are implicitly assuming $x_0 = 0$, which is consistent with our derivation of equation. However, the right-hand integral shows that, by saying e^x is the derivative of e^x, we are implicitly assuming $x_0 = -\infty$. Thus ranges of integration/differentiation we have tacitly assumed for these two definitions are different.

To get agreement between the interpolated binomial expansion method and the definition based on exponential functions we must return to equation, and replace the condition $x_0 = 0$ with the condition $x_0 = -\infty$". This is easy to do, because it simply amounts to setting the upper summation limit to infinity, i.e., we take the following formula for our generalized derivative

$$\frac{d^\nu}{dx^\nu} f(x) = \lim_{\varepsilon \to 0} \frac{1}{\varepsilon^\nu} \sum_{j=0}^{\infty} (-1)^j \frac{\Gamma(\nu+1)}{j!\,\Gamma(\nu+1-j)} f(x - j\varepsilon)$$

With this, we do indeed find that the νth derivative of e^{ax} is simply $(a)^\nu e^{ax}$, consistent with the purely exponential approach.

Still another approach to fractional calculus is to begin with a generalization of the formula for repeated integration. Suppose the function $f(x)$ has the specified values f_0, f_1, f_2, f_3, f_4, and f_5 at equally spaced intervals of width ε.

The integral of this function from $x = 0$ to 5ε can be approximated by ε times the cumulative sum of these values, and the integral of this new function is ε times the cumulative sum of those values, and so on. This is illustrated in the table below.

	f_0	f_0	f_0
	f_0+f_1	$2f_0+f_1$	$3f_0+f_1$
f_2	$f_0+f_1+f_2$	$3f_0+f_1+f_2$	$6f_0+f_1+f_2$
f_3	$f_0+f_1+f_2+f_3$	$4f_0+f_1+f_2+f_3$	$10f_0+f_1+f_2+f_3$
f_4	$f_0+f_1+f_2+f_3+f_4$	$5f_0+f_1+f_2+f_3+f_4$	$15f_0+f_1+f_2+f_3+f_4$
f_5	$f_0+f_1+f_2+f_3+f_4+f_5$	$6f_0+f_1+f_2+f_3+f_4+f_5$	$12f_0+f_1+f_2+f_3+f_4+f_5$

Thus the 1st, 2nd, and 3rd "integrations" yield the values

$$S_1 = f_0 + f_1 + f_2 + f_3 + f_4 + f_5$$

$$S_2 = (6-0)\,f_0 + (6-1)\,f_1 + (6-2)\,f_2 + (6-3)\,f_3 + (6-4)\,f_4 + (6-5)\,f_5$$

$$S_3 = \frac{(6+1)(6-0)}{2} f_0 + \frac{(6-0)(6-1)}{2} f_1 + \frac{(6-1)(6-2)}{2} f_2 \qquad \frac{(6-2)(6-3)}{2} f_5$$

As we divide the overall interval into more and more segments, ε becomes arbitrarily small, and so do the differences between the factors in any given term, so the successive integrations give

$$\frac{d^{-2}}{dx^{-2}} f(x) = \int_0^x \int_0^{u_1} f(u_1)\,du_2\,du_1 = \int_0^x (x-u)\,f(u)\,du$$

$$\frac{d^{-3}}{dx^{-3}} f(x) = \int_0^x \int_0^{u_1} \int_0^{u_2} f(u_1)\,du_3\,du_2\,du_1 = \frac{1}{2!} \int_0^x (x-u)\,f(u)\,du$$

and so on. Thus we have Cauchy's expression for repeated integrals, which we can express using the gamma function instead of factorials, as

$$\frac{d^{-\nu}}{dx^{-\nu}} f(x) = \frac{1}{\Gamma(\nu)_0} \int_0^x (x-u)^{\nu-1} f(u)\,du$$

The convergence properties of this formula are best when ν has a value between 0 and 1. There are two different ways in which this formula might be applied. For example, if we wish

Chapter 10

Applications of Integral

Equipped with the integral and able to calculate it for many functions, we now move on to some interesting applications, each arising from the notion of a limit of sums. The integral was first introduced with reference to the "area under the graph" of a function. We begin this section by applying this application to more general regions.

ARC LENGTH

In this section we are going to look at computing the arc length of a function. Because it's easy enough to derive the formulas that we'll use in this section we will derive one of them and leave the other to you to derive.

We want to determine the length of the continuous function $y = f(x)$ on the interval $[a ,b]$. Initially we'll need to estimate the length of the curve. We'll do this by dividing the interval up into n equal subintervals each of width and we'll denote the point on the curve at each point by Pi. We can then approximate the curve by a series of straight lines connecting the points. Here is a sketch of this situation for $n = 9$.

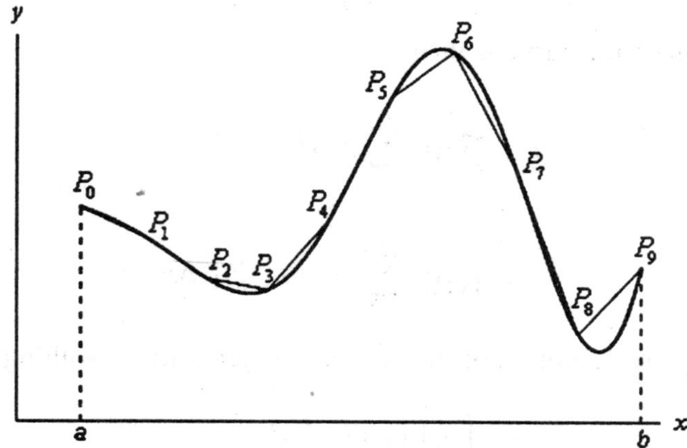

Now denote the length of each of these line segments by $|P_{i-1}P_i|$ and the length of the curve will then be approximately,

$$L \approx \sum_{i=1}^{n} | P_{i-1}P_i |$$

And we can get the exact length by taking n larger and larger. In other words, the exact length will be,

$$L \approx \lim_{n \to \infty} \sum_{i=1}^{n} | P_{i-1}P_i |$$

Now, let's get a better grasp on the length of each of these line segments. First, on each segment let's define $\Delta y_i = y_i - y_{i-1} = f(x_i) - f(x_{i-1})$. We can then compute directly the length of the line segments as follows.

$$| P_{i-1}P_i | = \sqrt{(x_i - x_{i-1})^2 + (y_i - y_{i-1})^2} = \sqrt{\Delta x^2 + \Delta y_i^2}$$

By the Mean Value Theorem we know that on the interval $[x_{i-1}, x_i]$ there is a point x_i^* so that,

$$f(x_i) - f(x_{i-1}) = f' \ (x_i^*)(x_i - x_{i-1})$$

$$\Delta\, yi = f' \ (x_i^*)\, \Delta x$$

Therefore, the length can now be written as,

$$| P_{i-1}P_i | = \sqrt{(x_i - x_{i-1})^2 + (y_i - y_{i-1})^2}$$

$$= \sqrt{\Delta x^2 + [f'(x_i^*)]^2 \Delta x^2}$$

$$= \sqrt{1 + [f'(x_i^*)]^2}\, \Delta x$$

The exact length of the curve is then,

$$L = \lim_{n \to \infty} \sum_{i=1}^{n} | P_{i-1}P_i |$$

$$= \lim_{n \to \infty} \sum_{i=1}^{n} \sqrt{1 + [f(x_i^*)]^2}\, \Delta x$$

However, using the definition of the definite integral, this is nothing more than,

$$L = \int_a^b \sqrt{1 + [f'(x)]^2}\, dx$$

A slightly more convenient notation (in my opinion anyway) is the following.

$$L = \int_a^b \sqrt{1+\left(\frac{dy}{dx}\right)^2}\,dx$$

In a similar fashion we can also derive a formula for $x = h(y)$ on $[c, d]$. This formula is,

$$L = \int_c^d \sqrt{1+[h'(y)]^2}\,dy \int_c^d \sqrt{1+\left(\frac{dx}{dy}\right)^2}\,dy$$

Again, the second form is probably a little more convenient.

Note the difference in the derivative under the square root! Don't get too confused. With one we differentiate with respect to x and with the other we differentiate with respect to y. One way to keep the two straight is to notice that the differential in the "denominator" of the derivative will match up with the differential in the integral. This is one of the reasons why the second form is a little more convenient.

Before we work any examples we need to make a small change in notation. Instead of having two formulas for the arc length of a function we are going to reduce it, in part, to a single formula.

From this point on we are going to use the following formula for the length of the curve.

ARC LENGTH FORMULA(S)

$$L = \int ds$$

Where,

$$ds = \sqrt{1+\left(\frac{dy}{dx}\right)^2}\,dx \qquad \text{if } y = f(x),\ a \le x \le b$$

$$ds = \sqrt{1+\left(\frac{dx}{dy}\right)^2}\,dy \qquad \text{if } x = h(y)\ c \le y \le d$$

Note that no limits where put on the integral as the limits will depend upon the ds that we're using. Using the first ds will require x limits of integration and using the second ds will require y limits of integration.

Thinking of the arc length formula as a single integral with different ways to define ds will be convenient when we run across arc lengths in future sections. Also, this ds notation will be a nice no :tion for the next section as well.

Now that we've derived the arc length formula let's work some examples.

Example: Determine the length of $y = \text{In}(\sec x)$ between $0 \le x \le \frac{\pi}{4}$.

Solution: In this case we'll need to use the first ds since the function is in the form $y = f(x)$. So, let's get the derivative out of the way.

$$\frac{dy}{dx} = \frac{\sec x \tan x}{\sec x} = \tan x \qquad \left(\frac{dy}{dx}\right)^2 = \tan^2 x$$

Let's also get the root out of the way since there is often simplification that can be done and there's no reason to do that inside the integral.

$$\sqrt{\left(\frac{dy}{dx}\right)^2} = \sqrt{1+\tan^2 x} = \sqrt{\sec^2 x} = |\sec x| = \sec x$$

Note that we could drop the absolute value bars here since secant is positive in the range given. The arc length is then,

$$L = \int_0^{\frac{\pi}{4}} \sec c \, dx$$

$$= \text{In } |\sec x + \tan x| \Big|_0^{\frac{\pi}{4}}$$

$$= \text{In } \left(\sqrt{2}+1\right)$$

Example: Determine the length of $x = \frac{2}{3}(y-1)^{\frac{3}{2}}$ between $1 \le y \le 4$.

Solution: There is a very common mistake that students make in problems of this type. Many students see that the function is in the form $x = h(y)$ and they immediately decide that it will be too difficult to work with it in that form so they solve for y to get the function into the form $y = f(x)$. While that can be done here it will lead to a messier integral for us to deal with.

Sometimes it's just easier to work with functions in the form $x = h(y)$. In fact, if you can work with functions in the form $y = f(x)$ then you can work with functions in the form $x = h(y)$. There really isn't a difference between the two so don't get excited about functions in the form $x = h(y)$.

Let's compute the derivative and the root.

$$\frac{dx}{dy} = (y-1)^{\frac{1}{2}} \qquad \Rightarrow \qquad \left(1+\frac{dx}{dy}\right)^2 = \sqrt{1+y-1} = \sqrt{y}$$

As you can see keeping the function in the form $x = h(y)$ is going to lead to a very easy integral. To see what would happen if we tried to work with the function in the form $y = f(x)$ see the next example.

Let's get the length.

$$L = \int_1^4 \sqrt{y}\,dy$$
$$= \frac{2}{3} y^{\frac{3}{2}} \Big|_1^4$$
$$= \frac{14}{3}$$

As noted in the last example we really do have a choice as to which ds we use. Provided we can get the function in the form required for a particular *ds* we can use it. However, as also noted above, there will often be a significant difference in difficulty in the resulting integrals. Let's take a quick look at what would happen in the previous example if we did put the function into the form $y = f(x)$.

Example: Redo the previous example using the function in the form $y = f(x)$ instead.

Solution: In this case the function and its derivative would be,

$$y = \left(\frac{3x}{2}\right)^{\frac{3}{2}} + 1 \qquad \frac{dy}{dx} = \left(\frac{3x}{2}\right)^{-\frac{3}{2}}$$

The root in the arc length formula would then be.

$$\sqrt{1 + \left(\frac{dy}{dx}\right)^2} = \sqrt{1 + \frac{1}{\left(\frac{3x}{2}\right)^{\frac{2}{3}}}} = \sqrt{\frac{\left(\frac{3x}{2}\right)^{\frac{2}{3}} + 1}{\left(\frac{3x}{2}\right)^{\frac{2}{3}}}} = \frac{\sqrt{\left(\frac{3x}{2}\right)^{\frac{2}{3}} + 1}}{\left(\frac{3x}{2}\right)^{\frac{1}{3}}}$$

he simplification work above was just to put the root into a form that will allow us to gral.

fore we write down the integral we'll also need to determine the limits. This uires x limits of integration and we've got y limits. They are easy enough to we know x as a function of y all we need to do is plug in the original y and get the x limits of integration. Doing this gives,

$$0 \le x \le \frac{2}{3}(3)^{\frac{3}{2}}$$

th, but there they are.

Let's now write down the integral that will give the length.

$$L = \int_0^{\frac{2}{3}(3)^{\frac{3}{2}}} \frac{\sqrt{\left(\frac{3x}{2}\right)^{\frac{2}{3}} + 1}}{\left(\frac{3x}{2}\right)^{\frac{1}{3}}} dx$$

That's a really unpleasant looking integral. It can be evaluated however using the following substitution.

$$u = \left(\frac{3x}{2}\right)^{\frac{2}{3}} + 1 \qquad du = \left(\frac{3x}{2}\right)^{-\frac{1}{3}} dx$$

$$x = 0 \qquad \Rightarrow \qquad u = 4$$

$$x = \frac{2}{3}(3)^{\frac{3}{2}} \qquad \Rightarrow \qquad u = 4$$

Using this substitution the integral becomes,

$$L = \int_1^4 \sqrt{u}\, du$$

$$= \frac{2}{3} u^{\frac{3}{2}} \Big|_1^4$$

$$= \frac{14}{3}$$

So, we got the same answer as in the previous example. Although that shouldn't really be all that surprising since we were dealing with the same curve.

From a technical standpoint the integral in the previous example was not that diffic[...] was just a Calculus I substitution. However, from a practical standpoint the inte[...] significantly more difficult than the integral we evaluated in Example 2. So, the m[...] story here is that we can use either formula (provided we can get the function [...] form of course) however one will often be significantly easier to actually eva[...]

Okay, let's work one more example.

Example: Determine the length of $x = \frac{1}{2}y^2$ for $0 \le x \le \frac{1}{2}$. Ass[...]

Solution: We'll use the second *ds* for this one as the functi[...] form for that one. Also, the other *ds* would again lead to a parti[...] derivative and root will then be,

$$\frac{dx}{dy} = y \qquad \Rightarrow \qquad \sqrt{1+\left(\frac{dx}{dy}\right)^2} = \sqrt{1+y^2}$$

Before writing down the length notice that we were given x limits and we will need y limits for this ds. With the assumption that y is positive there are easy enough to get. All we need to do is plug x into our equation and solve for y. Doing this gives,

$$0 \le y \le 1$$

The integral for the arc length is then,

$$L = \int_0^1 \sqrt{1+y^2}\,dy$$

This integral will require the following trig substitution.

$$y = \tan\theta \qquad dy = \sec^2\theta\,d\theta$$

$$y = 0 \quad \Rightarrow \quad 0 = \tan\theta \quad \Rightarrow \quad \theta = 0$$

$$y = 1 \quad \Rightarrow \quad 1 = \tan\theta \quad \Rightarrow \quad \theta = \frac{\pi}{4}$$

$$\sqrt{1+y^2} = \sqrt{1+\tan^2\theta} = \sqrt{\sec^2\theta} = |\sec\theta| = \sec\theta$$

$$\sqrt{1+y^2} = \sqrt{1+\tan^2\theta} = \sqrt{\sec^2\vartheta} = |\sec\theta| = \sec\theta$$

The length is then,

$$L = \int_0^{\frac{\pi}{4}} \sec^3\theta d\theta$$

$$= \frac{1}{2}(\sec\theta\tan\theta + \text{In}\,|\sec\theta + \tan\theta|)\Big|_0^{\frac{\pi}{4}}$$

$$= \frac{1}{2}\left(\sqrt{2} + \text{In}(1+\sqrt{2})\right)$$

SURFACE AREA

In this section we are going to look once again at solids of revolution. We first looked at them back in Calculus I when we found the volume of the solid of revolution. In this section we want to find the surface area of this region.

So, for the purposes of the derivation of the formula, let's look at rotating the continuous function $y = f(x)$ in the interval $[a, b]$ about the x-axis. Below is a sketch of a function and the solid of revolution we get by rotating the function about the x-axis.

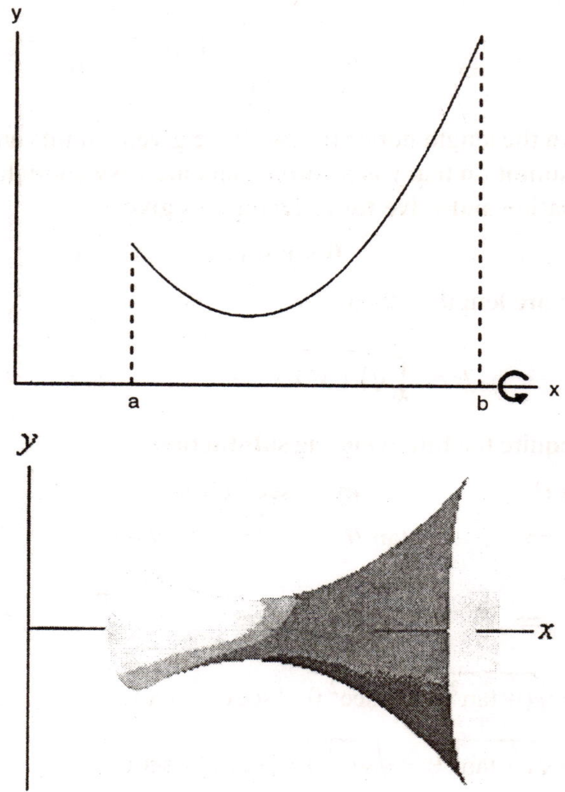

We can derive a formula for the surface area much as we derived the formula for arc length. We'll start by dividing the integral into n equal subintervals of width Δx. On each subinterval we will approximate the function with a straight line that agrees with the function at the endpoints of the each interval. Here is a sketch of that for our representative function using $n = 4$.

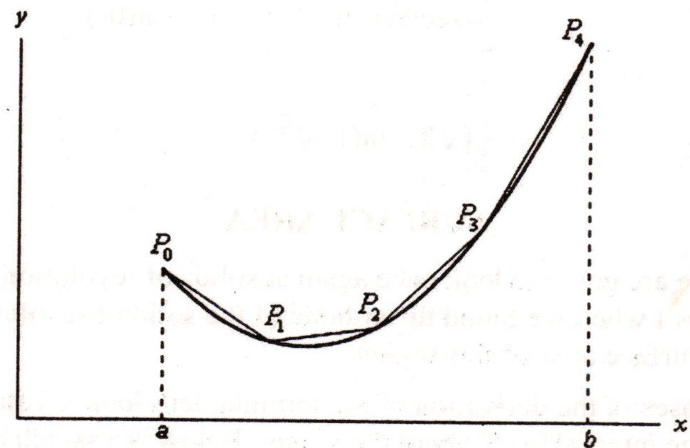

Now, rotate the approximations about the x-axis and we get the following solid.

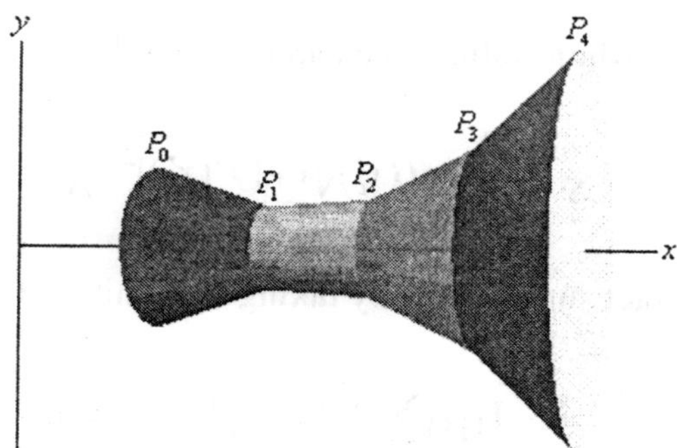

The approximation on each interval gives a distinct portion of the solid and to make this clear each portion is colored differently. Each of these portions are called frustums and we know how to find the surface area of frustums.

The surface area of a frustum is given by,

$$A = 2\pi r l$$

Where,

$$r = \frac{1}{2}(r_1 + r_2) \qquad r_1 = \text{Radius of right end}$$

$$r_2 = \text{Radius of left end}$$

And l is the length of the slant of the frustum.

For the frustum on the interval $[x_{i-1}, x_i]$ we have,

$r1 = f(x_i)$

$r_2 = f(xi-1)$

$l = |Pi-1\ Pi|$ (Length of the line segment connecting P_i and p_{i-1})

And we know from the previous section that,

$$|P_{i-1}P_i| = \sqrt{1+[f'(x_i^*)]^2}\ \Delta x \text{ Where } x_i^* \text{ is some point in } [x_{i-1}, x_i]$$

Before writing down the formula for the surface area we are going to assume that Δx is "small" and since $f(x)$ is continuous we can then assume that,

$$f(x_i) \approx f(x_i^*) \quad \text{and} \quad f(x_{i-1}) \approx f(x_i^*)$$

So, the surface area of the frustum on the interval $[x_{i-1}, x_i]$ is approximately,

$$Ai = 2\pi\left(\frac{f(x_i) + f(x_{i-1})}{2}\right)|P_{i-1}P_i|$$

$$\approx 2\pi f(x_i^*)\sqrt{1+[f'(x_i^*)]^2}\ \Delta x$$

The surface area of the whole solid is then approximately,

$$S \approx \sum_{i=1}^{n} 2\pi f(x_i^*)\sqrt{1+[f'(x_i^*)]^2}\ \Delta x$$

And we can get the exact surface area by taking the limit as n goes to infinity.

$$S = \lim_{n\to\infty}\sum_{i=1}^{n} 2\pi f(x_i^*)\sqrt{1+[f'(x_i^*)]^2}\ \Delta x$$

$$= \int_a^b 2\pi f(x)\sqrt{1+[f'(x)]^2}\ dx$$

If we wanted to we could also derive a similar formula for rotating $x = h(y)$ on $[c, d]$ about the y-axis. This would give the following formula.

$$S = \int_c^d 2\pi f(y)\sqrt{1+[h'(y)]^2}\ dy$$

These are not the "standard" formulas however. Notice that the roots in both of these formulas are nothing more than the two ds's we used in the previous section. Also, we will replace $f(x)$ with y and $h(y)$ with x. Doing this gives the following two formulas for the surface area.

SURFACE AREA FORMULAS

$S = \int 2\pi y ds$ **Rotation about x -axis**

$S = \int 2\pi x ds$ **Rotation about y-axis**

Where,

$$ds = \sqrt{1+\left(\frac{dy}{dx}\right)^2}\ dx \qquad \text{if } y = f(x),\ a \le x \le b$$

$$ds = \sqrt{1+\left(\frac{dx}{dy}\right)^2}\ dy \quad \text{if } x = h(y),\ c \le y \le d$$

There are a couple of things to note about these formulas. First, notice that the variable in

the integral itself is always the opposite variable from the one we're rotating about. Second, we are allowed to use either *ds* in either formula. This means that there are, in some way, four formulas here. We will choose the *ds* based upon which is the most convenient for a given function and problem.

Now let's work a couple of examples.

Example: Determine the surface area of the solid obtained by rotating $y = \sqrt{9-x^2}, -2 \le x \le 2$, about the *x*-axis.

Solution: The formula that we'll be using here is,

$$S = \int 2\pi y ds$$

Since we are rotating about the *x*-axis and we'll use the first *ds* in this case because our function is in the correct form for that *ds* and we won't gain anything by solving it for *x*.

Let's first get the derivative and the root taken care of.

$$\frac{dy}{dx} = \frac{1}{2}(9-x^2)^{-\frac{1}{2}}(-2x) = -\frac{x}{(9-x^2)^{\frac{1}{2}}}$$

$$\sqrt{1+\left(\frac{dy}{dx}\right)^2} = \sqrt{1+\frac{x^2}{9-x^2}} = \sqrt{\frac{9}{9-x^2}} = \frac{3}{\sqrt{9-x^2}}$$

Here's the integral for the surface area,

$$S = \int_{-2}^{2} 2\pi y \frac{3}{\sqrt{9-x^2}} dx$$

There is a problem however. The *dx* means that we shouldn't have any *y*'s in the integral. So, before evaluating the integral we'll need to substitute in for *y* as well.

The surface area is then,

$$S = \int_{-2}^{2} 2\pi \sqrt{9-x^2} \frac{3}{\sqrt{9-x^2}} dx$$

$$= \int_{-2}^{2} 6\pi \, dx$$

$$= 24\pi$$

Previously we made the comment that we could use either *ds* in the surface area formulas. Let's work an example in which using either *ds* won't create integrals that are too difficult to evaluate and so we can check both *ds*'s.

Example: Determine the surface area of the solid obtained by rotating $y = \sqrt[3]{x}, 1 \le y \le 2$, about the y-axis. Use both ds's to compute the surface area.

Solution: Note that we've been given the function set up for the first ds and limits that work for the second ds.

Solution 1: This solution will use the first ds listed above. We'll start with the derivative and root.

$$\frac{dy}{dx} = \frac{1}{3}x^{-\frac{2}{3}}$$

$$\sqrt{1+\left(\frac{dy}{dx}\right)} = \sqrt{1+\frac{1}{9x^{\frac{4}{3}}}} = \sqrt{\frac{9x^{\frac{4}{3}}+1}{3x^{\frac{2}{3}}}} = \frac{\sqrt{9x^{\frac{4}{3}}+1}}{3x^{\frac{2}{3}}}$$

We'll also need to get new limits. That isn't too bad however. All we need to do is plug in the given y's into our equation and solve to get that the range of x's is $1 \le x \le 8$. The integral for the surface area is then,

$$S = \int_1^8 2\pi x \frac{\sqrt{9x^{\frac{4}{3}}+1}}{3x^{\frac{2}{3}}}\, dx$$

$$= \frac{2\pi}{3}\int_1^8 x^{\frac{1}{3}}\sqrt{9x^{\frac{4}{3}}+1}\, dx$$

Note that this time we didn't need to substitute in for the x as we did in the previous example. In this case we picked up a dx from the ds and so we don't need to do a substitution for the x. In fact if we had substituted for x we would have put y's into integral which would have caused problems.

Using the substitution

$$u = 9x^{\frac{4}{3}}+1 \qquad du = 12x^{\frac{1}{3}}$$

The integral becomes,

$$S = \frac{\pi}{18}\int_{10}^{145}\sqrt{u}\, du$$

$$= \frac{\pi}{27}u^{\frac{3}{2}}\Big|_{10}^{145}$$

$$= \frac{\pi}{27}\left(145^{\frac{3}{2}} - 10^{\frac{3}{2}}\right) = 199.48$$

Solution 2: This time we'll use the second ds. So, we'll first need to solve the equation for x. We'll also go ahead and get the derivative and root while we're at it.

$$x = y^3 \quad \frac{dx}{dy} = 3y^2$$

$$\sqrt{1+\left(\frac{dx}{dy}\right)^2} = \sqrt{1+9y^4}$$

The surface area is then,

$$S = \int_1^2 2\pi x\sqrt{1+9y^4}\, dy$$

We used the original y limits this time because we picked up a dy from the ds. Also note that the presence of the dy means that this time, unlike the first solution, we'll need to substitute in for the x. Doing that gives,

$$S = \int_1^2 2\pi y^3\sqrt{1+9y^4}\, dy \qquad\qquad u = 1 + 9y^4$$

$$= \frac{\pi}{18}\int_{10}^{145}\sqrt{u}\, du$$

$$= \frac{\pi}{27}\left(145^{\frac{3}{2}} - 10^{\frac{3}{2}}\right) = 199.48$$

Note that after the substitution the integral was identical to the first solution and so the work was skipped. As this example has shown we can used either ds to get the surface area. It is important to point out as well that with one ds we had to do a substitution for the x and with the other we didn't. This will always work out that way.

Note as well that in the case of the last example it was just as easy to use either ds. That often won't be the case. In many examples only one of the ds will be convenient to work with so we'll always need to determine which ds is liable to be the easiest to work with before starting the problem.

CENTER OF MASS

In this section we are going to find the center of mass or centroid of a thin plate with uniform density ρ. The center of mass or centroid of a region is the point in which the region will be perfectly balanced horizontally if suspended from that point. So, let's suppose that the plate is the region bounded by the two curves $f(x)$ and $g(x)$ on the interval $[a,b]$. So, we want to find the center of mass of the region below.

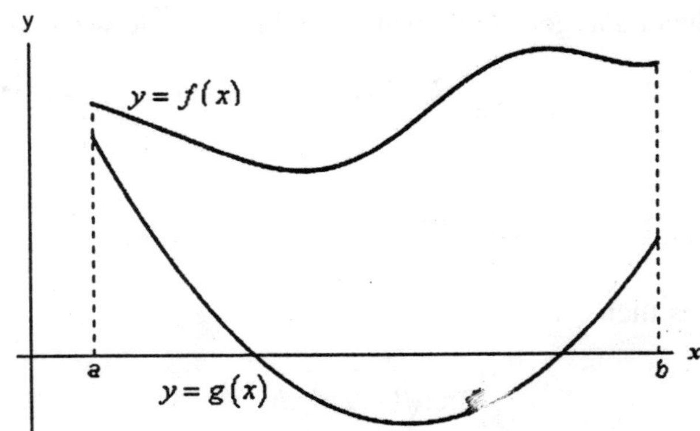

We'll first need the mass of this plate. The mass is,

$$M = \rho \text{ (Area of plate)}$$

$$= \rho \int_a^b f(x) - g(x)\, dx$$

Next we'll need the moments of the region. There are two moments, denoted by M_x and M_y. The moments measure the tendency of the region to rotate about the x and y-axis respectively. The moments are given by,

EQUATIONS OF MOMENTS

$$M_x = \rho \int_a^b \frac{1}{2}\left([f(x)]^2 - [g(x)]^2\right) dx$$

$$M_y = \rho \int_a^b x\left(f(x) - g(x)\right) dx$$

The coordinates of the center of mass, $(\bar{x}, \bar{y})$, are then,

CENTER OF MASS COORDINATES

$$\bar{x} = \frac{m_y}{m} = \frac{\int_a^b x\left(f(x) - g(x)\right) dx}{\int_a^b f(x) - g(x)\, dx} = \frac{1}{A}\int_a^b x\left(f(x) - g(x)\right) dx$$

$$\bar{y} = \frac{m_x}{m} = \frac{\int_a^b \frac{1}{2}\left([f(x)]^2 - [g(x)]^2\right) dx}{\int_a^b f(x) - g(x)\, dx} = \frac{1}{A}\int_a^b \frac{1}{2}\left([f(x)]^2 - [g(x)]^2\right) dx$$

Where,

$$A = \int_a^b f(x) - g(x)\, dx$$

Note that the density, ρ, of the plate cancels out and so isn't really needed.

Let's work a couple of examples.

Example: Determine the center of mass for the region bounded by $y = 2\sin(2x)$, $y = 0$ on the interval $\left[0, \frac{\pi}{2}\right]$

Solution: Here is a sketch of the region with the center of mass denoted with a dot.

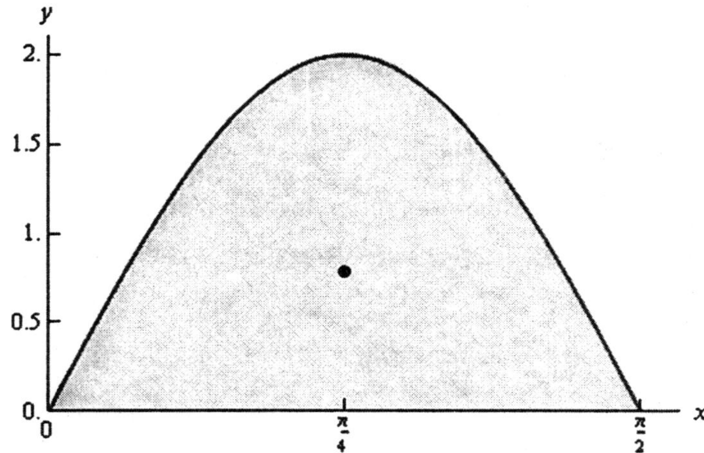

Let's first get the area of the region.

$$A = \int_0^{\frac{\pi}{2}} 2\sin(2x)\, dx$$

$$= -\cos(2x)\Big|_0^{\frac{\pi}{2}}$$

$$= 2$$

Now, the moments (without density since it will just drop out) are,

$$Mx = \int_0^{\frac{\pi}{2}} 2\sin^2(2x)\, dx \qquad My = \int_0^{\frac{\pi}{2}} 2x\sin(2x)\, dx$$

$$= \int_0^{\frac{\pi}{2}} 1 - \cos(4x)\, dx \qquad = -x\cos(2x)\Big|_0^{\frac{\pi}{2}} + \int_0^{\frac{\pi}{2}} \cos(2x)\, dx$$

$$= \left(x - \frac{1}{4} \sin (4x) \right) \Bigg|_0^{\frac{\pi}{4}} \qquad = -x \cos (2x) \Bigg|_0^{\frac{\pi}{2}} + \frac{1}{2} \sin(2x) \Bigg|_0^{\frac{\pi}{2}}$$

$$= \frac{\pi}{2} \qquad = \frac{\pi}{2}$$

The coordinates of the center of mass are then,

$$\bar{x} = \frac{\pi/2}{2} = \frac{\pi}{4}$$

$$\bar{y} = \frac{\pi/2}{2} = \frac{\pi}{4}$$

Again, note that we didn't put in the density since it will cancel out.

So, the center of mass for this region is $\left(\frac{\pi}{4}, \frac{\pi}{4} \right)$.

Example: Determine the center of mass for the region bounded by $y = x^3$ and $y = \sqrt{x}$.

Solution: The two curves intersect at $x = 0$ and $x = 1$ and here is a sketch of the region with the center of mass marked with a box.

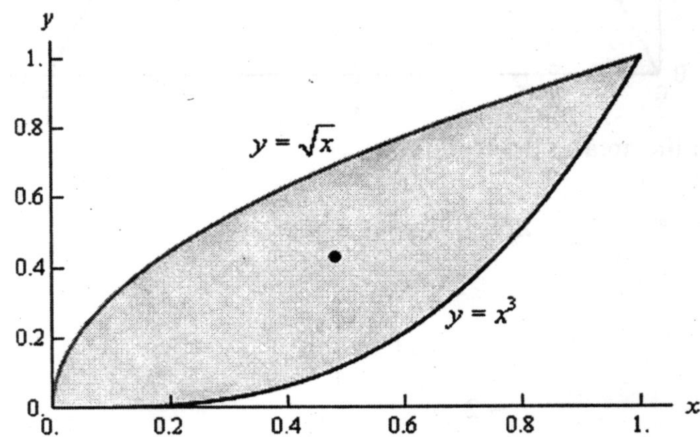

We'll first get the area of the region.

$$A = \int_0^1 \sqrt{x} - x^3 \, dx$$

$$= \left(\frac{2}{3} x^{\frac{3}{2}} - \frac{1}{4} x^4 \right) \Bigg|_0^1 = \frac{5}{12}$$

Now the moments, again without density, are

$$Mx = \int_0^1 \frac{1}{2}(x - x^6)dx \qquad My = \int_0^1 x(\sqrt{x} - x^3)dx$$

$$= \frac{1}{2}\left(\frac{1}{2}x^2 - \frac{1}{7}x^7\right)\Bigg|_0^1 \qquad = \int_0^1 x^{\frac{3}{2}} - x^4\, dx$$

$$= \frac{5}{28} \qquad = \left(\frac{2}{5}x^{\frac{5}{2}} - \frac{1}{5}x^5\right)\Bigg|_0^1$$

$$= \frac{1}{5}$$

The coordinates of the center of mass is then,

$$\bar{x} = \frac{1/5}{5/12} = \frac{12}{25}$$

$$\bar{y} = \frac{5/28}{5/12} = \frac{3}{7}$$

The coordinates of the center of mass are then, $\left(\frac{12}{25}, \frac{3}{7}\right)$.

HYDROSTATIC PRESSURE AND FORCE

In this section we are going to submerge a vertical plate in water and we want to know the force that is exerted on the plate due to the pressure of the water. This force is often called the hydrostatic force. There are two basic formulas that we'll be using here. First, if we are d meters below the surface then the hydrostatic pressure is given by,

$$P = \rho g d$$

Where, ρ is the density of the fluid and g is the gravitational acceleration. We are going to assume that the fluid in question is water and since we are going to be using the metric system these quantities become,

$$\rho = 1000 \text{ kg/m}^3 \qquad g = 9.81 \text{ m/s}^2$$

The second formula that we need is the following. Assume that a constant pressure P is acting on a surface with area A. Then the hydrostatic force that acts on the area is,

$$F = PA$$

Note that we won't be able to find the hydrostatic force on a vertical plate using this formula since the pressure will vary with depth and hence will not be constant as required by this formula. We will however need this for our work.

The best way to see how these problems work is to do an example or two.

Example: Determine the hydrostatic force on the following triangular plate that is submerged in water as shown.

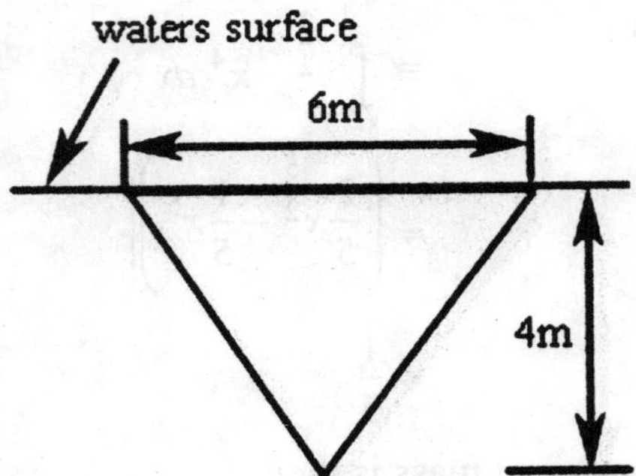

Solution: The first thing to do here is set up an axis system. So, let's redo the sketch above with the following axis system added in.

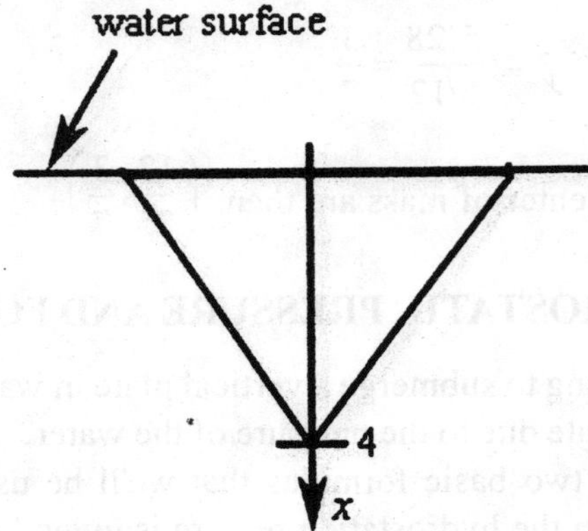

So, we are going to orient the x-axis so that positive x is downward, $x = 0$ corresponds to the water surface and $x = 4$ corresponds to the depth of the tip of the triangle.

Next we are break up the triangle into n horizontal strips each of equal width Δx and in each interval $[x_{i-1}, x_i]$ choose any point x_i^*. In order to make the computations easier we are going to make two assumptions about these strips. First, we will ignore the fact that the ends are actually going to be slanted and assume the strips are rectangular. If Δx is sufficiently small this will not affect our computations much. Second, we will assume that Δx is small enough that the hydrostatic pressure on each strip is essentially constant.

Below is a representative strip.

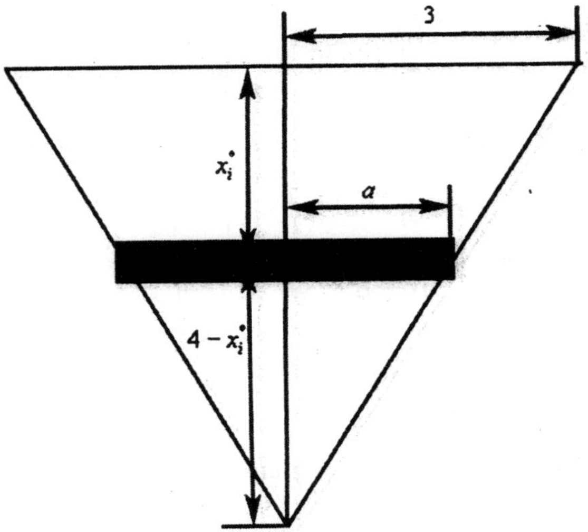

The height of this strip is Δx and the width is $2a$. We can use similar triangles to determine a as follows,

$$\frac{3}{4}=\frac{a}{4-x_i^{\bullet}} \qquad \Rightarrow \qquad a = 3-\frac{a}{4}x_i^{\bullet}$$

Now, since we are assuming the pressure on this strip is constant, the pressure is given by,

$$P_i = \rho g d = 1000(9.81)x_i^{\bullet} = 9810x_i^{\bullet}$$

And the hydrostatic force on each strip is,

$$F_i = P_i A = P_i(2a\Delta x) = 9810x_i^{\bullet}(2)\left(3-\frac{3}{4}x_i^{\bullet}\right)\Delta x = 19620x_i^{\bullet}\left(3-\frac{3}{4}x_i^{\bullet}\right)\Delta x$$

The approximate hydrostatic force on the plate is then the sum of the forces on all the strips or,

$$F \approx \sum_{i=1}^{n} 19620x_i^{\bullet}\left(3-\frac{3}{4}x_i^{\bullet}\right)\Delta x$$

Taking the limit will get the exact hydrostatic force,

$$F = \lim_{n\to\infty}\sum_{i=1}^{n} 19620x_i^{\bullet}\left(3-\frac{3}{4}x_i^{\bullet}\right)\Delta x$$

Using the definition of the definite integral this is nothing more than,

$$F = \int_0^4 19620\left(3x - \frac{3}{4}x^2\right)dx$$

The hydrostatic force is then,

$$F = \int_0^4 19620\left(3x - \frac{3}{4}x^2\right)dx$$

$$= 19620\left(\frac{3}{2}x^2 - \frac{1}{4}x^3\right)\Bigg|_0^4$$

$$= 156960N$$

Let's take a look at another example.

Example: Find the hydrostatic force on a circular plate of radius 2 that is submerged 6 meters in the water.

Solution: First, we're going to assume that the top of the circular plate is 6 meters under the water. Next, we will set up the axis system so that the origin of the axis system is at the center of the plate. Setting the axis system up in this way will greatly simplify our work.

Finally, we will again split up the plate into n horizontal strips each of width Δy and we'll choose a point y_i^* from each strip. We'll also assume that the strips are rectangular again to help with the computations. Here is a sketch of the setup.

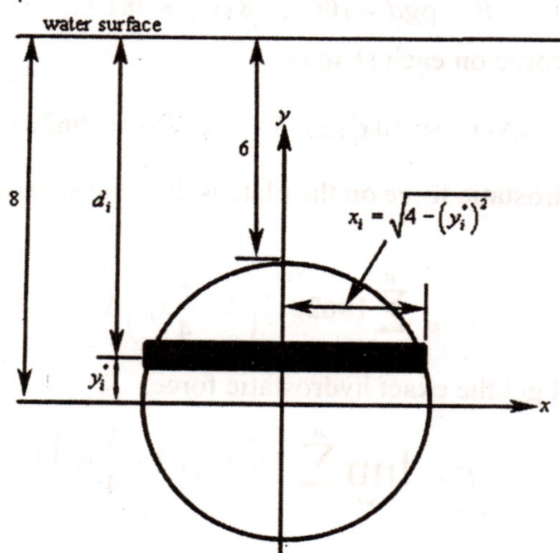

The depth below the water surface of each strip is,

$$d_i = 8 - y_i^*$$

And that in turn gives us the pressure on the strip,

$$P_i = \rho g d_i = 9810\,(8 - y_i^*)$$

The area of each strip is,

$$Ai = 2\sqrt{4-(y_i^*)^2}\,\Delta y$$

The hydrostatic force on each strip is,

$$F_i = P_i A_i = 9810(8 - y_i^*)(2)\sqrt{4-(y_i^*)^2}\,\Delta y$$

The total force on the plate is,

$$F = \lim_{n\to\infty}\sum_{i=1}^{n} 19620(8 - y_i^*)\sqrt{4-(y_i^*)^2}\,\Delta y$$

$$= 19620\int_{-2}^{2}(8 - y_i^*)\sqrt{4-y^2}\,dy$$

To do this integral we'll need to split it up into two integrals.

$$F = 19620\int_{-2}^{2} 8\sqrt{4-y^2}\,dy - 1920\int_{-2}^{2} y\sqrt{4-y^2}\,dy$$

The first integral requires the trig substitution $y = 2\sin\theta$ and the second integral needs the substitution $v = 4 - y^2$. After using these substitution we get,

$$F = 627840\int_{-\pi/2}^{\pi/2}\cos^2\theta\, d\theta + 9810\int_0^0 \sqrt{v\, dv}$$

$$= 313920\int_{-\pi/2}^{\pi/2} 1 + \cos(2\theta)\, d\theta + 0$$

$$= 313920\left(\theta + \frac{1}{2}\sin(2\theta)\right)\Bigg|_{\frac{\pi}{2}}^{\frac{\pi}{2}}$$

$$= 313920\pi$$

Note that after the substitution we know the second integral will be zero because the upper and lower limit is the same.

PROBABILITY

In this last application of integrals that we'll be looking at we're going to look at probability. Before actually getting into the applications we need to get a couple of definitions out of the way.

Suppose that we wanted to look at the age of a person, the height of a person, the amount of time spent waiting in line, or maybe the lifetime of a battery. Each of these quantities have values that will range over an interval of integers. Because of this these are called continuous random variables. Continuous random variables are often represented by X.

Every continuous random variable, X, has a probability density function, $f(x)$. Probability density functions satisfy the following conditions.

1. $f(x) \geq 0$ for all x.

2. $\int_{-\infty}^{\infty} f(x)dx = 1$

Probability density functions can be used to determine the probability that a continuous random variable lies between two values, say a and b. This probability is denotedby $P(a \leq X \leq b)$ and is given by,

$$P(a \leq X \leq b) = \int_a^b f(x)dx$$

Let's take a look at an example of this.

Example: Let $f(x) = \frac{x^3}{5000}(10 - x)$ for $0 \leq x \leq 10$ and $f(x) = 0$ for all other values of x.

Answer: Each of the following questions about this function.

1. Show that $f(x)$ is a probability density function.
2. Find $P(1 \leq X \leq 4)$
3. Find $P(x \leq 6)$

Solution:

1. Show that $f(x)$ is a probability density function.

First note that in the range $0 \leq x \leq 10$ is clearly positive and outside of this range we've defined it to be zero.

So, to show this is a probability density function we'll need to show that $\int_{-\infty}^{\infty} f(x)\,dx = 1$.

$$\int_{-\infty}^{\infty} f(x)\,dx = \int_0^{10} \frac{x^3}{5000}(10 - x)\,dx$$

$$= \left(\frac{x^4}{2000} - \frac{x^5}{25000}\right)\Bigg|_0^{10}$$

$$= 1$$

Note the change in limits on the integral. The function is only non-zero in these ranges and so the integral can be reduced down to only the interval where the function is not zero.

2. Find $P\ (1 \leq X \leq 4)$

In this case we need to evaluate the following integral.

$$P\ (1 \leq X \leq 4) = \int_1^4 \frac{x^3}{5000}(10 - x)dx$$

$$= \left(\frac{x^4}{2000} - \frac{x^5}{25000}\right)\Bigg|_1^4$$

$$= 0.08658$$

So the probability of X being between 1 and 4 is 8.658%.

3. Find $P\ (x \leq 6)$

Note that in this case $P\ (x \geq 6)$ is equivalent to $P\ (6 \leq X \leq 10)$ since 10 is the largest value that X can be. So the probability that X is greater than or equal to 6 is,

$$P\ (X \geq 6) = \int_6^{10} \frac{x^3}{5000}(10 - x)\, dx$$

$$= \left(\frac{x^4}{2000} - \frac{x^5}{25000}\right)\Bigg|_6^{10}$$

$$= 0.\ 66304$$

This probability is then 66.304%.

Probability density functions can also be used to determine the mean of a continuous random variable. The mean is given by,

$$\mu = \int_{-\infty}^{\infty} \frac{x^3}{5000} xf(x)\, dx$$

Let's work one more example.

Example: It has been determined that the probability density function for the wait in line at a counter is given by,

$$f(t) = \begin{cases} 0 & \text{if } t < 0 \\ 0.\text{l}e^{-\frac{t}{10}} & \text{if } t \geq 0 \end{cases}$$

Where t is the number of minutes spent waiting in line. Answer each of the following questions about this probability density function.

1. Verify that this is in fact a probability density function.
2. Determine the probability that a person will wait in line for at least 6 minutes.

3. Determine the mean wait in line.

Solution:

1. Verify that this is in fact a probability density function.

This function is clearly positive or zero and so there's not much to do here other than compute the integral.

$$\int_{-\infty}^{\infty} f(t)dt = \int_{0}^{\infty} 0.\text{l}e^{-\frac{t}{10}}dt$$

$$= \lim_{u\to\infty} \int_{0}^{n} 0.\text{l}e^{-\frac{t}{10}}dt$$

$$= \lim_{u\to\infty} \left(-e^{\frac{t}{10}}\right)\Bigg|_{0}^{n}$$

$$= \lim_{u\to\infty} \left(-e^{\frac{t}{10}}\right) = 1$$

So it is a probability density function.

2. Determine the probability that a person will wait in line for at least 6 minutes.

The probability that we're looking for here is $P\ (x \geq 6)$.

$$P\ (X^3\ 6) = \int_{6}^{\infty} 0.\text{l}e^{-\frac{t}{10}}dt$$

$$= \lim_{u\to\infty} \int_{6}^{u} 0.\text{l}e^{-\frac{t}{10}}dt$$

$$= \lim_{u\to\infty} \left(-e^{-\frac{t}{10}}\right)\Bigg|_{6}^{n}$$

$$= \lim_{u\to\infty} \left(e^{\frac{6}{10}} - e^{-\frac{n}{10}}\right) = e^{-\frac{3}{5}} = 0.548812$$

So the probability that a person will wait in line for more than 6 minutes is 54.8811%.

3. Determine the mean wait in line.

Here's the mean wait time.

$$\mu = \int_{-\infty}^{\infty} t\,f(t)dt$$

$$= \int_0^{\infty} 1te^{\frac{t}{10}} dt$$

$$= \lim_{u \to \infty} \int_0^{n} 1te^{\frac{t}{10}} dt \qquad \text{integrating by parts...}$$

$$= \lim_{u \to \infty} \left(-t + 10)e^{-\frac{t}{10}} \right) \Big|_0^u$$

$$= \lim_{u \to \infty} \left(10 - (u + 10)e^{-\frac{u}{10}} \right) = 10$$

So, it looks like the average wait time is 10 minutes.